# Roots of Modern Technology

Siegfried Wendt

# Roots of Modern Technology

An Elegant Survey of the Basic Mathematical and Scientific Concepts

Prof. Dr.-Ing. Siegfried Wendt
Albertstr. 1
67655 Kaiserslautern
Germany
E-mail: Siegfried.Wendt@hpi.uni-potsdam.de

ISBN 978-3-642-12061-9        e-ISBN 978-3-642-12062-6

DOI 10.1007/978-3-642-12062-6

Library of Congress Control Number: 2010926862

© 2010 Springer-Verlag Berlin Heidelberg

This work is subject to copyright. All rights are reserved, whether the whole or part of the material is concerned, specifically the rights of translation, reprinting, reuse of illustrations, recitation, broadcasting, reproduction on microfilm or in any other way, and storage in data banks. Duplication of this publication or parts thereof is permitted only under the provisions of the German Copyright Law of September 9, 1965, in its current version, and permission for use must always be obtained from Springer. Violations are liable to prosecution under the German Copyright Law.

The use of general descriptive names, registered names, trademarks, etc. in this publication does not imply, even in the absence of a specific statement, that such names are exempt from the relevant protective laws and regulations and therefore free for general use.

*Typesetting:* Data supplied by the authors

*Production & Cover Design:* Scientific Publishing Services Pvt. Ltd., Chennai, India

Printed on acid-free paper

9 8 7 6 5 4 3 2 1

springer.com

# Contents

**1 Explaining Modern Technology** ........................................................... 1
What Socrates Would Ask Me .................................................................. 2
Omitting Irrelevant Subjects Is an Art ....................................................... 5
No One should Be Afraid of Formulas ..................................................... 7

## Part I: Fundamentals of Mathematics and Logic

**2 Mathematicians Are Humans Like You and Me – They Count and Arrange** ................................................................................................. 13
What a Number "Sees" When It Looks into a Mirror ............................. 14
Sets Are Everywhere ............................................................................... 25
Functions Tell Us How to Get Results .................................................... 31
"Come Closer!" Is What Limits Want .................................................... 40
An Eye for an Eye and a Tooth for a Tooth – That's the Principle of Equations ................................................................................................. 44

**3 Mathematicians Are Nothing Special – They Draw and Compare** ...... 53
How Mr. Euclid's Ideas Have Grown Up ................................................ 53
How the Fraction "Zero Divided by Zero" and the Product "Infinity Times Zero" Are Related ........................................................................ 64
Relations Which We Can Deduce, but Not Really Understand .............. 74

**4 When It Helps to Ignore Any Meaning** ................................................ 81
Where Discretionary Powers Are Not Allowed ...................................... 81
Games Which Can Be Played without Thinking .................................... 83
How Logical Thinking Can Be Replaced by Pattern Recognition ......... 87
Detours Which Are Shorter Than the Direct Route ................................ 98
How We Can Enter into Four- or Higher-Dimensional Spaces Using Simple Steps .......................................................................................... 100

**5 About the Methods for Computing the Future** .................................. 109
Attempts to Reduce Expectations to Numbers ..................................... 110
How We Can Calculate the Number of Possible Cases ........................ 112
What You Can Do If You Don't Want to Know All the Details ........... 118
How to Handle the Situation When the Cases Are No Longer Countable .............................................................................................. 125
Statistics Are More Than Just Listing the Results of Counts ............... 129

| 6 | **What Talking and Writing Have in Common** ............................... 131 |
|---|---|
| | How Speech and Writing Are Interrelated.................................................132 |
| | What Grammar Has to Do with the Meaning of Texts ..............................133 |
| | How to Control Conversations in Order to Make Sure All Participants Get a Fair Chance to Say Something ..........................................................141 |

**Part II: Fundamentals of Natural Sciences**

| 7 | **What the Moon Has to Do with Mechanical Engineering**.................147 |
|---|---|
| | What Galileo Galilei Could Teach Us without Upsetting the Pope ...........148 |
| | What Sir Isaac Newton Found Out about Forces and Moving Bodies on Earth and in the Sky..............................................................................153 |

| 8 | **How Albert Einstein Disregarded Common Sense**..........................173 |
|---|---|
| | How Meters and Clocks Were "Relativized" and the Speed of Light Was Made the Standard Reference ............................................................173 |
| | How the Beautiful World of Mr. Newton Got Bended ..............................189 |

| 9 | **How a Few Frog Legs Triggered the Origin of Electrical Engineering**............................................................................................ 207 |
|---|---|
| | The Tremendous Consequences of Accidental and Simple Observations ...............................................................................................208 |
| | How Mr. Maxwell Transferred His Ideas from the Bath Tub to Free Space...........................................................................................................216 |
| | How the Feasibility of High Voltage and Radio Waves Became Evident without Experimenting .................................................................227 |
| | What We Get by Multiplying or Dividing Volts, Amperes and Similar Things..........................................................................................................233 |

| 10 | **Small, Smaller, Smallest – How the Components of Matter Were Found**................................................................................................... 241 |
|---|---|
| | How the Age-Old Assumption That Matter Is Composed of Atoms became Experimentally Relevant ..............................................................242 |
| | What Can Be Deduced from the Assumption That Gases Are Small Balls Flying Around ...................................................................248 |
| | How Particles Which Had Been Called "Indivisible" Broke Apart............258 |

| 11 | **How the Difference between Particles and Waves Disappeared**..........267 |
|---|---|
| | How Waves Can Be Forced to Show Us That They Really Are Waves .........................................................................................................267 |
| | How It became Necessary to Consider Rays of Light and Heat as Flying Packets of Energy............................................................................271 |
| | A Theory Which Could Be Confirmed, but Stayed Inconceivable.............281 |
| | Phenomena Which Even Einstein Thought to Be Impossible ...................302 |

## 12   How "Recipes" in the Cells of Living Organisms Were Found and Can Be Rewritten ..................................................................309
How Organization and Life Are Connected ................................309
How the Living became "Technological Matter" .......................316
Like the Mother, Like the Father - How Inheritance Works.......317
How New Recipes Can Be Smuggled into Living Cells..............338
How to Provide Evidence Confirming "Who It Was" .................341

## Part III: Fundamentals of Engineering

## 13   Why Engineers Are "Playing with Models"................................347
What Engineers Are Needed for ................................................347
A Look into the Toy Box of Engineers.......................................352
How the Sine Function Makes the Jobs of Engineers Easier......377

## 14   Everything becomes Digital – Really Everything?........................389
What Zeros and Ones Have to Do with Digital Systems............389
Why Engineers Want to Digitize as much as Possible ...............398
Computer Hardware: How Digital Systems Which Execute Programs Are Built .................................................................411
Computer Software: How Programmers Can Tell Their Computers What They Expect Them to Do ................................421
An Engineering Job Which Is Not Yet Adequately Done ..........434

**Concluding Remarks**..................................................................437

**Acknowledgments**......................................................................439

**References**..................................................................................441

**Name Index**................................................................................443

**Subject Index**.............................................................................445

# Chapter 1
# Explaining Modern Technology

A couple of years ago, on a subway ride in Berlin, I overheard a conversation between two students, and I remember one of them saying, "Last semester I took Dr. Anderson's course on computer electronics. But I should have stayed in bed instead of getting up early each Tuesday morning in order to attend his eight o'clock lecture, since nobody learned anything from this course. This semester, I am attending the lectures of Dr. Heymann, and she is the best professor I ever had. Her explanations are so illustrative that even the most complex subjects have become absolutely clear to me." This made me think back to my own years in high school and college, and I remembered my math teacher as the one whom I still would grade an A plus, while at the same time, I remembered some other teachers who did a rather poor job. I am convinced that almost all of my readers have experienced extremely good teachers and others who weren't worth the salary they were paid.

During my over thirty years of teaching engineering courses, I certainly tried to be a good teacher. But it was not until I was approaching retirement that I explicitly considered the question about how to become a good teacher. Although I don't have the complete answer to this question, at least I am convinced that there is a specific condition that must be satisfied under all circumstances: whenever a teacher enters the classroom, he must consciously consider the fact that his way of viewing the world may differ substantially from the views of the students. This makes him aware that he must try to view the world through the eyes of his students.

As a professor of engineering sciences, I am, of course, mainly interested in the different ways a person may view the world with respect to its technological aspects. Thus, it happened that I not only considered the differences between my view and the views of my students, but I also asked myself what the differences were between my view and the views of my wife and my three year old grandson. This made me aware of the fact that, for a little child, all technical devices belong to the natural world. A child doesn't wonder about the buttons which can be pushed to turn the light on or off, or that there is a cell phone which can be used to talk to Grandma. The child doesn't ask where the cell phone comes from or why a car can move so much faster than a running horse. For the child, cell phones and cars are as natural as apples and horses. We might say that for a little child, either everything is a miracle or nothing is a miracle.

When my philosophical considerations had reached this point, it only required a short further step for me to come up with the question of how someone who had died hundreds or thousands of years before our time, would view our world. Obviously, my thinking had been influenced by a novel I had read some years ago, and a movie I had seen at about the same time. The German novel "Pachmayr" [SPO] and the French movie "Hibernatus" were created around 1970, and both works involved complications which occur when a person who died long ago suddenly comes to life in the present time, and still remembers everything about his former life. In the movie, the time gap was one hundred years, while in the novel it was almost five hundred years. Thus, it seemed quite natural to me to ask myself how a person from the middle ages or from classical antiquity would react if he would be confronted with a world where there are electric lights, pills for headaches, food freezers, microwave ovens, airplanes, television sets, automobiles, mobile phones and computers. Instead of considering this question with respect to an abstract person, I wanted to think of a real person who had lived in the past. Finally, I chose the Greek philosopher Socrates (469-399 BC) because he is the perfect example of someone who continued asking critical questions until he either reached the fundamental essence of the subject, or realized that the limits of knowledge had been reached for the person being asked. Thus, the following chapters are my answer to the question, "How could I help Socrates understand our present world?"

## What Socrates Would Ask Me

In part, Socrates would be in the same situation as a newborn child who must learn about the world. The difference between the situations for the newborn and Socrates is that Socrates would be able to recall the rich experiences of his former life. The only things Socrates must learn are how the present world differs from that of his former life.

An important guideline for my teaching efforts always has been the saying, "A picture is worth a thousand words." Therefore, as you will see, a major part of the text of this book is just a sequence of comments about diagrams. For example, consider Fig. 1.1 where rectangles represent six areas of knowledge which I placed in levels, one on top of the other like the floors of a building.

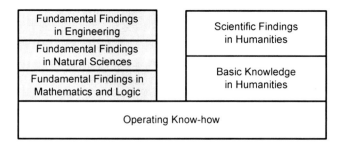

**Fig. 1.1** Layers of knowledge

The different types of knowledge shown must be acquired from bottom to top. This means that a newborn child must first acquire the knowledge which I called "Operating Know-how." This is the kind of knowledge needed to operate complex technical equipment. Actually, for our Socrates as for any newborn child, the unknown world is like a new machine which they must learn to use. This learning does not require any understanding since it consists only of remembering causal relationships. Thus a child not only learns what happens when certain buttons are pressed – the light goes on, the door bell rings or the garage door opens – but it will also soon know in which drawer grandma keeps the candy. Learning one's native language also means acquiring operating know-how, since this is nothing but remembering the causal relationships between certain acoustic patterns and types of human behavior. One characteristic of all kinds of operational know-how is that it completely loses its value when its owner is placed in a new environment. There was a time when I often had to travel for professional reasons, and in those days it frequently happened that I could not immediately drive away from the airport in a rental car because the buttons and switches for the wipers and the headlights were not where I expected them to be. Each time in these situations I had to acquire new operating know-how which was valuable only as long as I drove that particular make and model of car. Ask yourself how much of what you actually know is valuable only when you are in certain environments. Even knowledge about the location of the nearest toilet is of that type.

Despite the tremendous amount of operating know-how Socrates would need in order to survive in our present world, this is not the kind of knowledge he would ask me to provide. Nor would his questions be related to the two areas of knowledge which are on the right hand side in Fig. 1.1, since these subjects are just what Socrates had experienced in his former lifetime, namely politics, sociology, economics and arts. He was involved in wars in his time, and so he would not be surprised to learn that since then many more wars have been fought, people have had to leave their countries and move to other regions, and empires have grown and vanished. He had known sculptors, painters and poets and would consider it normal that in the meantime many new artists have created new works, and that aesthetic criteria have undergone evolutionary changes. Certainly it would be very interesting for him to learn that Athens is no longer the center of the universe, that the earth is a globe, that Columbus had discovered America, that Marco Polo had traveled to China and that today there are precise maps of the entire earth. But probably all of these changes would not cause him to say, "I don't understand it." The same can be said about the fact that letterpress printing was invented and that our children learn to read and write in elementary school. Although historians say that Socrates never produced any written text, Socrates certainly would appreciate the fact that today it is quite easy to obtain a fundamental education in humanities just by reading a few books. Today there are even books on the market with titles such as "Everything an Educated Person Should Know." But even after Socrates had read these books and had acquired all that knowledge, he would still be dissatisfied and see the need to ask for my help. He would come and say, "All this knowledge does not help me in the least to understand the strange phenomena I regularly encounter almost all the time, and wherever I go. You press a button and a huge stadium gets bright; then you press the same button and the stadium gets dark again. You press other buttons and the church bells in the tower start to ring or heavy doors swing open. Objects looking like extremely long houses on wheels with hundreds of people sitting inside move at high speeds on rails, although no apparent reason for the movement can be discerned. And, of course, the devices and systems which you call television and mobile telephones remain absolute mysteries to me. On the other hand, I have been with you long enough to be sure that mankind is still completely human and has not turned into a crowd of demigods. Therefore, I am looking for someone who is willing and able to clearly explain to me the fundamental findings which enabled humans to create such mysterious devices and systems."

In my present situation as a retired professor of engineering, nothing could be more gratifying to me than to be asked by Socrates to give him these explanations. Erwin Chargaff whom, in many respects, I consider a model of a very wise man, once wrote [CH 1]: "A real explanation requires two persons, one who explains and another one who understands." From my long experience as a university professor, I must add another requirement: "Understanding an explanation requires a person who wants to understand it." In the case of Socrates I can be sure that he wants to understand and that he would not refuse to make the necessary effort. The effort I must impose for the reader is not more than the careful study of the 440 pages of this book. I certainly would have preferred to reach my goal with fewer pages, and I really tried hard to do this. But I had to guide Mr. Socrates and my other readers from bottom to top over the three stacked grey areas of knowledge on the left hand side of Fig. 1.1. This stack tells us that the knowledge about fundamental engineering findings can be acquired only after one has gained a certain insight into the fundamental findings in natural sciences, and that these require knowledge about fundamental findings in mathematics and logic. Therefore, this book is structured as a sequence of three groups of chapters with each group corresponding exactly to one of the grey areas of knowledge in Fig. 1.1.

You will not find any information about the internal details of technical products in this book. The purpose of this entire book is to provide the knowledge which leads the reader to believe that these products can be conceived and built. No one would ever have invested a minute in the development of a technical product if he or she had not been convinced about its possible success. This conviction about possible success has always been based on the level of knowledge a person had reached. From a particular level of knowledge, they could look down and see all the findings or results which would help them solve all the problems which might turn up in the course of the development process. Someone who begins the development of a technical product has a vague idea of the overall solution that can be implemented if certain types of problems can be solved. Solving any of these problems does not require any findings outside of the plateaus of knowledge which have already been reached and can be seen from the place where the developer is standing. In most cases, solving all these problems still requires hard work. Sometimes good luck is needed to come up with a reasonable solution. Never does any magic come into play. Thus, it is my goal to lead my readers to that small set of plateaus of knowledge, the plateaus on which the technicians walk when they are looking for solutions of problems with which they are confronted when developing new products. While walking around on these plateaus is part of the job of professionals, I shall not take my readers on such excursions. Instead, I shall help them climb the rock walls which lead to these plateaus. In the course of history, each of these rock walls has been conquered for the first time. But since then, the walls have been climbed often, and hooks which make it easier to go up have been hammered into the rock. There are no cable cars leading up to the plateaus of knowledge, but with the help of an experienced mountain guide, even inexperienced mountaineers can reach the plateaus with reasonable effort.

## Omitting Irrelevant Subjects Is an Art

When the physicist Richard Feynman (1918-1988) was given the Nobel Prize, a journalist asked him to summarize in three sentences what he got the prize for. This request certainly was for an explanation far below an acceptable limit of brevity, and Feynman answered, "If I could explain this in three sentences, it certainly would not be worth the Nobel Prize. But I could explain the essence of my work to your educated readers on two pages."

Although I did not win the Nobel Prize, I am somehow in a similar situation since I am expected to present my subject in a minimum number of pages. I have omitted everything that could be omitted without reducing the comprehensibility and the intellectual depth of the presentation. Since this book is restricted to findings which are fundamental for our present technical systems, I could omit all findings from mathematics and natural sciences which have not yet been applied to any technical products. Therefore this book does not contain the theories of cosmology which explain the origin of the world. If you are interested in the so-called "big bang" theory, you must look in other books. By the way, did you know that Albert Einstein commented on the idea of the big bang with the words, "This theory is as absurd as having an encyclopedia originate from the explosion of a printing shop?" Among the subjects I omitted in this book is the theory of strings, which is an attempt to combine cosmology with quantum theory, and Darwin's theory of evolution which is an explanation of the origin of living species.

But there are subjects which I omitted even though they are technologically relevant – for instance acoustics and geometric optics. My decisions not to discuss them are based on my judgment that, in these cases, the corresponding knowledge plateaus can be reached without a mountain guide. I also omitted most of the historical details of the processes which finally resulted in the knowledge we have today. The way in which I will guide you up the rock walls will always lead directly to the plateau we want to reach. Certainly it would have been nice to show you all the dead end roads which have been tried in the past. For example, consider the definition of electricity which a student of the German philosopher Georg Wilhelm Friedrich Hegel (1770-1831) once wrote [HEG]:

> "Electricity is the pure purpose of the shape which liberates itself from it, which begins to cancel its indifference, since electricity is the immediate emergence, or the existence which does not yet emerge from or be conditional upon the shape, or is not yet the dissolution of the shape itself, but is the superficial process wherein the differences leave their shape but still are dependent on it and are not yet independent with it."

Today, we can only laugh about this, and it remains absolutely mysterious how someone could ever dream up such nonsense. Perhaps Professor Hegel said something reasonable, but a student taking notes confused all of it. Certainly we can take this definition as a hint about how difficult it was in those days to gain a fundamental insight about the basis for today's electrical engineering.

When I discussed my intention to write this book with friends and relatives, some of them spontaneously expressed their doubts about whether I could succeed at all during a time when "the total knowledge of mankind doubles every few years." Why am I not afraid of this doubling? It is simply because it is not a doubling of our knowledge, but only of the available information, which can be acquired by searching and reading. Admittedly, people very often use the term "knowledge of mankind." But this is not a correct concept, because knowledge is only what one person knows. In order to be able to get access to and use the enormous wealth of information stored in libraries and on the internet, a person must possess some knowledge, because someone who knows nothing cannot ask questions. We cannot expand the range of our understanding unless we have already understood some fundamentals. Do you know why, for example, the sky is blue on a sunny day? Actually, I don't know this myself, but I do know exactly where I could look or whom I could ask to get a comprehensive answer if the need arose.

I recently read the following statement of the Austrian philosopher Konrad Paul Liessmann in the German journal "Research and Teaching" [LI]:

"What a person can or should know today is no longer determined by some standard theories of education, but mainly by the market which constantly changes. This knowledge can be produced and acquired rapidly, but just as rapidly be forgotten."

Nothing of what he said in this statement applies to the fundamental findings I present to you in this book. These findings and their presentation could not be produced rapidly; they required a laborious process of hard work. They cannot be acquired rapidly, since in spite of all the support of an experienced mountain guide, a lot of effort is still needed to climb up the steep walls of understanding. However, those who invest this effort and reach the plateaus of knowledge shall never forget the essential findings, and shall be glad to be able to refer to them for the rest of their lives. That which is discussed in this book will not be affected by a constantly changing market.

## No One Should Be Afraid of Formulas

In 1905, Albert Einstein climbed Mount Sinai. When he arrived at the top, he was shrouded by a cloud and heard a voice saying to him, "Albert, take this slab of stone, carry it down the mountain, and then read what I wrote on it and explain it to your people." When Albert Einstein came down from the mountain, he read "$E=mc^2$," and then he explained this formula to the people. Did it happen this way? We all know that this never happened. Why then does almost every author of a science book for non-professionals introduce this formula as if it had come from heaven? The reason for this can be found in the preface to the book "A Brief History of Time" by the English physicist Stephen Hawking [HA 1]. There he wrote:

"I was told that each formula in the book would halve the number of its readers. Therefore I decided to refrain from including any formulas. But finally, I made one exception, $E=mc^2$."

Publishers of books on science for non-professionals apparently believe that such books do not sell well to a broad public if they contain mathematical formulas. Why do publishers have this belief? Unfortunately, mathematics has acquired the image that it can be understood only by an extremely small minority of people who are gifted in a very exotic way, and that therefore it is quite natural for mathematics to remain closed for the rest of mankind. The German philosopher Arthur Schopenhauer (1788-1860) contributed to establishing this image of mathematics by writing [SCH]: "The talent for mathematics is a very specific and unique one which is not at all parallel to the other faculties of a human head, and indeed has nothing in common with them." And in the book "Lies in Educational Politics," the German author Werner Fuld [FU] wrote: "Does not Schopenhauer's statement call to mind our years in school, and especially those schoolmates from whom we could gratefully copy each math paper, but whom, on the other hand, we considered rather dumb if not even mentally deficient?" Obviously, the author Fuld must have had some problems with his lessons in math, and he sought to get applause from those of his readers with similar experiences. By his statements, he reinforces a prejudice which can be easily disproved. Of course, I myself do know some extremely unworldly mathematicians whom Mr. Fuld presumably would call "mentally deficient." I would be more specific about such mathematicians and say, "They cannot communicate adequately." But it is a severe logical sin to generalize from a small number of pathological cases. Think of the many engineers who move with great competence through many fields of mathematics; Mr. Fuld would have great difficulty in finding such pathological cases among them.

How about the belief that a rare and exotic talent was required for understanding mathematics and having fun dealing with it? Here the situation is the same as in all kinds of arts: a rare and special talent is required for creating a valuable work of art, but almost everybody is gifted enough to understand the work and enjoy it. Only a very few people can be good composers or writers, but many can enjoy listening to the music and reading the novels. If you think mathematics is extremely difficult and requires a rare talent which you don't have, this might be true concerning the creation of mathematical works. But if you don't understand a mathematical work presented to you, this might not be a consequence of not having a particular talent. Very often it will be the consequence of the fact that the "performing artist" was not gifted enough to communicate the work to others. The Spanish philosopher Ortega y Gasset (1883-1955) once wrote [OYG 1]:

> "Mathematicians exaggerate the difficulty of their subject a bit. If it appears so incomprehensible today, it is because the necessary energy has not been applied to simplifying its teaching."

Although I think that Ortega's statement is correct, I also think that there is something special about dealing with mathematics. It is not primarily a question of a good memory, but it requires the motivation to struggle for understanding. Throughout this entire book, I shall refer to the rock walls and plateaus of knowledge, where acquiring the understanding of mathematical findings corresponds to conquering a rock wall leading to a plateau. In contrast to this, acquiring knowledge about history, literature or art corresponds to walking around in a plane or on a plateau where the hikers need only a good memory to help them remember all the places they have been. But these hikers will never experience the moment of joy which causes the mountaineers to shout "Hooray!" when they reach the top of the mountain.

Attempts to teach findings in natural science and engineering without using mathematics are similar to attempts to teach the evolution of styles in painting to a blind person. Formulas are just statements in a special type of language which help to state facts about abstract structures as briefly and precisely as possible. Any attempt to state these facts by exclusively using natural language would result in bulky and confusing phrases which nobody would understand except for their authors. Formulas are needed especially for defining concepts about higher levels of abstractions which otherwise could not be defined at all. For instance, there would not be a relativity theory, or any type of electrical engineering, because they depend heavily on abstract concepts like imaginary numbers and four-dimensional spaces which can be defined only by using formulas. These concepts are completely formal and even Einstein and all the other geniuses in physics and mathematics depend on formulas to deal with these concepts.

In the course of decades of academic teaching, I became strongly convinced that I finally know why so many people are scared of formulas: they believe that mathematicians, in contrast to normal people, have a sixth sense for formulas. They don't believe that what the teacher tells them about a certain formula is all that could be said about it. They believe that there is something "behind the formula" to which they don't have any access. Throughout the entire book, I keep emphasizing that nobody needs a sixth sense to understand a formula, because there is never anything hidden behind it.

In about 1925, the Spanish philosopher Ortega y Gasset, whom I mentioned above, wrote an essay entitled "Mission of the University". In this essay, I found the following paragraph which hopefully motivates you to go through all the remaining chapters of this book [OYG 2]:

> "Consider the gentleman who professes to be a doctor, a magistrate, a general, a philologist, or a bishop, i.e., a person who belongs to the leadership class of society. If he is ignorant about what the physical universe is today, he is a perfect barbarian, no matter how well he may know his laws, or his medicine or his Holy Fathers. It is certain that all the other things he does in life, including parts of his own profession which transcend its proper academic boundaries, will turn out unfortunately. His political ideas and actions will be inept. He will bring an atmosphere of unreality and cramped narrowness to his family life, and this will warp the upbringing of his children. With his friends, he will emit thoughts that are monstrosities and opinions that are a torrent of drivel and bluff."

If you now follow me through the coming chapters and finally climb all the rock walls to the plateaus of knowledge, you certainly will not belong to the kind of people Ortega was talking about.

Now it's time to tackle the first wall. Let's go!

# Part I
# Fundamentals of Mathematics and Logic

# Chapter 2
# Mathematicians Are Humans Like You and Me – They Count and Arrange

"Math – no, thanks!" said my daughter and the majority of all people I've met during the course of my life. How about you – do you also belong to those who haven't liked mathematics since they encountered it in high school? Let's assume we are not talking about mathematics but about a lamb roast, and you say, "I have never liked a lamb roast." Do you think a gourmet cook would accept this just as it was said? No, he would suspect that your aversion to a lamb roast was the consequence of the fact that you never have been served an optimally prepared lamb roast. Now let me persuade you to taste the meal of mathematics which I have prepared especially for you.

In this chapter, I shall present a view of mathematics "from high above" as if mathematics were a continent over which we could fly in a satellite. Think of Europe which actually can be seen from a satellite. It is evident that from such a far distance we cannot see any details, but only the rough topography of the continent. Sometimes we want to look a little bit closer which in our analogy means that we then use binoculars.

The textbooks for the mathematics courses from my years in high school have long disappeared, but I still remember that on their covers were printed certain strange words which at the time nobody explained to me, words like Arithmetic, Algebra and Analysis. Only the word Geometry referred to the world I knew. In our analogy, these words on the covers of math books correspond to the names of the countries printed on the maps of Europe. When we study a map of Europe, we are interested in the structure of the continent, that is, how the outlines are shaped, where the borders between the countries are located, which way the rivers flow and where the mountains and the big cities are located. In this situation, we do not ask what Europe is good for and how it could be applied. In the same way, we first have to get to know the structure of the "math continent." We should not yet ask what it could be used for. Of course, we already know that we had no reason to study this continent except that it had become fundamental for physics and technology.

## What a Number "Sees" When It Looks into a Mirror

The mathematician Leopold Kronecker (1823-1891) said, "The natural numbers have been created by the dear God and everything else is the work of humans." In the following paragraphs, we shall reconstruct this human act of creation. For each day in the creation of the world, the bible includes the closing statement: "And God saw that it was good." I hope that we can write at the end of our creation story: "And we saw how wonderfully and logically one result implied the next."

The *natural numbers* are so natural that even small children quite intuitively grasp the corresponding concepts. The grandchild goes up the stairs holding grandma's hand and counts: "One, two, three, ...." The natural numbers serve to count, and that means that a position in a sequence is assigned to each element of a set, and that the size of the set is found at the same time. Of course, the steps of a staircase have their positions in a sequence before we count them. By counting, we only assign names to these positions, for after there is a sixth step there is a seventh, etc. The name which we assign to the last position is the name of the number of steps in the staircase. When we count, we pass through all positions of a sequence, and therefore the most natural way of counting consists of sequentially saying the names of the numbers. We can remember the sequence of these names as if they were words in a rhythmic song: "One, two, three, four, five, six, seven – all the angels look down from heaven." It is true, though, that we also count things which don't have a natural sequential order: think of apples in a basket. In such a case, we produce an artificial order which is destroyed when we are done counting. For example, we pour the apples from the basket onto the table and then put them, one after the other, back into the basket "singing" our numbers song.

Although the natural numbers and their application in counting are the nucleus of mathematics, the true concept of mathematics was not born until addition, subtraction, multiplication and division were performed as abstract operations. They are called the four basic arithmetic operations where the word "arithmetic" refers to the Greek word "arithmos" for number. People always used to illustrate these operations by thinking of containers which hold countable objects, for example apples in baskets. *Addition* means pouring the contents from one container into a second container on top of its contents. The inverse of this operation is called *subtraction*, and this means that a part or all of the contents of one container is removed and put into another container which originally was empty. Of course, we cannot take out more apples than had been in the basket originally. Only magicians in a circus can demonstrate that they can take 8 balls out of a bag which originally seemed to contain no more than 5 balls. Subtraction is the first operation which motivates us to create a new number, "*Zero*," which is the number of apples remaining in a basket after we have removed all of the apples which were there originally. Zero can occur only as a result of computing and not when we count. Therefore, it does not belong to the natural numbers, but to those numbers which Kronecker said were not created "by the dear God."

In addition and subtraction, we deal only with objects of the same type, for example, apples or steps. In *multiplication* and *division*, however, objects of two different types are counted, for example baskets and apples. Multiplication is illustrated by the idea that we originally have 'b', a certain number of baskets, each containing an equal number 'a' of apples. As an example, think of 5 baskets each containing 8 apples. By pouring the contents of all of these baskets into one basket, we obtain the so-called product, namely 40 apples. Another way to illustrate a product is with a rectangular floor which is covered by many smaller rectangularly shaped tiles, and where we wish to find the total number of tiles. Two numbers must be known, namely how many tiles fit into one row and how many tiles fit into one column. From these two numbers, we can get the total number of tiles by multiplication.

Division is based on the idea that the apples in one basket must be equally distributed into a given number of originally empty baskets. If we are not allowed to cut apples into parts it will quite often occur that an equal distribution is impossible. This can be done in the case of 3 baskets if 15 apples are to be distributed, but not if we start with 17 apples.

While addition and multiplication can be performed on any pair of natural numbers, this is not true for subtraction and division. This is a consequence of the restriction that the result of the operation must be a natural number or zero. Now we lift this restriction and allow the objects to be distributed to be cut into parts if necessary. At this point we no longer think of apples but of sticks of butter, since these can be cut easily and precisely. When such a stick is cut into three parts of equal length, it is said that each part has one third the length of the original stick. "One third" is formally written 1/3. Such a form consisting of two numbers separated by a slash is called a *fraction*. The number in front of the slash is called the numerator; it stands for the number of objects to be distributed. The number behind the slash is called the denominator; it stands for the number of containers or receivers among which the objects are to be distributed. In our example, we wanted to distribute 17 sticks equally among 3 receivers. First, we perform the easy part of the task and distribute 15 sticks with each receiver getting 5 of these. Then we cut each of the two remaining sticks into three parts of equal length and distribute the resulting six thirds among the three baskets with each basket receiving two of the thirds. The operation of distributing 17 sticks among 3 receivers can be written formally as 17/3 = 5 + 2/3.

The above statement which says that the division of 17 by 3 cannot be performed is now revised by introducing a new type of number. If it is possible to cut objects into parts of equal size, it is certainly possible to distribute any natural number of such objects among any natural number of receivers. By definition, the number of objects one receiver gets is said to be a number though it may not be natural. Thus, 1/3 and 17/3 are numbers of a new type, called "fractions," created by us. The set of numbers which we discussed up to now is the union of the natural numbers, the zero and those fractions which do not result in natural numbers.

We eliminated the impossibility of division for certain pairs of numbers by allowing objects to be cut into parts of equal size. Now we eliminate the impossibility of subtraction for certain pairs of numbers by a formal trick. At this point, we cannot see a way to subtract 8 from 5 since this would mean that we can take 8 objects out of a container which originally contained only 5 objects. As long as we are restricted to operations in the real world, we have no way of overcoming this impossibility. But in mathematics it has sometimes been very common to leave the real world and enter the so-called *formal* world. In this formal world, objects and operations are defined without any reference to reality. The formal *objects* are "created" just by giving them names, and the formal *operations* are defined just by writing down which formal objects should be the result of the formal operation on other formal objects. The first formal objects which I shall now introduce are the so-called *negative numbers*. Imagine that we could place the numbers which we already have in front of a mirror. All these numbers are ordered according to their "weight," i.e., for each pair of different numbers we can always say which of the two numbers is greater. Therefore, we can imagine a straight line onto which we place the numbers ordered according to their weight. This line begins with the position of the zero, and from there on the natural numbers follow at equal distances. The fractions will be placed between the positions of these numbers. Now we imagine we could put this straight line in front of a mirror in such a way that the zero position touches the mirror, and that the line stands at a right angle (90 degrees) to the mirror surface. Now we see this straight line twice, both in front of and behind the mirror surface. Thus, each number in front of the mirror has its partner in the mirror. Since the zero sits on the mirror surface, it is the only number which does not have a separate partner. Obviously, we must now introduce a way to make clear whether we are talking about a number in front of the mirror or about its partner behind the mirror. For example, we could identify the partner of five by calling it "mirrored five." Or, at least in the case of writing, we could use black ink for the numbers in front of the mirror and red ink for their partners in the mirror. This use of colors is actually used when people denote the financial condition of a company by writing monetary numbers in black or red. This indicates that a company is making either a profit or a loss. Mathematicians use the *minus sign* to identify the numbers in the mirror (-1, -2, -3, etc.). In this way they indicate that such a number can be obtained only as the result of a subtraction where more objects are taken out of a container than have previously been in it. Such numbers are called *negative numbers*.

I have very often heard the statement, "Negative numbers don't mean anything to me since I cannot relate them to any picture from the real world." People say this with the same kind of regret as if they were talking about a personal weakness: "I cannot make attractive drawings." They believe that mathematicians relate much more meaning to the concept of negative numbers than a "normal person." But this belief is not justified. The description above prescribed what picture you should use to relate to the concept of negative numbers, and this is exactly the picture all mathematicians have. A negative number is a mirrored positive number – and nothing more. Once you can accept this idea of a number sitting in front of a mirror and seeing its reflection, you are well-prepared to accept another type of reflection which will be introduced later.

The reasons for "creating" fractions and negative numbers came up when we found that a particular arithmetic operation could not be performed on all pairs of known (natural) numbers unless we allowed the results to be numbers which were of a type different from those we already knew. Consequently, we must now ask whether arithmetic operations with numbers of the new types (fractions and negative numbers) again force us to create new types of numbers. There is no space or reason here for presenting all the cases which have to be considered in order to answer this question. The answer is that no new types need to be introduced. Figure 2.1 represents the results obtained in our process of creating numbers. Besides the natural numbers originally given, there are four new types of numbers, namely the zero, the positive fractions, the negative whole numbers and the negative fractions. All these numbers together are called *rational numbers* which comes from the Latin word 'ratio'. In the figure, the four arithmetic operations are symbolized by rectangles, each of which has two arrows entering and one arrow leaving. The two arriving arrows symbolize the two numbers which are the inputs of the operation, and the leaving arrow symbolizes the result. In each case, the result is a rational number. Division is the only operation which does not allow the input to be any pair of rational numbers; there is the restriction that zero not be allowed to be the denominator. It does not make any sense to request that a certain number of objects be distributed without specifying the number of receivers for the distribution.

Although the results shown in Fig. 2.1 do not imply any need for finding new types of numbers, the rational numbers actually do not yet represent the end of the process of creating numbers. The old Greek mathematicians found that there are certain relations between the lengths of lines in geometric shapes which are not based on rational numbers. The two best-known examples are the relation between the diagonal and the edge of a square, and the relation between the circumference

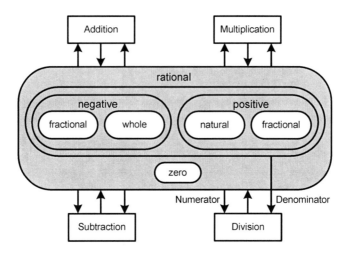

**Fig. 2.1** The rational numbers and the four arithmetic operations

and the diameter of a circle. It is impossible to express these relations exactly as fractions, although they can be approximated arbitrarily closely by fractions. Numbers of this type are called *irrational* numbers. Rather good approximations are 707/500 for the relation of the square and 3927/1250 for the relation of the circle. All over the world, the relation between the circumference and the diameter of a circle is symbolized by the Greek letter $\pi$ ("pi"). The set of numbers which we get by gathering the rational numbers and the irrational numbers together is called the set of *real numbers*.

There is a fundamental difference between the square's diagonal relation and the circle's number $\pi$. In the case of the diagonal relation d, a so-called arithmetic equation exists which exactly defines this number. Although the concept of equations will be presented in a later section of this book, you will easily understand the following equation: d∗d=2. This equation says that the multiplication of d with itself has the result 2. In the case of the number $\pi$ an arithmetic equation defining the number does not exist. The types of numbers which cannot be defined by an arithmetic equation are called *transcendental* numbers. The prefix "trans" indicates that something beyond certain limits is considered. The transcendental numbers lie beyond any considerations which start with natural numbers and lead to a definition of the number we are interested in. In later sections we shall encounter this type of number again.

Before describing the last part of the process of creating numbers, we stay for a while on the playground of the real numbers. In this sentence, I used the word "playground" deliberately in order to make it clear to you that many mathematical findings have been found by playing around and not by searching for solutions to serious problems. Naturally, children beginning to play do not question what their play might be good for, and this attitude is also often very true for mathematicians. We could even think that the following biblical quotation was meant for mathematicians: "Unless you become like little children, you shall not enter into the kingdom of heaven." (Matthew 18, 3). Therefore, let's begin to play.

| Name of the operation | Reduction to known operations | Formal representation |
|---|---|---|
| Multiplication | Addition of multiple equal summands: Product = 2 + 2 + 2 + 2 + 2 | Product = 10 = 5 ∗ 2 |
| Division (Inversion of Multiplication) | Search for the Factor Q in the Multiplication: 10 = 5 ∗ Q | Quotient = 2 = 10:5 = $\frac{10}{5}$ |
| Power | Multiplication of multiple equal factors: Power = 2 ∗ 2 ∗ 2 ∗ 2 ∗ 2 | Power = 32 = $2^5$ |
| Root (First Inversion of Power) | Search for the Base R in the Power: 32 = $R^5$ | Fifth root of 32 = 2 = $\sqrt[5]{32}$ |
| Logarithm (Second Inversion of Power) | Search for the Exponent L in the Power: 32 = $2^L$ | Logarithm of 32 to the Base 2 = 5 = $\log_2 32$ |

**Fig. 2.2** Powers, roots and logarithms

The table in Fig. 2.2 summarizes the results of our playing around with numbers and operations, and leads us to new concepts and operations. We see that the division operation is the inverse of the multiplication operation, and that multiplication gives us the idea to introduce the concept of *power*. By formally replacing the plus sign in the definition of the multiplication by the multiplication symbol, we obtain the definition of the power relation. In the example, we write the power operation as $2^5$ where the 2 is called the *base* and the 5 is called the *exponent*. These words refer to facts from our every day lives: the position of the basement is underneath the rest of the house, and the position of something which is exposed is above in order to make sure that it can be seen by everybody. Multiplication has only one type of inversion because the order of the two factors can be reversed without having an effect on the result. In contrast to this, the power relation has two types of inversions because reversing base and exponent may change the result: $2^5=32$ and $5^2=25$. The first type of inversion of the power relation provides the concept of *roots*. The symbol used in formulas for the operation "root" is an abstraction from the letter r which makes us think of the word root. The second type of inversion of the power relation provides the concept of the "*logarithm*". The name of this operation comes from the two Greek words, "logos" and "arithmos".

The new operations introduced in Fig. 2.2, namely the power, the root and the logarithm help us to dream up the next steps in our play activities. In Fig. 2.2 we restricted ourselves to using only natural numbers. Now we ask whether it makes sense to ask for the results of these operations if their operands are no longer natural numbers, but negative numbers or fractions. If we define a power $b^n$ as a chain of n numbers which all have the value b and are connected by multiplication symbols, the expressions $b^0$, $b^{-3}$ or $b^{4.5}$ cannot be interpreted as meaningful powers. But Fig. 2.3 shows that it really makes sense to define powers with exponents which are negative numbers or fractions.

$$\frac{2^7}{2^4} = \frac{\cancel{2}\cdot\cancel{2}\cdot\cancel{2}\cdot\cancel{2}\cdot 2\cdot 2\cdot 2}{\cancel{2}\cdot\cancel{2}\cdot\cancel{2}\cdot\cancel{2}} = 2^3$$

Conclusions: $\quad \dfrac{b^m}{b^n} = b^{m-n} \qquad b^0 = 1 \qquad b^{-m} = \dfrac{1}{b^m}$

$(3\cdot 3)\cdot(3\cdot 3)\cdot(3\cdot 3)\cdot(3\cdot 3)\cdot(3\cdot 3) = (3^2)^5 = 3^{10}$
$(3\cdot 3\cdot 3\cdot 3\cdot 3)\cdot(3\cdot 3\cdot 3\cdot 3\cdot 3) = (3^5)^2 = 3^{10}$

Conclusions: $\quad (b^m)^n = (b^n)^m = b^{m\cdot n} \qquad \left(b^{1/m}\right)^m = b^1 = b$

$$b^{1/m} = \sqrt[m]{b}$$

**Fig. 2.3** Powers whose exponents are not natural numbers

From Fig. 2.3 follows $b^0=1$, $b^{-3}=1/b^3$ and $b^{4.5}=b^{9/2}=(b^{1/2})^9=(\sqrt{b})^9$. The values of these powers are real numbers, i.e., they are elements of the set of numbers which we have already found. But we are forced to create a new type of number when we ask for the square root of a negative number, i.e., when we ask for a number x which satisfies the equation $x^2 = -1$. This number x cannot be a real number since the square of a real number is always a positive number, even if the original number is negative: $(-1)*(-1) = +1$. In a textbook written in 1768 by the brilliant mathematician Leonhard Euler (1707-1783), I found the amusing statement [EU]: "When we have to compute the square root of a negative number, we certainly are in an embarrassing situation." We now remember the situation when we used a mirror to create the negative numbers. A method which was successful once could be successful again. Therefore, we will take all of the real numbers and place them in front of a mirror, but we cannot proceed in the same way as we did when we created the negative numbers. Otherwise, we would not be able to distinguish between the negative numbers and the new type of numbers. When creating the negative numbers, the straight line with the numbers was placed at a right angle to the surface of the mirror. Now the line with the numbers to be mirrored must be placed at a different angle. A reasonable choice turns out to be 45 degrees.

As in the previous case, the zero is the number which touches the surface of the mirror. In this way, we get a partner in the mirror for every real number in front of it except for the 0. Since the numbers in front of the mirror are called real numbers, the numbers in the mirror could have been called unreal numbers, but Leonhard Euler suggested the name *"imaginary numbers"* and introduced the letter i to identify the number whose square is -1. That which I pointed out when we created the negative numbers by using a mirror can be repeated here: there is nothing strange or miraculous about imaginary numbers, they are just real numbers in a mirror, and the only property we assign to them is that their square is a negative real number: i∗i=-1, 2i∗2i=-4, 3i∗3i=-9, etc. You should not think that mathematicians know more about imaginary numbers than what I just told you.

However, the process of creating numbers is not finished with the creation of the imaginary numbers, for now, we must check what types of numbers we get as results from arithmetic operations where at least one of the operands is imaginary. The different possible combinations are shown in the table in Fig. 2.4. The thickly-framed field stands out as the one with a result which is neither real nor imaginary. When the pair of operands for an addition or a subtraction contains one real and one imaginary number, the result is a number of a type which does not belong to the types we already know. This means that again we are forced to create a new type of number. These new numbers are called *complex numbers*.

|  | Addition or Subtraction | Multiplication or Division |
|---|---|---|
| Both operands are imaginary. | Imaginary result:<br>5i + 2i = 7i<br>5i - 7i = -2i | Real result:<br>5i ∗ 2i = -10<br>10i : 5i = 2 |
| One operand is real, the other one is imaginary. | Complex result:<br>5 + 2i = 5 + 2i<br>5 - 7i = 5 - 7i | Imaginary result:<br>5i ∗2 = 10i<br>10i : 5 = 2i<br>10 : 5i = -2i |

**Fig 2.4** Arithmetic operations with imaginary numbers

A complex number is always the sum of its real part and its imaginary part. These two parts stay separate and are not merged in the result in contrast to a normal addition where the two operands can no longer be seen in the result. In normal addition, the two operands +7 and (-4) are merged and don't show up in the result, +3. In contrast to this, in a complex number the real part +7 and the imaginary part 3i keep their identity and stay visible. The real part and the imaginary part are like a husband and wife who do not disappear by being married.

Although it was Leonhard Euler who introduced the symbol i for the square root of -1, he was not the first one who had the idea of "creating" numbers which are square roots of negative numbers. This idea was born about 200 years before Euler began to use such numbers. In 1545, the Italian mathematician Gerolamo Cardano (1501-1576) published a book that introduced the use of complex numbers for solving certain mathematical problems.

Creating one type of number after the other might become boring if you are not a fanatic mathematician. Therefore, you probably will be pleased to hear that now we really have reached the end of this number creation process. The result of an arithmetic operation will always be of one of the number types we created so far if the operands are of these types, too. Addition and subtraction are the simplest cases since here the operation can be performed on the real parts and the imaginary parts independently: (4+3i)+(12+5i)=(16+8i). In the case of multiplication we have to do what is shown in Fig. 2.5.

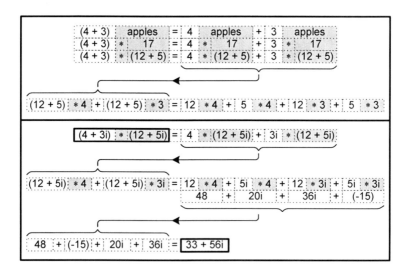

**Fig 2.5** Example showing the multiplication of complex numbers

In the upper section of this figure, the method of multiplying two sums is developed. In each of the first three rows, you see the same pattern of white and grey fields. The first row shows that this pattern is not restricted to multiplication. It illustrates the possibility of expanding the expression "four plus three apples" to the longer expression "four apples plus three apples". The next two rows show that the apples can be replaced by any factor following the multiplication symbol. In the third row, this factor is a sum, and thus we have a product of two sums on the left side of the equation. By reordering the right side of this equation, we obtain the expression on the left side of the fourth row. Here, the method of expanding a white-grey-pattern can be applied twice, and this finally results in four products, the factors of which are no longer sums, but numbers.

In the lower section of Fig. 2.5, the method from the upper section is applied to a product of two sums, each of which has one real and one imaginary summand. You see that the product of these two complex numbers is again a complex number, 33+56i.

Now we have reached the point where I can show you that still another number creator, often unnoticed, was sometimes involved behind the scenes of our number-creation activities. First we look at Fig. 2.6. On the left hand side, all the types of numbers which we know by now are represented in a container view of the kind already used in Fig. 2.1. In this view, you should think of containers into which you are looking from above. There is one big container which contains smaller containers. Certainly, it would be more natural for you to consider if these were not containers for numbers but for flour, sugar and rice. In our case, the containers are for the zero, the positive real numbers, etc. On the right of this container view, the same numbers are represented as points in a plane. The points on the horizontal line represent the real numbers with the negative numbers on the left, the zero in the middle, and the positive numbers on the right. The points on the vertical line below and above the zero represent the imaginary numbers. This

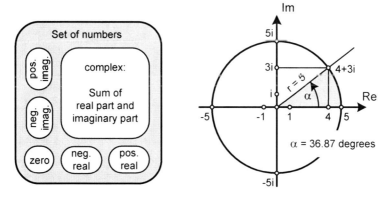

**Fig 2.6**   Different types of numbers and their representations as points in a plane

vertical line is a copy of the horizontal line turned by 90 degrees. The points in the plane which are neither on the horizontal line nor on the vertical line represent the complex numbers. As an example, I selected and marked the number 4+3i. Instead of identifying this number by providing its real part and its imaginary part, I could have used an alternative identification by providing the radius 5 and the angle 36.87 degrees. The horizontal line of real numbers, the line of the radius and the circle define the grey area which has the shape of a piece of pie. The particular form of identification to choose depends on the actual situation. Please note that a complex number is not a point in a plane, but can be represented as such a point.

By applying this representation of complex numbers to the numbers in Fig. 2.5, we get Fig. 2.7. Even if I had not explicitly expressed the relations in the thickly framed fields, you presumably would have soon noticed that the radius of the product is equal to the product of the radiuses of the factors, and that the angle of the product is equal to the sum of the angles of the factors. Isn't this amazing? When complex numbers were created as sums by adding the real parts and the imaginary parts of the numbers, radiuses and angles were not considered at all. Obviously, we were able to create something containing hidden laws of great importance, although we were absolutely unaware of these laws during the creation process. That's what I meant when I said that another number creator was involved behind the scenes in our number creation activities.

|  | Re + Im | Radius | Angle |
|---|---|---|---|
| 1. Factor | 4 + 3i | 5 | 36.87 degrees |
| 2. Factor | 12 + 5i | 13 | 22.62 degrees |
| Product | 33 + 56i | 65 = 5*13 | 59.49 degrees = (36.87+22.62) degrees |

**Fig 2.7**  Radiuses and angles for the multiplication in Fig. 2.5

This new insight now makes it easy for us to define the division of two complex numbers as the inverse of their multiplication. Since in the multiplication operation, two radiuses are multiplied and two angles are added, the division operation can be performed by dividing radiuses and subtracting angles. This new insight also indicates how powers and roots of complex numbers can be computed: we have to compute powers or roots of radiuses and we have to multiply or divide angles. But certainly, this is only well-defined for real exponents. It is easy to write $2^i$, but it is not at all easy to find out whether this power expression defines a number or whether it does not make any sense at all. The richness of our language makes it possible not only to write down reasonable expressions or statements, but also to construct grammatically-correct sequences of words which are without any reasonable meaning. As an example, consider the expression "the natural number between 17 and 18". This can be written or said, but it has no meaning since it contradicts the definition of the natural numbers. Thus, the expression $2^i$ could be nonsense or it could define a number. What I have presented to you so far is totally insufficient for determining which is true. You'll have to wait until the end of Chapter 3. Then all the results we need for finding the answer will have been presented. However, it might be interesting for you to see the answer even if you cannot understand how it is obtained: $2^i$ equals the complex number 0.769+0.639i.

I think that now, at the end of the story of the creation of numbers, you will agree that we can say, "We saw how wonderfully and logically one result implied the next."

## Sets Are Everywhere

I still very well remember the time when the so-called set theory was introduced as a subject in elementary school teaching. All of a sudden, parents realized that they no longer could help their children with their homework. Adult education centers reacted by offering evening courses for parents to make them familiar with the modern way to teach mathematics, called the "new math." One of my classmates from high school became an elementary school teacher, and every now and then we still meet. At one of these meetings, I mentioned set theory and asked her to give me her view of it. I still remembered that mathematics had not been her favorite subject in high school. But I also knew that being an expert in mathematics is not a requirement for understanding the basic concepts of set theory if they are adequately explained. Obviously, she had not been lucky enough to be taught by a good teacher at her college of education. From what she enthusiastically told me, I easily realized that she had not really understood the sense and the purpose of set theory. In the following section, I will tell you what her college teacher should have told my classmate.

The mathematical concept of a *set* corresponds to the idea of a container which contains nothing, little or much, where the set is what's in the container. In the previous section, I told you the story about the creation of the numbers, and there, in a few cases, I used the term "set" in its mathematical sense. When we look again at figures 2.1 and 2.6, we can imagine looking from above into containers which contain numbers. For example, there is a container for the zero, another one for the positive rational numbers, etc. Of course, the containers for numbers are not real containers like baskets or boxes. Only specific objects can lie in specific containers, like apples in baskets. If the objects are abstract, like numbers, their containers must be abstract, too. Most sets we are interested in are sets of objects of the same type – like sets of apples or sets of numbers. It doesn't make much sense to imagine a container which contains two apples, the fraction 3/5 and the moon. Our interest in sets is restricted to cases where the elements in a set share some common characteristics which are worth talking about. For instance, we talked a lot about the characteristics of complex numbers.

The objects which belong to a set are said to be the *elements* of the set. When we consider a well-defined set and some arbitrary object, it makes sense to ask whether or not the object is an element of the set. For example, we may ask whether the number $\pi$ is an element of the set of rational numbers, and the answer is no. The symbol $\in$ is used in formulas to state the fact that a given object, x, is an element of a given set, S. The formula for this is written $x \in S$. The shape of the symbol $\in$ resembles the capital letter E to which we can associate the word "enclosed". Accordingly, we can express the fact that x is not contained in S by writing $x \notin S$.

The use of the concept of sets lies in two quite different areas which I call the "language area" and the "infinity area". In the language area, a small number of words with very strict meanings have been selected in order to make communication clearer and unambiguous. It was the language area which motivated the introduction of set theory into elementary school teaching. Mathematicians, however, think the language area is rather trivial. For them, it's the infinity area which makes set theory interesting. We shall now spend some time in the language area and, only at the end, shall we have a short look at the infinity area.

It is appropriate for us to enter the language area like a child in the first grade. At first, we have to determine a *universe* which means only that we have to agree upon the type of objects we are going to talk about. Once we have determined our universe, we will not talk about other types of objects until we explicitly change the universe. Our first universe is shown in Fig. 2.8. It contains 18 geometric shapes which differ in size, shape and color. If we don't want to talk about the universe as a whole, but only about certain selected elements, we may draw container borders within the universe and associate the selected elements with their containers. In Fig. 2.8 you see a border which encloses all white elements, and another border which encloses all triangles. The set of elements within such a border is called a *subset* of the universe. Once we have determined two subsets, we can ask two questions concerning these subsets. For the first question, we ask for the *intersection* of the two subsets which is the set of those elements which are contained in both subsets at the same time. In Fig. 2.8, the intersection shown is the set of all white triangles. Instead of asking for the intersection, we can ask for the *union* of two given subsets which is the set of all elements which are contained in at least one of the two subsets. If their intersection is empty, two subsets are said to be *disjoint*. In this case, there are no elements which are contained in both subsets.

Sets 27

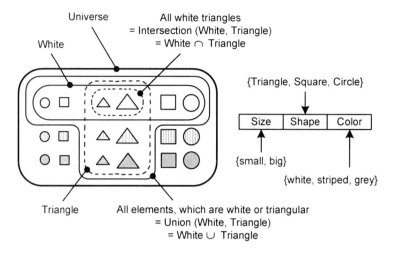

**Fig 2.8** Illustration for the concepts of sets

When we select a subset S from a universe U, we cannot avoid selecting a second subset at the same time. It is the set which contains all elements which have not been selected in the first place. This set of all non-selected elements is called the *complement* of the original subset S. In our examples in Fig. 2.8, the complement of the set of all triangles is the set of all non-triangles, and the complement of the set of all white shapes is the set of all non-white shapes. By definition, a set and its complement are always disjoint, since they cannot share any elements. The union of a set and its complement is always the entire universe.

The formula $S_1 \cap S_2$ is read as "the intersection of the two sets $S_1$ and $S_2$." You may imagine the symbol $\cap$ as a bridge which connects two river banks. The formula for an intersection means that an object belongs to both sets. The formula $S_1 \cup S_2$ is read as "the union of the two sets $S_1$ and $S_2$." You may imagine the symbol $\cup$ to be a container into which objects are thrown from both sides. The formula for a union means that an object may come from either set.

With this, our visit to the elementary school ends, and we return to the world of adults. Now we no longer play with shapes but with so-called "tuples." The term tuple does not belong to our everyday language, but is used only in mathematics. Nevertheless, as you will soon see, there is no special theory behind it. A *tuple* is just an ordered set of positions or containers where each position is filled with an element from a set for that specific position. Many situations in our everyday life can be looked at as being tuples. Think of the position of a conductor in front of his orchestra, and think of the music stand where his score is placed. Each pair consisting of a conductor and a score is a tuple with two positions. The set from which an actual element for a given position can be selected is called the *domain* for this position. Every time when you have a finite set of positions and the associated domains, you know the set of all possible tuples using these positions. On the right hand side of Fig. 2.8 you see three ordered positions (size, shape, color), where the domains for each of these positions are assigned as finite sets. While the elements of the domains are listed within curly brackets { , , , }, the tuples are written using round brackets ( , , , ). In the case of sets, the order in which the elements are listed is irrelevant. The curly brackets can be seen as borders of a container into which the elements are thrown. Thus, the formula {a, b, c} means exactly the same set as {b, c, a}. In contrast to this, the round brackets can be seen as borders of a form containing specific fields which have to be filled. For instance, the first field could be used for the age of a person, and the second field for the age of the spouse. Thus, the tuple (53, 56) is not the same as (56, 53).

In most cases, we don't consider elements which can be put on a table and grouped as in Fig. 2.8. Therefore, we no longer make drawings of our sets, but we define them using formulas.

As I mentioned earlier, formulas are nothing but adequate abbreviations for situations which we could have expressed in natural language. You previously saw formulas in the section where the creation of the numbers was considered, but those formulas were of a type which people usually don't think of as formulas. Nevertheless, the arithmetic expression (3+4)*5 is a formula, since it is an abbreviation of the much longer text "the result which we get when we first add the two numbers 3 and 4 and then multiply this sum by the number 5". An example of an expression which everybody will call a formula is the following: $(x, y)$ with $y=(x+2)*x-3$. This stands for "all ordered pairs of two numbers where the second number is obtained by adding 2 to the first number, then multiplying this sum by the first number and finally subtracting 3 from the product". Here, the letters $x$ and $y$ stand for the longer expressions "first number" and "second number", i.e., the letters are substitutes for actual numbers whose values are left open until the formula is applied. Such substitutes in formulas are called *variables*. Although in most cases formulas do contain variables, this is not a requirement for an expression to be a formula.

The formulas defining the sets from Fig. 2.8 are given in Fig. 2.9. In the formulas in Fig. 2.9, you find two symbols, × and | , with which you probably are not familiar. The symbol × is the multiplication symbol in so-called Cartesian products where the two factors are sets. "Oh no, not again!" you might now possibly object. "We just learned how to multiply complex numbers, and now you expect us to multiply sets." But again, you shall see that in spite of the strange wording, the idea behind it is quite simple. The term "Cartesian product" is named after the French philosopher René Descartes [DES] (1596-1650). The *Cartesian product* of n sets is defined as the set of all tuples with n positions, where the domains of these are the factors of the product. Our universe in Fig. 2.8 is the Cartesian product of the domains for the three positions size, shape and color. The domain of the size has two elements, that of the shape has three elements, and that of the color also has three elements. Therefore, the Cartesian product must contain 2*3*3=18 tuples. If you count the elements in Fig. 2.8, you will find 18 elements.

The first formula in Fig. 2.8 which contains the symbol | is read as "The set 'white' is defined as the set of all elements x from the actual universe for which it is true that their color is white." In this text, the symbol | stands for the section "from the actual universe, for which it is true that..."

| | |
|---|---|
| **Domains of the Attributes:** | Size    = {small, big} <br> Shape = {Triangle, Square, Circle} <br> Color   = {white, striped, grey} |
| **Universe** | = Size × Shape × Color <br> = { (small, Triangle, white),  . . .  (big, Circle, grey) } |
| **White**   = { x \| x has the color white. } <br> **Triangle** = { x \| x is a Triangle. } ||
| **Intersection:** White ∩ Triangle = { x \| (x ∈ White) AND (x ∈ Triangle) } <br> **Union:**        White ∪ Triangle = { x \| (x ∈ White)   OR   (x ∈ Triangle) } ||

**Fig. 2.9** Formulas for the sets in Fig. 2.8

We stayed long enough in the language area for you to realize how the concepts of set theory can help to communicate clearly and unambiguously. These concepts have been created for no other purpose. We now move on to the second area where set theory is useful. I called this the infinity area. Because a set could have infinite size, it is no longer reasonable to talk about "the number of elements in a set," but to use the term "*power*" instead. The power of the set in Fig. 2.8 is 18. All the sets you learned about in the story of the creation of numbers have infinite power. From the beginning of mankind, humans have struggled with their fate. This fate forces them to think about the infinite but, at the same time, prevents them from really understanding it. My son was still in kindergarten when he asked me, "What is greater than infinity?"

The mathematician George Cantor (1845-1918) is called the father of set theory. He used this theory mainly to prove that there are different powers of infinity. When mathematicians argue about infinity, they always use the infinity of the set of natural numbers as a reference for infinite powers; this might be called the "natural infinity." Realizing that there is no biggest natural number is a fundamental consequence of having understood the concept of counting. Cantor asked himself how he could compare the powers of infinite sets. He came up with an answer to this question by first answering it for finite sets and then transferring the answer to infinite sets. Assume you could not count, but you wanted to compare the powers of two finite sets. What would you do? Imagine a set of cups and a set of saucers. You would make pairs by placing cups on saucers until no more pairs could be formed. At the end of this process, you would have reached one of three possible situations: either you found a saucer for each cup, or there are cups or saucers left over. In each case, you can say whether the two sets have equal powers or which one has a bigger power.

Now we can transfer this method to compare two infinite sets. As an example, we consider two sets, one being the set of all natural numbers, and the other being the set of all positive even numbers. Intuitively, you might say that the set of all positive even numbers has half the power of the set of all natural numbers, since only every second natural number is even. But this is not true according to Cantor's definition since, as in the case of cups and saucers, you can make pairs which you can write down as tuples: (1, 2), (2, 4), (3, 6), (4, 8) and so on. You see that for each natural number or for each positive even number, you find its partner in the pair. You shall not reach an end in the process of making pairs because you are working with infinite sets. According to Cantor's definition, the powers of the two sets considered are equal. Whenever it is possible to order the elements of an infinite set as a countable sequence, this set has the same power as the set of all natural numbers.

Consider Fig. 2.10. Each circle stands for a pair of two natural numbers. These circles are connected by a directed path beginning at the circle for the pair (1, 1). By this path, all circles are ordered in a countable sequence with the count number contained inside each circle. Thus the set of all pairs of natural numbers has the same power as the set of all natural numbers alone. It could even be shown that the same result is obtained when tuples with more than two positions are considered.

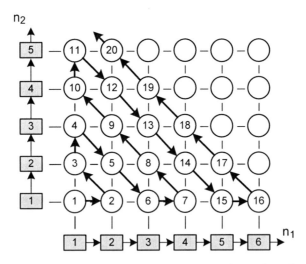

**Fig. 2.10**  Enumeration of all ordered pairs of natural numbers

We have not yet considered a set whose power is larger than natural infinity. Recall that each positive rational number can be written as a fraction n/d where both the numerator n and the denominator d are natural numbers. Therefore the set of all positive rational numbers is a subset of the set of all pairs of natural numbers and cannot have a power greater than natural infinity. But Cantor showed that the power of the set of all real numbers is greater than natural infinity. This means that he proved that the real numbers cannot be ordered completely as a countable sequence. I won't show you how he did it. If you want to know, you will have to ask an expert. Further, Cantor was even able to show that there are sets which have powers greater than the power of the set of all real numbers.

## Functions Tell Us How to Get Results

In everyday language, the noun and the verb "function" are used frequently. "I just don't function early in the morning." "Our TV set doesn't function any more." "I have no idea what the function of this device might be." These phrases talk about the purpose of a system which might not be fulfilled correctly or could be unknown. From this everyday meaning of the word function, one cannot conclude how this word is interpreted in mathematics. A mathematical function is a definite assignment by which each element of an input set gets a partner in an output set. Instead of talking about the pair (input element, output element), mathematicians prefer to say (argument, result). Here, the word argument is used with a quite different meaning than usual where arguments are a type of discussions. The input and output sets of a function may be the same set or two different sets. Let's consider the simple example of doubling natural numbers. The input set of this function is the set of all natural numbers, and the output set is the set of all even natural numbers. In this case, the output set is a subset of the input set. The function of doubling is reversible since to each element of the output set belongs exactly one input element. By halving the output element, we get the corresponding input element. A simple example of a non-reversible function is "omitting the sign." The input set is the set of all real numbers, and the output set is the set of all non-negative real numbers. Each non-zero result is assigned to two input elements, e.g., if the output result is 5, the input element could be -5 or +5

In most texts where the concept of mathematical functions is introduced, the input and output sets of the examples are sets of numbers. From this, most people conclude that it is part of the definition of mathematical functions that they determine how the resulting number is computed from input numbers. Therefore, I emphasize that the concept of mathematical functions does not refer to numbers at all. Neither the input set nor the output set must be a set of numbers. Although it is true that in many cases of interest these sets are number sets, this does not mean that the concept of functions is connected to the concept of numbers. The fact that many people wear glasses does not justify the assumption that all human beings wear glasses. Anything concrete or abstract which is unambiguously connected to something may be considered as an input element of a function. For instance, think of the children in first grade in a certain elementary school. Many things are connected unambiguously to these children, e.g., the date and location of their birth, their parents, etc. Therefore, we may say that the relation between the children and their parents is a function where each child is an input element and each pair of parents is an output element. If a child has no sisters or brothers in the class, the function is reversible since then it is possible to start with a pair of parents and unambiguously get to the corresponding child.

Until now, all the functions considered as examples related one set to one other set, and therefore they had only one input position and one output position. But this is not necessarily the case. A function may have multiple input positions and multiple output positions where each has its own domain. As an example, we consider the function

$(CEO, CFO, CIO)$ = Top Management (*Corporation*, *Point in Time*).

The form in which this is written is the standard form for functions in mathematical formulas: the name of the function, Top Management, is followed by the list of input positions enclosed in round brackets. This indicates the fact that the input elements are in a tuple. The input positions are written using italic letters in order to emphasize the difference between positions and what actually fills these positions. The position *Corporation* may be filled by "General Electric Co." or "General Motors Co.". In contrast to this, "Top Management" is the name of the function, and this must be distinguished from the result positions which are listed on the left side of the equation symbol. Here, too, the position names are written using italic letters. Possible entries for the position $CEO$ might be "Henry Ford" or "James P. Goodwin". Whenever the value of the input tuple is given, the value of the output tuple is determined.

Of special interest are those functions with only one output position and possibly multiple input positions where the same domain applies to all these positions. A very simple example of such a function is the Addition function,

*sum* = Addition (*first summand, second summand*).

Our next example is introduced in Fig. 2.11 which shows a round table with three chairs. We assume that these chairs cannot be moved. This could be the situation in the restaurant of a cruise ship where tables and chairs are fastened to the floor in order to avoid accidents in case of rough seas. The right side of the figure lists the six possibilities for how three persons A, B and C could be seated in the chairs at

Functions 33

this table. We now assume that the captain can tell these persons to change their seating order. There are six different movement orders the captain could give, including the order "Stay seated!" Using the two domains, "seating plans" (like A=3, B=1, C=2) and "orders" like "move clockwise!", each having six elements, we can define some functions which are presented in Fig. 2.12

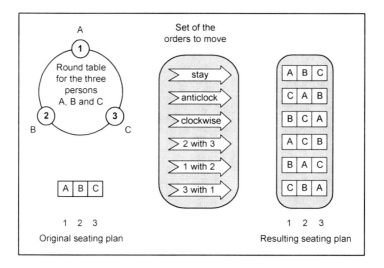

**Fig. 2.11**   Example about "seating plans"

We now concentrate on the "order combination function" which has the same formal structure as the addition function: there are two input positions and one output position, and all three positions have the same domain, the set of the six movement orders. The order combination function specifies the results when the captain gives two orders sequentially. The results are defined by the table in the lower part of the figure. The shaded fields contain the results which are not changed by interchanging the two input orders. If, for instance, the captain first says: "Move anticlockwise!" and then says: "Move clockwise!", the final seating plan is the same as if he had said: "Stay seated!" Not all fields in Fig. 2.12 are shaded since in many cases the result depends on which of the two orders was given first. Such a dependency on the order of the input elements is not new to us, since we already have seen it in subtraction and division. When the result of a function with two input positions is independent of the order of the input elements in all cases, this function is said to be *"commutative"*. The best known examples of commutative functions are the addition and multiplication functions.

Six "move-functions" with one argument:
*new seating plan* = **stay seated** (*original seating plan*)
*new seating plan* = **move counterclockwise** (*original seating plan*)
⋮
*new seating plan* = **exchange seats 3 and 1** (*original seating plan*)

One "move-function" with two arguments:
*new seating plan* = **execute** (*move-order, original seating plan*)

One "order combination function" with two arguments:
*equivalent order* = **sequence of** (*first order, second order*)

|  |  | second order |  |  |  |  |  |
|---|---|---|---|---|---|---|---|
|  |  | stay | anti-clock | clock | 2and3 | 1and2 | 3and1 |
| first order | stay | stay | anti-clock | clock | 2and3 | 1and2 | 3and1 |
|  | anti-clock | anti-clock | clock | stay | 1and2 | 3and1 | 2and3 |
|  | clock | clock | stay | anti-clock | 3and1 | 2and3 | 1and2 |
|  | 2and3 | 2and3 | 3and1 | 1and2 | stay | clock | anti-clock |
|  | 1and2 | 1and2 | 2and3 | 3and1 | anti-clock | stay | clock |
|  | 3and1 | 3and1 | 1and2 | 2and3 | clock | anti-clock | stay |

**Fig. 2.12** Alternative functions for Fig. 2.11

Playing around with functions leads to structures which might be called "Function chains". In such a structure, the result of an "inner function" is used as an input element of an "outer function". In the example of Fig. 2.12 the possibility of creating chains is quite obvious. Consider the example

3and1 (*original seating plan*) = 2and3 ( clockwise (*original seating plan*) )

Here, the original seating plan is first changed by all three persons moving clockwise. Then the resulting seating situation is changed by interchanging the two persons on chairs 2 and 3. The final seating situation is the same as if the original seating plan had only been changed once by interchanging the two persons on chairs 1 and 3.

The next function we consider is called a *"polynomial"*. This name has no everyday meaning, but is used only in mathematics. Nevertheless, polynomials have

Functions

a rather simple functional structure. They are defined by chaining multiplications and additions in a specific way. An example is presented in Fig. 2.13. A polynomial function P(x) has only the single input position x. The highest exponent which occurs in a polynomial determines the so-called *degree* of the polynomial. Thus, the example in Fig. 2.13 is a 3$^{rd}$ degree polynomial. The numbers multiplying the powers of x are called the coefficients of the polynomial; they can be seen as weights of the powers of x.

The Polynomial written as a **sum** is determined by its coefficients:

The Polynomial written as a **product** is determined by its zeros and the coefficient of the highest power:

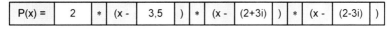

**Fig. 2.13**  Alternative representations of a polynomial

Mathematicians have come up with many questions which could be asked concerning polynomials. They found that a polynomial can be expressed in two alternative forms, either as a sum of weighted powers of x, or (in factored form) as a product based on those values of x for which the result of the polynomial is zero. In factored form, each factor enclosed in brackets is a term where a number is subtracted from x. When x gets the value of this number, the factor and with it the entire product becomes zero. This value of x is called a *"zero"* of P(x). The number of these factors is always equal to the degree of the polynomial. Depending on the coefficients, zeros can be complex numbers. If all coefficients are real numbers, complex zeros can only occur in pairs where the real parts of both partners in a pair are equal, and the imaginary parts differ only in sign. This is the case in the example in Fig. 2.13. It is also possible for a zero to occur more than once in the product form. That's enough for you to know about polynomials. The main thing is that polynomials and their zeros belong together like cars and their wheels.

We are still sitting in our space ship flying over the mathematics continent and actually looking down on function country. Maybe you have read that we could see the Great Wall of China from a spaceship with only the naked eye. Of similar importance is the function we consider next; it cannot be missed when we look down on function country. When I introduced complex numbers, I said that it is only a question of appropriateness whether a complex number is represented by its real part and its imaginary part, or by its radius and its angle (see Fig. 2.6). From that statement we can conclude that there must exist two pairs of functions which transform the one form of representation into the other and vice versa. When the real part and the imaginary part are given as inputs, two functions provide the radius and the angle as their results, and when the radius and the angle are given, two other functions provide the real part and the imaginary part. These pairs of functions are presented in Fig. 2.14.

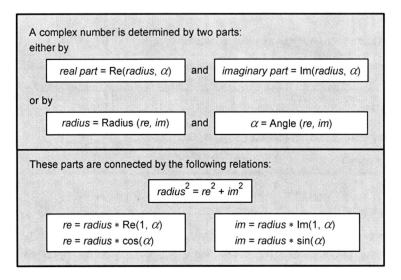

**Fig. 2. 14**   Functional relations between the parts determining a complex number

Again looking at Fig. 2.6, we consider the complex number (4 + 3i) which has radius 5 and angle 36.87 degrees. One-fifth of this complex number will have the radius 1 and still have the angle 36.87 degrees, since the division by a real number does not change the angle. Whenever we start with a complex number which has radius 1, we can obtain any other complex number which has the same angle by multiplying the real part and the imaginary part by the desired radius. This is expressed in the lower part of Fig. 2.14 for the complex number $(re) + i\,(im)$ with angle $\alpha$. By restricting the position *radius* in the two functions Re(*radius, angle*) and Im(*radius, angle*) to the value 1, two new functions are defined which have only one input position each. These two functions are called "sine" and "cosine", and they are abbreviated by sin and cos, respectively. These two functions are called transcendental functions; this means that the result for a given input cannot be obtained by a finite arithmetic computation. But at least we can draw diagrams which represent these functions. Fig. 2.15 shows the graph of the sine function and how it is obtained. The circle on the left has the radius 1 and, correspondingly, its circumference is $2\pi$. This circumference appears again as a horizontal straight line on the right – only it is shortened by applying a certain scaling. Thus, each point

Functions

on the circle corresponds unambiguously to a point on the horizontal line. Therefore, each point on the horizontal line has an imaginary part assigned to it, and its length can be taken from the circle on the left. The curve is the line which connects all the ends of the vertical lines having the length of the corresponding imaginary parts. The graph of the cosine function has the same form, but compared to the sine function, it is shifted left by $\pi/2$, i.e., its maximum is located at $\alpha=0$ and is repeated at $\alpha=2\pi$, while its minimum is located at $\alpha=\pi$.

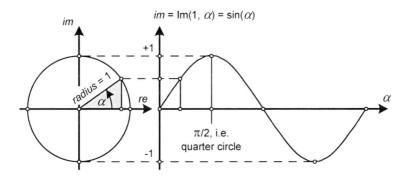

**Fig. 2. 15**   The function $\sin(\alpha)$

Although there is a particular connection between complex numbers and the functions sine and cosine – more details about this will be presented in Chapter 3 - these functions can also be defined without any reference to complex numbers. In that case, the definition refers to a right-angled or right triangle. The longest side of such a triangle is called its *"hypotenuse"*. It is situated opposite the right angle. The other two sides are called *"legs"*; and they enclose the right angle. Usually, the letter $c$ is chosen as the variable for the length of the hypotenuse, and the letters $a$ and $b$ are used for the lengths of the legs. Now, we assume that the angle $\alpha$ is enclosed by the hypotenuse and the leg which has the length $b$. Then $\sin(\alpha)$ is defined to be the fraction $a/c$, and $\cos(\alpha)$ is defined to be the fraction $b/c$. If we apply these definitions to the grey triangle in Fig. 2.15, the radius is the hypotenuse while the horizontal real part and the vertical imaginary part are the two legs. Thus, we have $c=radius$, $a=im$ and $b=re$, and therefore $\sin(\alpha)=im/radius$ and $\cos(\alpha)=re/radius$ which corresponds to the definitions in Fig. 2.14.

An interesting area in function country is called the *"recursive function definition"*. It is well known and emphasized by high school teachers that one should never use a term that is defined in its own definition, since otherwise the definition would be cyclic. At first look, the recursive definition of a function seems to be cyclic since, as a consequence of its being recursive, the name of the function appears in its definition. But a second look shows that the definition is not cyclic, but a spiral with a well-defined end point.

The concept of recursive function definitions can best be introduced by presenting a vivid example. In Fig. 2.16 you see the elements of the game "Towers of Hanoi". On the foundation, three thin pillars are erected at equal distances, and on these pillars can be placed circular discs with holes in their centers. There are four discs with different diameters. In the left upper corner of the figure, you see a tower built with the four discs.

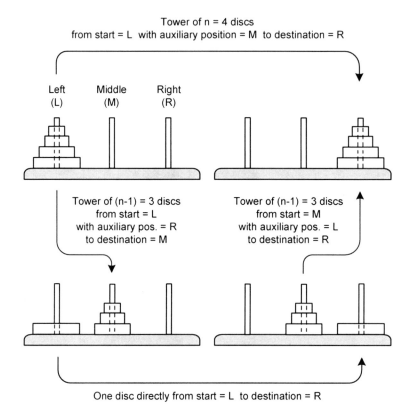

**Fig. 2. 16** The game "The towers of Hanoi"

The game is played according to the following rules. At the beginning, all discs are on the left pillar where they are ordered according to their size, the biggest disc sitting directly on the foundation. At the end of the process, the same tower of discs in the same order must be located on the right pillar. The discs have to be moved one at a time, and never shall a disc be placed onto a smaller disc. Usually the game is played with seven discs. For each given number of discs n, there exists one optimal sequence of moves which requires the minimum number of moves: $2^n-1$. We can define a function which results in the optimal sequence of moves. In the definition of this function, we do not assume that the original tower is sitting at the left pillar and must be transferred to the right pillar. We leave open which pillars shall be the start, the destination and the auxiliary, i.e., the function will have these positions as input positions which actually have to be filled. By looking at the recursive function definition in the lower part of Fig. 2.17, you can see that this is an appropriate decision. This definition is based on the assumption that it would be easy to find the optimal sequence of moves for n discs if we already knew the optimal sequence for (n-1) discs. Then we would first transfer a tower of (n-1) discs from the start to the auxiliary position, which leaves the biggest disc alone on the start. This disc then can be moved directly to the destination. And finally, we could move the tower of (n-1) discs, one at a time, from its auxiliary position to the destination using the optimal sequence for (n-1) discs. This process is illustrated in Fig. 2.16.

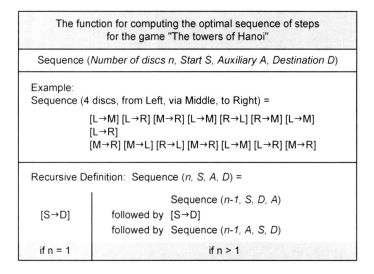

Fig. 2. 17   Recursive definition of the "Hanoi-sequence-function"

That the function definition in Fig. 2.17 is not cyclic is seen by distinguishing the two cases n=1 and n>1. If the original tower contains only a single disk – i.e., n=1 – the result of the function is determined directly, and this case will always be reached by a finite number of recursions. For each time when the number of discs is greater than one, this number is reduced by one which finally will lead us to the case n=1.

When I decided to present the examples in Fig. 2.11 and 2.16, I had in mind to show you that you can find functions everywhere in your daily life – you only have to adapt your eyes appropriately. However, as I said earlier, most of the functions mathematicians work with are not taken from daily life. There is a special mathematical discipline called "function theory" where functions on the domain of complex numbers are the subject. In Chapter 3, I shall introduce differentiation and integration which are functions whose domains are also functions. A function whose domains are also functions has already been presented to you. You can see this in Fig. 2.12.

## "Come Closer!" Is What Limits Want

The politician says, "I shall not comment on the outcome of the election before I have had time to analyze it." The chemist says, "Before one can begin the synthesis of chemical products, one first must learn how to analyze." And the health advisor says, "The best therapy for you would be psychoanalysis." When people use the word *"analysis"* they have the idea that something must be taken apart. There is a field within mathematics which also is called analysis, and here it is infinity which has to be taken apart. But now, the problems of infinity are not those which George Cantor tried to solve by comparing powers of infinite sets. Now the question is about how we can get from infinite to finite results. What I mean by this colloquial characteristic can best be explained by describing a paradox which was presented by the Greek philosopher Zenon of Elea (about 450 BC). Principally, the argument can be applied to any race, but Zenon chose the race between the hero Achilles and a turtle, since in this case the difference of speed is so extreme that the absurdity of the paradox is evident beyond all doubt. The paradox seems to prove that Achilles can never catch up with the turtle when it starts with a lead. The reasoning goes as follows: First, Achilles has to run the distance of the lead granted to the turtle. When he reaches the point from where the turtle started, the turtle is no longer there but is at a point some distance away. Now, Achilles has to run this distance, but when he reaches the end of it, the turtle again is no longer there, but again is at some distance away. Thus the distance Achilles has to run is cut into an infinite sequence of sections which get shorter and shorter, but since their number is infinite, the question arises about whether or not Achilles can ever catch up with the turtle. Since the days of Zenon, mathematicians have been very creative, and they have found a way to add up an infinite number of summands and get a finite result.

At the top of Fig.2.18 you find a formula describing the details about this paradox. In this formula, $v_A$ and $v_T$ are the speeds of Achilles and the turtle, respectively. Each summand in brackets represents the relative time Achilles needs to run a specific section of the infinite sequence of sections. The first summand, namely the 1, stands for the time Achilles needs to run the length of the original lead of the turtle. I chose a lead of 50 meters, and I assumed that Achilles runs at a speed of ten meters per second which is the typical speed of a runner in the one-hundred meter race in the Olympics. The speed of the turtle is assumed to be one hundredth of the speed of Achilles ($q = v_T/v_A = 1/100$).

# Limits

$$\text{Catch-up-time} = \frac{\text{Lead}}{v_A} * \left(1 + \left(\frac{v_T}{v_A}\right) + \left(\frac{v_T}{v_A}\right)^2 + \left(\frac{v_T}{v_A}\right)^3 + \left(\frac{v_T}{v_A}\right)^4 + \cdots \right)$$

$$\text{Catch-up-time} = \frac{50 \text{ m}}{10 \text{ m/s}} * \left(1 + \left(\frac{1}{100}\right) + \left(\frac{1}{100}\right)^2 + \left(\frac{1}{100}\right)^3 + \left(\frac{1}{100}\right)^4 + \cdots \right)$$

$$\text{Sum}(q, n) = 1 + q + q^2 + q^3 + \ldots + q^n$$

$$q * \text{Sum}(q, n) = \quad\quad q + q^2 + q^3 + \ldots + q^n + q^{n+1}$$

$$\text{Sum}(q, n) - q * \text{Sum}(q, n) = 1 \quad\quad\quad\quad\quad\quad - q^{n+1}$$

$$(1 - q) * \text{Sum}(q, n) = 1 \quad\quad\quad\quad\quad\quad - q^{n+1}$$

$$\text{Sum}(q, n) = \frac{1 - q^{n+1}}{1 - q}$$

For $0 < q < 1$: $\quad \text{Sum}(q, \infty) = \lim_{n \to \infty} \frac{1 - q^{n+1}}{1 - q} = \frac{1}{1 - q}$

$$\text{Catch-up-time} = 5 \text{ s} * \frac{1}{1 - 0.01} = \frac{5 \text{ s}}{0.99} = 5.05050505\ldots \text{ s}$$

**Fig. 2.18** Using an infinite sum to calculate the catch-up time for the Achilles-turtle race

In the middle of Fig. 2.18 you see the equations that show the trick which made it possible to compute the sum of an infinite number of summands. First, we assume that we were interested only in the sum of the first ($n+1$) summands where the last summand is $q^n$. This sum, Sum($q, n$), is a function with two input positions. Now we multiply both sides of the equation for this sum by the factor $q$ which is the base of the powers to be added. In the case of the race, we have $q=1/100$. By writing the product appropriately under the original sum, we see that the sum and the product share a long section in the middle and differ only at the left and the right ends. This makes it possible to get rid of the "unmathematical three little dots" (used to represent unwritten intermediate terms) by subtracting the lower line from the upper line. Dividing both sides of the equation by $1-q$ gives us the formula for computing the function Sum($q, n$) for finite values of $n$. It is given in the second to last line in the middle of Fig. 2.18.

Here, my wife says, "Look, that's just the reason why I don't like math. I have no difficulty in understanding your reasoning, but never in my life would I have come up with this trick. I don't like a scientific discipline where you constantly have to come up with new tricks, and where you are called stupid if you don't find the tricks." I appreciate that my wife can express her aversion against mathematics so clearly. I can't be sure whether I myself would have found the tricky solution shown in Fig. 2.18, but certainly, I will not call you stupid if you don't come up with such tricks. The only thing I expect is that you can easily follow the reasoning when it is adequately explained.

We still have a short distance to go before we reach our final goal, the formula for computing the sum of an infinite sequence of summands. The number $n$ of summands is still an input position in our function for the sum of finite length. Now we check what happens if we make $n$ greater and greater. With growing $n$, the power $q^{n+1}$ gets smaller and smaller when $q$ is less than one. Think of our example of the race where $q$ is 1/100. In this case, $q^{n+1}$ has already dropped to one millionth when $n$ reaches the value of only two. As long as $n$ has a finite value, the power $q^{n+1}$ has a non-zero value, but this value can be made as close to zero as one might like just by making $n$ big enough. This is a characteristic of any so-called *limit value*, that it is possible to get as close to it as one likes just by making $n$ big enough. Mathematically speaking, $q^{n+1}$ goes towards zero as $n$ goes towards infinity. In formulas, limit values are symbolized by "lim", the first three letters of the word limit. In our example, the limit value of the sum enclosed in brackets is 100/99. The time Achilles needs to catch up with the turtle is found to be 500/99 seconds. He needs 5 seconds to reach the position the turtle started from, and then he needs only 5/99 seconds to finally catch up.

The next limit I shall present to you plays such a dominant role in mathematics that any one who does not know this limit will not be taken seriously by mathematicians. I am talking about the concept of steady growth. When they hear the word growth, many people do not think first of the growth of invested capital or money. Most will think of the growth of trees or children. In the case of growing capital, there are predefined points in time when the interest is added to the capital, and this results in a stepwise growth. Natural growth, however, is always steady growth since there are no steps. The limit situation we are aiming at is obtained by starting with stepwise growth and making the equal time intervals between the steps shorter and shorter. The diagram on the left side of Fig. 2.19 shows one step of the growth of capital C with a growth rate of 100 percent. Presumably, you think this is an exorbitant rate, but this is true only if this capital is money and the time interval is rather short. In Fig. 2.19, however, no assumptions are made concerning the type of capital and the duration of the time interval T. When both the growth rate and the step width are reduced to one-fourth of what they are on the left side, the growth process is shown in the diagram on the right side. While at the end of the time interval T, the capital in the left diagram has grown by a factor of only two, the capital in the right diagram has grown by a factor of 2.4414 = [ ( (1+0.25)*(1+0.25) )*(1+0.25) ]*(1+0.25) = $(1+0.25)^4$. This is an application of the formula $(1+p)^n$ for compound interest with an interest rate of $p=0.25$ and $n=4$ time intervals being used.

# Limits

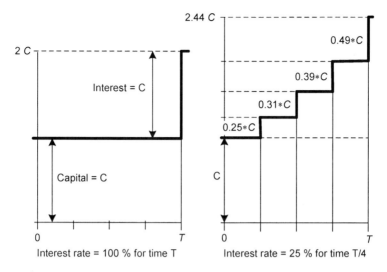

**Fig. 2. 19** Stepwise growth at different rates

Now we look at what we get when we increase the number of intervals *n* within the total time *T* and, at the same time, reduce the growth rate per interval to the value of 1/*n*. Here we get the limit which is shown in Fig. 2.20. This limit is symbolized by the small letter e in honor of the great mathematician Leonhard Euler (1707-1783). By setting both the initial capital *C* and the time *T* to one, we get the standardized function of steady growth, f(*x*)=e$^x$, which is called the *"exponential function"* and is shown in Fig. 2.20.

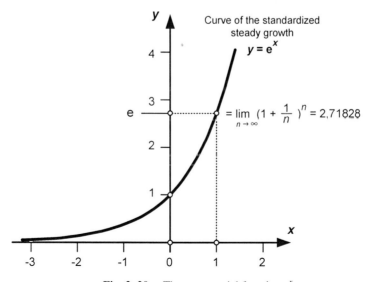

**Fig. 2. 20** The exponential function e$^x$

## An Eye for an Eye and a Tooth for a Tooth – That's the Principle of Equations

Around the year 800 AD, the Arab scientist Al-Hwarizmi, who lived in Bagdad, wrote a booklet on solving equations. One of the words occurring in the title of this booklet is *al-gabr* which means to set, complete or re-establish. This word is the root of the word "algebra" which is the name of the mathematical discipline dealing with equations. I still remember my time in high school when somebody told me that algebra meant computing with letters. It was not a teacher who said this, since a teacher would have given a more precise and complete explanation. Although computing with letters is not a completely wrong definition, it misses one important point. Algebra is the discipline of generating sequences of equations or inequalities. An *equation* is a proposition saying that terms on both sides of the equals sign identify the same thing. An *inequality* is a proposition saying that the two things identified by the terms on both sides of the relation symbol are not equal and are related to each other in the way specified by the symbol. The relation symbols for inequalities which occur most often are $\neq, <, >, \leq$ and $\geq$. From here on, the presentation is restricted to equations because once equations have been understood, this understanding can be transferred easily to inequalities.

Let's have a look at a first example of an equation:

The president of the United States in 1864 = Abraham Lincoln

Although the terms on the left and right sides of the equals sign look quite different, they identify the same person. Identification in this context means providing enough information for unambiguous specification of an individual element from a concrete or abstract universe. Though many letters occur in the equation above, they are not used in the sense meant by the statement, "Algebra is computing with letters." This statement means that letters are used as so-called "*variables*" which stand for elements and which will be specified later. In the equation

(Sum of the summands $(2k-1)$ for $k$ from 1 through $n$ ) = $n^2$,

two variables $k$ and $n$ occur. What this equation says can be said in natural language: "For any natural number $n$, the sum of the first $n$ odd numbers equals the square of $n$." For instance, if we choose $n$ to be five, we get $1+3+5+7+9=25=5^2$. In the natural language form of the equation, the term "odd numbers" can be used since it can be assumed that everybody knows what that is. But in the formula, the concept of odd numbers has to be expressed by the term

(2k-1) where k is used for counting the odd numbers. For example, if k is set to be five, we get the fifth odd number as (2*5-1)=9.

Since this equation is true for any natural number $n$, the equation states a so-called mathematical law. But variables are needed not only to express mathematical laws; they are also useful for describing problems where certain individual elements have to be found. In these cases, the standard letter for the variable is $x$, and it represents an unknown value or an unknown element. Maybe you have already heard in a movie or read in a novel about a Mr. X. In this case, the variable X stands for the name of an unknown person who is to be found. The example I chose to explain the use of $x$ is a simple brain-teaser: "Today is Anna's birthday. I am not telling you what her present age is, but it is hidden in the following statement: When Anna's age is five times the age which she had three years ago, she will be exactly twice as old as she is today. What is her present age?" In this case, the variable $x$ is used to represent Anna's present age and, using this, the text in the brain-teaser can be easily transformed into the formal equation of Fig. 2.21.

| When Anna will have five times | the age which she had three years ago |  | she will be twice as old | as she is today |
|---|---|---|---|---|
| 5 * | ( x - 3 ) | = | 2 * | x |

**Fig. 2. 21**  Transforming a brain-teaser into an equation

In this figure, I arranged the text and the formula in such a way that you can see just how the formula was created. The equation in Fig. 2.21 is not a mathematical law, but just a statement about Anna's present age. Please note that the number which is identified by the terms on the two sides of the equation do not equal Anna's age, but equal twice her age. There is only one value for $x$ which makes this equation a true statement, and this value is 5. That means that 5 is the solution of the equation. And how can this solution be found? I mentioned earlier that algebra is the discipline of generating sequences of equations. This generation process begins with a first equation being given, and then the next equation is obtained from it. The process hopefully ends with the equation which we were looking for. What is the equation we are looking for in our example of Fig. 2.21? This final equation obviously is $x=5$, since once we find it we know Anna's present age. The sequence of equations which begins with the equation in Fig. 2.21 and ends with $x=5$, is shown in Fig. 2.22.

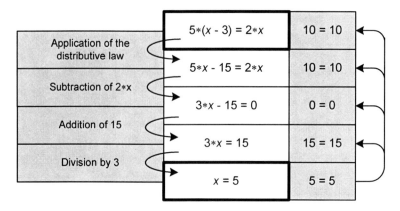

**Fig. 2. 22**   Steps for solving an equation

The basic rule for generating the next equation from a given one is very simple: any operation may be performed as long as it is performed on both sides of the equation at the same time. The operation used may change that which it operates upon, but since the same change occurs on both sides of the equal sign, the equation remains correct. Look how the sequence of equations in Fig. 2.22 is generated. The first operation uses the distributive law. This law gives us the right to transform the product of a number with a sum into a sum of products. This operation leaves the result of the computation unchanged, and therefore it needs to be performed only on the left hand side of the equation. Each of the next three operations changes that which is operated upon and therefore each must be performed on both sides of the equation at the same time. The choice of the sequence of operations was guided by the desired form for the final equation. Once we know the value of $x$, we can determine the numerical values for the two sides of the equations. These values are indicated by the arrows on the right side which lead from the bottom up, whereas the equations were generated from the top down.

Here again it may happen that some of my readers – and maybe my wife, too – will say: "How would I know which operations to perform on an equation in order to generate a sequence of equations which ends with the appropriate final equation?" Actually, because our first equation is of a very simple type, there exists a well-defined set of rules concerning how to choose the appropriate operations. But I don't see the need to present these rules, particularly because there are only a few types of equations which can be solved following such strict rules.

It often happens that there is not just a single unknown element $x$, but that the number of unknowns is greater than one. In these cases, a single equation is not sufficient to provide all the information needed to completely specify the unknowns. It can be proven that the number of equations which are needed must equal the number of unknowns. In Fig. 2.23 you see two examples of sets of equations, each with two unknowns, $x_1$ and $x_2$. On the left hand side of the figure, there are two equations and their solution is given at the bottom. On the right hand side, again there are two unknowns, but here three equations are given. In this case, no solution exists, since the three equations are contradictory. In other words, no combination of values for $x_1$ and $x_2$ will satisfy all three equations.

| $x_1 + 2x_2 = 29$<br>$6x_1 - x_2 = 18$ | $x_1 - 4x_2 = 1$<br>$3x_1 + 2x_2 = 17$<br>$2x_1 - 3x_2 = 10$ |
|---|---|
| Solution:<br>$x_1=5$ and $x_2=12$ | No solution exists. |

**Fig. 2.23** Examples of equation systems with two unknowns

When information is provided as specifications for some unknown variable, there are three possible cases. In the first case no value, of possible values in its universe, exists which fits the given information. In the second case, there is exactly one value, and in the third case there are more values than one that fit the specified information. Think of a robbery where someone claims to have seen the robber. If he then describes the person as someone 8 feet tall and weighing less than 45 pounds, this specification probably does not fit anybody in the world. If, however, he describes the person as someone about 6 feet tall and weighing at most 180 pounds, his specification will include millions of people. In the best case, he can specify the robber so accurately that his description matches the properties of only one living person. Equations with unknowns are nothing more than information provided in order to find specific values whose properties match the given specification.

Up until about 200 years ago, mathematicians dealt only with equations based on numbers. But in Figs. 2.11 and 2.16, you were introduced to functions which have nothing to do with numbers. In modern algebra, equations are no longer restricted to numbers. Modern algebra actually does not care at all what type of elements is used on both sides of an equation. The interest is focused on characteristic properties of the functions and the consequences thereof. Perhaps you remember that earlier we looked at a certain type of function having two input positions with one common domain for both the input and output positions. Functions of this type are called *closed*. Mathematicians asked themselves what characteristic properties of such functions could be found without considering their domain. Now consider Fig. 2.24. The first two rows show that a domain must be defined for which a closed operation can be considered. In the last three columns, three specific domains and a closed operation for each of them are presented for illustrating the abstract statements. When no specific operation is considered, the operator symbol □ is used for an operation which has no intuitive concrete meaning. We need such a symbol for writing the operation in the form $f(a, b) = a \square b$ which is called the *infix* form because the operator symbol is standing in between the two operands. Obviously, the infix form can be used only for functions having two input positions.

| The universe, i.e. the set of possible operands | The set of all whole numbers | The set of all positive rational numbers | The set of the six orders to move in Fig. 2.11 |
|---|---|---|---|
| The combining operator □, i.e. the function with two arguments | Addition + | Multiplication * | Sequence ⇨ |
| General independence from the order of the two operands (commutative law), i.e. (a □ b) = (b □ a) | yes | yes | no |
| General independence from the order of computing (associative law), i.e. (a □ b) □ c = a □ (b □ c) | yes | yes | yes |
| Existence of a neutral operand v, such that for each operand a it is true that (v □ a) = (a □ v) = a | yes $v = 0$ | yes $v = 1$ | yes $v$ = stay |
| General inversability, i.e. for each operand a exists a partner $a_{inv}$, such that (a □ $a_{inv}$) = ($a_{inv}$ □ a) = v | yes $a_{inv} = -a$ | yes $a_{inv} = \frac{1}{a}$ | yes s. Fig. 2.25 |

**Fig. 2. 24** Definition and examples of algebraic groups

The so-called *commutative* law is presented in the third row of Fig. 2.24. Here, the question asked is whether the result of the operation always stays the same when the two operands are interchanged. Certainly you learned early in elementary school that this general interchangeability of the two operands exists for addition and multiplication. But this is not true for the operation considered in the rightmost column.

Equations 49

The so-called *associative* law is presented in the fourth row. Whenever the letters "soci" occur in a word – think of social security, association or high society – it is about communities sharing something. Now consider the case of the associative law in a community of three operands. In this case, it must be decided which operation is to be performed first, either the operation with the first two operands or the operation with the last two operands. If in all cases the result does not depend on the order of the operations, it is said that the associative law holds for this closed operation with its domain.

The next structural property which a closed operation may have is the existence of a so-called *neutral operand*, $\nu$. What this means is shown in the fifth row of Fig. 2.24. The Greek letter $\nu$ is chosen to symbolize the neutral operand, since this corresponds to the Latin letter n which is the first letter of the word neutral. This special operand $\nu$ is called neutral since, for any operand $a$ from the domain of operands, the operation which combines $a$ with $\nu$ has the result $a$. The best known neutral operands are the numbers 0 and 1 for addition and multiplication, respectively. You all know that adding 0 to any number $a$ gives the result $a$, and that multiplying any number $a$ by 1 also gives the result $a$.

For a closed operation with a neutral operand $\nu$, we can ask if all elements in the domain of operands have a corresponding *inverse* partner. What this means is shown in the sixth row of Fig. 2.24. By definition, the result of an operation which combines an operand $a$ with its inverse partner $a_{inv}$ is the neutral element $\nu$. Again, addition and multiplication are the best known examples of closed operations with general invertability. In the case of addition, inversion means changing the sign, since adding any number $a$ to its inverse partner ($-a$) results in the neutral element 0. In the case of multiplication, inversion of a number $a$ means computing its reciprocal $1/a$, since multiplying $a$ by $1/a$ results in the neutral element 1.

While the domains for addition and multiplication are sets of numbers, the domain of the third operation considered in Fig. 2.24 is the set of the six move orders introduced in Fig. 2.11. In this example, the overall closed operation comes from the compression of two consecutive orders into a single order which has the same effect. For this operation, the associative law holds, but the commutative law does not. Here the neutral element is the order "Stay seated!". The inversion table in Fig. 2.25 can be derived from the function table given in Fig. 2.12. In this case, the inverse partners of some operands $a$ are identically $a$. For instance, the inverse of the order "The persons on the chairs 2 and 3 must interchange their seats!" is exactly the same order, because when this order is given twice, it has the same effect as if the neutral order "Stay seated!" had been given.

Any closed operation which has the three structural properties described within the thick rectangular frame in Fig. 2.24 is called an *algebraic group*. The group of people sitting around a table as shown in Fig. 2.11, who are ordered by someone to move (according to the set of six move orders), is an example of an algebraic group.

| $a$ | stay | anticlock | clock | 1 and 2 | 2 and 3 | 3 and 1 |
|---|---|---|---|---|---|---|
| $a_{inv}$ | stay | clock | anticlock | 1 and 2 | 2 and 3 | 3 and 1 |

**Fig. 2. 25**  Inversion of the orders in Fig. 2.11

When the concept of an algebraic group was presented to me for the first time, I was rather surprised to learn that such an abstract concept can be applied to many different areas. Each law which can be derived from these abstract properties holds equally for addition, multiplication and changing seating orders. In addition, it will apply to all operations found in the future which have the group properties. Figs. 2.26 and 2.27 show an example of an interesting law which can be derived from the group properties defined in Fig. 2.24. From the tables in Fig. 2.26 and 2.27, you can immediately conclude that our subject belongs to algebra, since these tables show sequences of equations. In both figures, we begin with the equation at the top and move down sequentially. Fig. 2.26 shows the derivation of the inversion law for chains of two operands. This law, shown in the last row, says that there are two alternative ways of inverting the chain which both provide the same result. Either we first compute the result of the chain which we then invert, or we first invert the operands and then compute the result of the chain in reverse order compared to the original chain. The first equation in Fig. 2.26 is taken from the definition of general invertability in Fig. 2.24, and the transformations downward are nothing more than applications of group properties, namely the associative law and the definitions of the neutral element and general invertability.

In Fig. 2.27, the inversion law is extended to chains of more than two operands. In this derivation, I use the result from Fig. 2.26. Besides that, the derivation uses only the associative law. The final result of the derivation is found in the shaded row at the bottom. We now check what the result means in the case of our three examples in Fig. 2.24. In the case of addition, it means that a sum of multiple summands can be inverted either by first computing the sum and then inverting its sign, or by first inverting the signs of all the summands and then computing the sum. There is no need to reverse the sequence of the summands, since addition is a commutative operation. In the case of multiplication, the law says that the reciprocal of a product of multiple factors can be obtained either by first computing the product and then computing its reciprocal, or by first computing the reciprocals of all factors and then computing the product. Like addition, multiplication is a commutative operation, and therefore there is no need to reverse the order of the factors. For the last example, the operation is the compression of a sequence of two orders into an equivalent one, and this operation is not commutative. Therefore, in this case, the reversal of the sequence of orders according to the inversion law is relevant.

# Equations

| | | |
|---:|:---:|:---:|
| a □ inv(a) | = | v |
| (a □ v) □ inv(a) | = | v |
| (a □ (b □ inv(b)) □ inv(a) | = | v |
| ((a □ b) □ inv(b)) □ inv(a) | = | v |
| inv(a □ b) □ ((a □ b) □ inv(b)) □ inv(a) | = | inv(a □ b) □ v |
| ((inv(a □ b) □ (a □ b)) □ inv(b)) □ inv(a) | = | inv(a □ b) |
| (v □ inv(b)) □ inv(a) | = | inv(a □ b) |
| inv(b) □ inv(a) | = | inv(a □ b) |

**Fig. 2. 26** Derivation of the inversion law for two operands

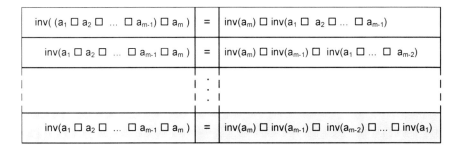

**Fig. 2. 27** Derivation of the inversion law for a chain of more than two operands

At first, I could not imagine what a great variety of laws could be derived starting from nothing but the group properties. But eventually it became clear to me why there is a special discipline within mathematics called "group theory".

A group needs only one closed operation, but there are other algebraic structures which are based on two closed operations – think of the combination of addition and multiplication. Then, not only must the structural properties of the individual functions be considered, but also the characteristics of the combination of the two. An example of a law which says something about the combination of two closed operations is the so-called distributive law which we already applied in Fig. 2.22. Two important algebraic structures which are defined using two closed operations are called *fields* and *lattices*. I shall not tell you what the definitions of these structures are. I only mention them because I want you to know where they belong, in case you read or hear about them.

# Chapter 3
# Mathematicians Are Nothing Special – They Draw and Compare

In the title of Chapter 2, mathematicians are said to be humans who count and arrange. Now this is supplemented by the statement that they also draw and compare. Although some drawings were used in Chapter 2 to illustrate numbers as relations concerning distances between points, now numbers no longer stand at the beginning of our considerations. Instead, we now start from points in drawings or in space.

## How Mr. Euclid's Ideas Have Grown Up

Geometry is fun, since here everything is visual and there are no formulas. Undoubtedly, this is true when a ruler and a pair of compasses are used for drawing shapes in a plane. In these shapes, there are angles which can be measured and distances between points which also can be measured, and the results of these measurements can be related to each other. Lots of laws about such relations have been discovered and are explained in textbooks on geometry, e.g., the law that the sum of the three interior angles of any triangle drawn on a plane is 180 degrees, i.e., equal to two right ($90^0$) angles. In classical antiquity, this type of geometry was brought to a high level of maturity. Today it is called "Euclidean geometry," referring to the Greek mathematician Euclid (365-300 BC). Today, we no longer need to climb steep walls to reach the plateau of Euclidean geometry. It might even seem that there is a cable car leading up to that plateau. But this impression obscures a very difficult problem which most people don't see. Since everybody believes they know exactly what a straight line or a plane is, nobody gets the idea that it might be difficult to define these objects with mathematical rigor. In the very beginning of his famous text "Elements", Euclid made the following statements [LO]:

- A point is what has no parts.
- A line is what has length, but no width.
- The extreme ends of a line are points.
- A straight line is a line which lies in between its points in a homogeneous way.
- A plane is an object which lies in between its straight lines in a homogeneous way.

Euclid evidently assumed that everybody knew what length and width are, and what it means for points to lie in a homogeneous way. In Chapter 4, where the concept of axioms is introduced, we shall resume the discussion of this problem.

The law which is presumably the most important in Euclidean geometry is called "the law of Pythagoras," although historians agree on the fact that this law had already been discovered by Arabian or Egyptian mathematicians before Pythagoras (570-510 BC) was born. This law describes an interesting property of so-called right triangles, i.e., triangles with one angle of 90 degrees. On the left side of Fig. 3.1 you see the classical shape which illustrates this law. All three edges of the triangle shown are used as edges of squares. Looking at such a drawing, someone must have come up with the idea that the area of the big square equals the sum of the areas of the two smaller squares. When I saw this drawing for the first time in high school, I certainly did not see that there might be such a simple relationship between the three squares - I had to get this information from the math teacher. But believing what the teacher tells you is not sufficient in mathematics. You have to be convinced by a so-called proof. A proof is a sequence of "evident conclusions" the correctness of which cannot be doubted. Since the time of Euclid, many different proofs of the law of Pythagoras have been found. From all these proofs, the one I like best is based on the two drawings in the center and on the right side of Fig. 3.1. This proof is simple and vivid. The drawings show that the area of the square with the edge (a+b) can be filled using four copies of the right triangle together with either the big grey square (center drawing) or the two small grey squares (right drawing). From this it follows that the sum of the areas of the two small grey squares, $a^2+b^2$, must be equal to the area of the big grey square, $c^2$.

Shapes can be drawn not only in planes, but also on surfaces of solid bodies. Since the earth is a sphere, it is quite natural that shapes drawn on spheres were

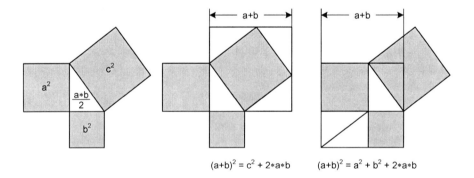

**Fig. 3.1** Verifying the law of Pythagoras

# Coordinate Systems and Matrices

studied rather early and with great interest. Many laws which hold for shapes in planes no longer hold for shapes on the surface of spheres. As an example, let's look at the sum of the three angles of a triangle. If the triangle is drawn in a plane, the sum of the three angles is 180 degrees. Now assume we draw a triangle on the surface of the earth with the first point at the North Pole. From there we draw a straight line southward until we reach the equator; here is the second point. The second edge is drawn eastward on the equator, with the third point on the equator at a distance of one fourth of the equator. From there the third edge goes back to the North Pole. In this triangle, all three edges have the same length, and if such a triangle had been drawn in a plane, the three angles would each be 60 degrees. But all three angles of our triangle on the sphere are 90 degree angles.

When talking about geometry, we should, at least, mention topology. The term topology refers to the Greek word *topos* for location. While in conventional geometry we are interested in angles and distances between points, these are of no interest in topology. In topology, we ask whether it is possible to continuously distort a shape or a solid body in such a way that it is exactly the same as a second given shape or body. This can be illustrated using the assumption that the shape is drawn on a thin elastic foil or that the body consists of an elastic material which can be distorted continuously without breaking apart. The distortion may change the area of the shape or the volume of the body. Then it is possible to start with a circle and make a square out of it, or to start with a cube and make a sphere out of it. But it is impossible to start with a combination of two squares, one of which is drawn inside the other one, and make a single circle from this combination.

Topology as an explicit discipline of mathematics began in about 1850. But it goes back to a question which had previously been asked by the ancient Greeks: "How is it possible that points constitute a space?" Geometric objects have length, area or volume, but a point doesn't have any of these. If geometric objects consist only of points, how is it possible that these objects have substance although their constituents do not? The error in this argument lies in the assumption that the distance between two points is created by the points in between. Distance cannot be created by points, but is an elementary property of pairs of points. Think of the friendship of two persons. Friendship is not a property of an individual, but a relationship between the two individuals.

The actual definition of the concept of a topological space is so abstract that it is questionable whether topology should be considered an aspect of geometry.

We now leave traditional geometry which is characterized by drawing, measuring and comparing. We leave by a bridge which was built around 1640, the time of the 30 Years War in Europe. It takes us back into the world of numbers and arithmetic operations. The building of this bridge really must be considered a great success in conquering the mountains of mathematics, since drawing shapes and measuring distances and angles seem to have nothing in common with arithmetic operations.

The main barrier which had prevented the building of this bridge before then was a problem which is illustrated on the left hand side of Fig. 3.2. In this diagram, the product a∗b appears twice, namely both as the area of the shaded rectangle and as the length of the long vertical line on the left. The fact that the same value can be interpreted alternatively as an area or as a length is possible only if this value is neither an area nor a length, but just a number. Assume that a is the number 3 and b is the number 1.6. Then the product is the number 4.8. Certainly, 4.8 cm is something different from 4.8 cm$^2$, but in both cases there is the same number, 4.8. It always has been quite clear that 3 apples are something different than 3 pears, and that the number 3 represents only what the two sets have in common, namely the result of counting the elements. We have to think in the same way when we look at the left diagram in Fig. 3.2. When I made this diagram, I had to choose a scale, i.e., I had to decide which length should correspond to the number 1. Usually, this length is called the unit length. Once the unit length is given, the unit area follows from it, since it is just the area of a square with edges of unit length.

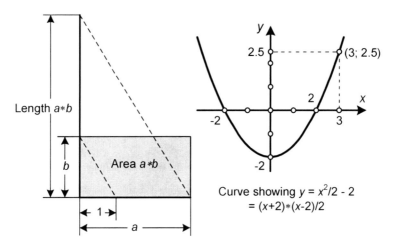

**Fig. 3.2** Geometry and numbers: the coordinate system

Although I didn't mention it explicitly at the time, we crossed the bridge between the world of drawings and the world of numbers once before. This occurred in connection with Fig. 2.6 where the complex numbers were illustrated as points in a plane. There, too, I had to choose a unit length. The points in the plane in Fig. 3.2 are not representations of complex numbers, but of pairs of real numbers (x, y) which we call the coordinates of the points. The number x in the

Coordinate Systems and Matrices    57

first position of the tuple corresponds to a distance on the horizontal axis, and the number y in the second position corresponds to a distance on the vertical axis. Thus the pair (3; 2.5) belongs to the point in the diagram on the right side of Fig. 3.2. The relation which assigns a pair of numbers to each point in the plane is called the *coordinate system*. If we do not restrict ourselves to points in planes, but deal with points in three-dimensional space, we need three coordinates (x, y, z). The z-axis usually is perpendicular to the plane which is determined by the two axes for x and y.

All geometric shapes we are interested in can be drawn on the x-y plane. If we restrict ourselves to the use of a ruler and a pair of compasses only, many shapes are excluded, although they might be very interesting. But since the plane is not only the plane for drawings but for representing pairs of numbers, we are no longer restricted to defining shapes by drawing them, but we can define shapes using formulas. The curve in the right diagram of Fig. 3.2 is a so-called parabola, and I defined it by the formula $y=x^2/2 - 2$. To any given value $x$, a corresponding value $y$ can be computed using this formula, and each pair (x, y) for which this relationship holds defines a point of the curve. This concept of describing a curve in a plane by a formula $y=f(x)$, where the pairs (x, y) correspond to the points of the curve, is so commonly used today that I previously used it in Figs. 2.15 and 2.20 to illustrate the sine function sin(x) and the exponential function $e^x$ without previous explanation. The formulas which until now were looked at only as arithmetic relations now may be looked at as shapes having geometric properties. Although this seems to be so simple to us today, it was a tremendous achievement when the first mathematicians climbed that mountain.

It is possible to do geometry using pairs of numbers, and this is shown in the example in Fig.3.3. Each corner of the triangle is described by a pair of numbers. From these we get the pairs of numbers assigned to the edges by subtracting the coordinates of one end from the coordinates at the other end. These subtractions cannot be done without ordering the points, and this is indicated by the arrows on the edges. The pairs assigned to the edges are the basis for computing both the lengths of the edges and the angles between them. The lengths are obtained using the law of Pythagoras as it is shown in Fig. 3.3. The angles cannot be obtained by

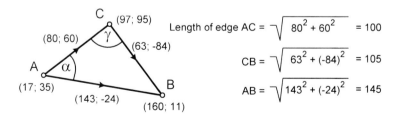

**Fig. 3.3** Computing line lengths with coordinates

arithmetic operations, but at least we can get their cosines. The cosine function was introduced in Fig. 2.15 where the relationships between the four aspects of a complex number – real part, imaginary part, radius and angle – were considered. The method which provides the cosines of the angles of a triangle on the basis of the pairs of numbers assigned to the edges is formal and rather simple. It is an application of a computing formalism which is used in many different areas of mathematics. This formalism is called *vector-* and *matrix-multiplication*.

I shall introduce this formalism using an example which has nothing at all to do with geometry. This may help you to understand that this formalism is universal and that its use is not restricted to a narrow mathematical field. Assume that you have to solve the following problem. Some customers want to buy different quantities of a variety of items from a single provider, and for this they have to select a provider from a list. Each provider has his own price list. The question is which provider should be chosen by each customer to make sure that the customer gets the lowest total cost for his list of supplies. In the example in Fig. 3.4, there are four customers A, B, C and D and two providers I and II. There are three types of items 1, 2 and 3. Looking at the list of customer C, you see that he wants to buy three pieces of item 1, four pieces of item 2 and two pieces of item 3. The price list of provider II shows that he charges five currency units per piece of item 1, four units per piece of item 2 and three units per piece of item 3. The diagram shows an arrangement of three rectangles which are divided into rows and columns, and at each intersection of a row and a column is a square field containing a number. The numbers in the rectangles at the left and at the top define the original problem. These are the price lists of the providers and the supply lists of the customers. Such rectangles filled with numbers are called *matrices*. In the special case that a matrix has only one row or one column, it is called a *vector*, and if both the number of rows and the number of columns are one, the matrix is called a *scalar*. The numbers in the matrix in the lower right corner, the result matrix, represent the amount of the bill the customer of the corresponding column has to pay if he buys his supplies from the provider of the corresponding row. I marked the minimal amounts for each customer by enclosing them in circles. Each amount is computed as a sum of three products. For example, the amount 37 which is in the shaded field is obtained as the result of 5*3+4*4+3*2. The numbers used in this computation are taken from the shaded row of the left matrix and the shaded column of the upper matrix.

I characterized the three matrices using terms we know from arithmetic multiplication, namely first factor, second factor and product. If all three matrices are scalars, the formalism shown actually corresponds exactly to the multiplication of two numbers. The 90 degree curves, which in the example correspond to the different items, help us to find the pairs of numbers which have to be multiplied.

# Coordinate Systems and Matrices

The number of products which have to be added to get the content of one field in the result matrix corresponds to the number of these curves. Therefore, two matrices can be multiplied only if the number of columns of the first factor matrix equals the number of rows of the second factor matrix, because otherwise we could not draw curves where each has a well-defined connection to both sides. The size of the result matrix is determined by the number of rows of the first factor and by the number of columns of the second factor.

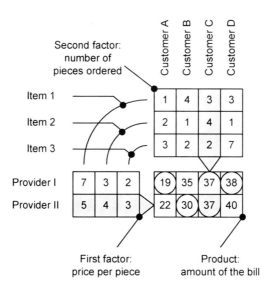

**Fig. 3.4** Pattern for matrix multiplication

The commercial example in Fig. 3.4 was chosen to demonstrate that the use of matrix multiplication is not restricted to geometry. An interpretation of the formalism is not possible without considering its application. In the following section, either the factors or the result will be vectors in planes or in three-dimensional space. The problem which led us to the concept of matrix multiplication was the question of how we could compute the values of the angles or at least the cosines of these angles in the triangle in Fig. 3.3. Fig. 3.5 shows how these cosines are obtained: the vectors of the two edges which enclose the angle are multiplied in such a way that the result is a scalar. This requires that the vector which is used as the first factor is arranged as a *row-vector* while the second factor must be a *column-vector*. When two vectors are multiplied in this way, the result is a scalar, and therefore this way of multiplying two vectors is called computing the *scalar product*. The scalar product always equals the product of the lengths of the two edges and the cosine of the enclosed angle. Isn't that amazing? I certainly could prove this to you, but I don't see the need to present this proof.

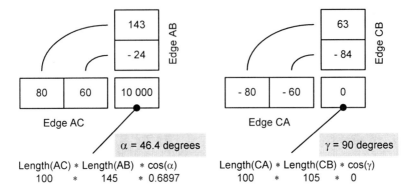

**Fig. 3.5** Computing the scalar products of the triangle edges in Fig. 3.3

When we describe points by tuples of coordinates, the values of these coordinates depend on how we placed the coordinate system in the plane or the space. In Fig. 3.6, the point P in the upper right corner is described in two different coordinate systems. These two systems are both rectangular and share the intersection of the x- and the y-axes. Since one of these coordinate systems is obtained from the other one by rotation, the subscripts o and r are used to identify the two systems; they refer to the words *original* and *rotated*.

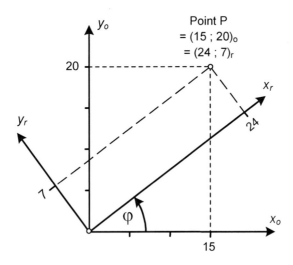

**Fig. 3.6** Specifying a point in two different coordinate systems

Coordinate Systems and Matrices 61

In the o-system the coordinates of the point P are (15, 20), and in the r-system they are (24, 7). Now the question is how to obtain the r-coordinates from the o-coordinates and vice-versa, once the rotational angle φ is given. This can be achieved easily by a matrix multiplication as shown in Fig. 3.7. The numbers in the matrices are obtained from the angle φ by taking its sine and cosine. Here I just repeat what I said earlier: I could prove this to you, but I don't see the need to present this proof. You wouldn't learn much from it.

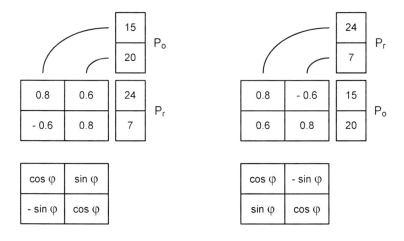

**Fig. 3.7**  Matrix multiplications for the example in Fig. 3.6

Next, I shall show you that matrices not only can be multiplied but that it is possible to compute the reciprocal of a matrix which is the basis for division. Once we have the reciprocal 1/B of a matrix B, we can get the quotient of two matrices A/B by computing A*(1/B). But this requires that we know what the "*unit matrix*" looks like. The two matrices in Fig. 3.7 will help us to obtain the unit matrix since they were introduced to transform the coordinates in both directions: $M_1*P_o=P_r$ and $M_2*P_r=P_o$. We can combine these two equations and get $M_2*(M_1*P_o)=P_o$, and from this follows $M_2*M_1=1$, the unit matrix. This means that $M_2$ is the reciprocal of $M_1$ and vice-versa. In Fig. 3.8 the two matrices from Fig. 3.7 are multiplied and the product is the unit matrix. The *unit matrix* of dimension n is a square matrix with n rows and n columns where all cells are filled with zeros except the cells in the diagonal leading from the upper left to the lower right; these are filled with ones. If the unit matrix of dimension n is multiplied by any square matrix M of the same dimension, the product will be identical to M.

Matrix multiplication is a rather simple formal procedure where certain numbers must be multiplied according to a formal pattern and then their products must be added. This computation can be executed even by someone who has no

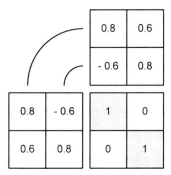

**Fig. 3.8** Verifying the reciprocity of two matrices from Fig. 3.7

idea about what the purpose of this procedure is. Now you have seen not only one but two completely different areas where matrix multiplication is extremely helpful. You must not feel bad at the thought that you would never have come up with such a great idea – I didn't invent it either. It is the same situation which we encounter quite often. It needs a genius to find the concept, but once it has been found, it can be understood and applied by quite normal people.

What I am going to show you now could have been found only by a genius, something we observed previously. Suppose you were expected to come up with a method to solve the problem which is represented at the top of Fig. 3.9 where the two edges a and b enclose the angle φ. These edges are described by their pairs of coordinates, and from these the method provides the coordinates of the vector which is perpendicular to the plane determined by a and b, and whose length equals the area of the shaded parallelogram.

Many years ago, someone taught me the procedure for finding this vector, and I was very amazed to learn how many geometric problems can be solved formally by using matrix computation. In the lower left of Fig. 3.9 you see the formal pattern, and in the lower right this pattern is applied to a simple example. In a first step, the coordinates of the edge a are used to fill certain cells of a square matrix. Each coordinate appears twice in this matrix, and its sign is inverted once. The cells of the diagonal leading from the upper left corner to the lower right are filled with zeros. These rules for filling the cells of a matrix are applicable only if the matrix is three-dimensional. In the second step of the procedure, this matrix is multiplied by the coordinates of the edge b. The result of this multiplication is the vector asked for – what a miracle! Since the result of the procedure is a vector, this type of multiplication is called vector-multiplication which emphasizes the fact that there is a scalar multiplication, too. But I prefer to call it "perpendicular product" since this refers to the fact that the product is perpendicular to the plane determined by the two factors.

Coordinate Systems and Matrices                                                    63

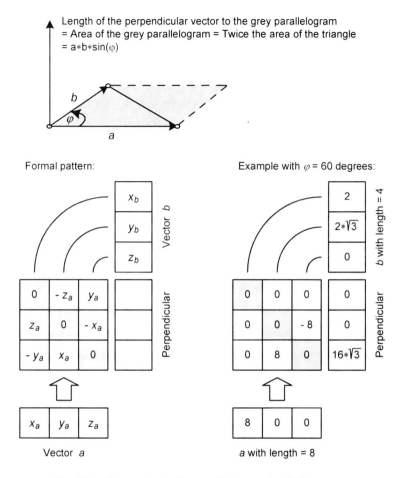

**Fig. 3.9** Computing the "perpendicular product" of two vectors

I chose the numbers for the example in the lower right so that the relevant facts can be easily seen. The triangle, specified by the two edges a and b, was chosen to be half of a triangle with three edges of equal length, that length being 8. The edge a has this full length, but b has only half of it. The angle enclosed by a and b must then be 60 degrees. The plane determined by the two edges a and b is the x-y plane since the z-coordinates of both a and b are zero. It is easy to see that the resulting vector is perpendicular to the x-y plane since only its z-coordinate has a non-zero value. And its length is as required, since it equals the area of a triangle with all angles of 60 degrees and all edges of length 8.

In my comments about Fig. 3.9, I did not mention a question which you may already have asked yourself. There are two possibilities for positioning a vector perpendicular to a given plane. The vector shown in Fig. 3.9 has an upward direction, but it would also have been a correct solution of the problem if the resulting vector had pointed downwards. Somewhere in my procedure, I had to decide which side of the plane the resulting vector would point. This decision was made by selecting the order of the factors. The procedure illustrated in Fig. 3.9 shows that the first factor a is used in a different way than the second factor b. Therefore, it is no wonder that the product depends on the order of the factors. If I had reversed the order, the resulting vector would have pointed downwards. In Fig. 3.9, the chosen order of the factors is illustrated by the arrow at the angle φ. How the direction of the resulting vector and the order of the factor edges are related is determined by the so-called "*right hand rule*": Put your right hand on the plane in such a way that your small finger is on the plane and is curved in the direction of the arrow that circles around φ. Then your thumb will point in the direction of the resulting vector.

While the perpendicular product is restricted to three-dimensional vectors, the scalar product is not restricted to any dimension. The scalar product requires only that the two factors have the same dimension.

Without the concepts of vector- and matrix-multiplication, it would not be possible to solve geometric problems with the help of computers. While Mr. Euclid could draw shapes and could find right angles, or see how certain distances between points are related just by looking at these shapes, a computer cannot look and reason about shapes. Therefore, computers always use coordinate systems to solve geometric problems because, in this case, points and edges can be described in the form of pairs or triples of numbers. Instead of looking and measuring, the computer adds and multiplies and compares numbers, since that's exactly what computers do very well.

## How the Fraction "Zero Divided by Zero" and the Product "Infinity Times Zero" Are Related

When I introduced the term "topology" I said that it might be questionable whether this topic should be considered a part of geometry. Correspondingly, the question might be asked about whether the differential and integral calculus, which will be introduced now, should not be considered topics within geometry, since the fundamental concepts of calculus cannot be defined without referring to illustrations showing graphs of curves in a plane. Calculus was developed shortly after the concept of coordinate systems had been found, and this made it possible to represent functions as curves in a plane. When mathematicians got interested in certain properties of these curves, they had to look for new methods which could provide the answers to their questions. They soon found it desirable to simultaneously consider two curves which represent two functions f(x) and s(x),

Differentiation and Integration 65

where s(x) is the slope function of f(x). What this means is illustrated in Fig. 3.10. The upper part of this figure shows a graph of the curve for a function f(x), and the curve for its partner function s(x) is shown below.

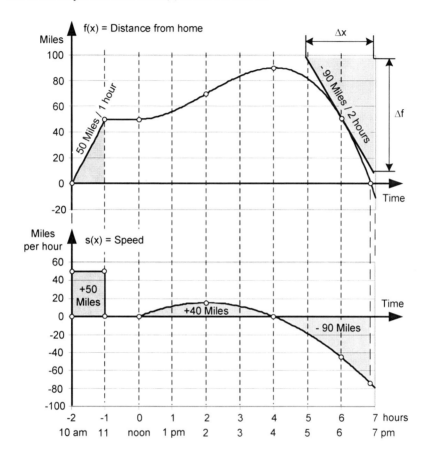

**Fig. 3.10** Example of the relationship between slope and area

In order to illustrate that the special type of relationship between the two functions f(x) and s(x) considered here is of great practical relevance, I shall not introduce this relationship by looking at two abstract functions, but by looking at two functions with everyday interpretations. Let's assume that a family goes on a one day excursion and leaves home in the morning at 10 a.m. The mother drives for one hour at a speed of 50 miles per hour. At 11 a.m., they arrive at a nice lake where they stay for one hour until noon in order to relax and swim in the clear water. Then, at noon, they continue their ride, but now the son is driving. He is not yet an experienced driver, so therefore he drives more slowly. This is just what the family wants, since they are now passing through very pretty countryside. The son accelerates very slowly until he reaches the maximum speed of 15 miles per hour at 2 p.m. At that time, they are 20 miles away from the lake and 70 miles from home. Now the son begins reducing his speed until he comes to a stop at 4 p.m. At that time, the family has reached their maximum distance from home, 90 miles, and they decide to return. Now the father takes the wheel. They want to be back home no later than 7 p.m. which leaves them 3 hours for a distance of 90 miles. Instead of going at a constant speed of 30 miles per hour, the father constantly increases his speed to a maximum of 73.3 miles per hour which is reached 9 minutes before 7 p.m., the time of their arrival back home. At 6 p.m., the speed was 45 miles per hour.

Instead of reading the story I just told you, you could get the same information by studying the graphs shown in Fig. 3.10. The upper graph shows how the family's distance from home changes over time, and the lower graph shows how their speed changes over time. The formal relationship between these two curves is indicated by the grey shaded areas. The grey triangles in the upper curve are used to compute the slope of the curve at given points. The slope at a point of a curve tells us how steep the tangent to the curve is at that point, and whether it leads up or down. The slope is given by the fraction $\Delta f/\Delta x$ where $\Delta f$ is the length of the vertical edge and $\Delta x$ is the length of the horizontal edge of the actual grey triangle. The value of this fraction does not depend on the actual size of the triangle, since if $\Delta x$ is changed by a factor, $\Delta f$ will be changed by the same factor. In our example, $\Delta f$ is measured in miles and $\Delta x$ in hours, and therefore the slope $\Delta f/\Delta x$ is a speed, measured in miles per hour. If the slope leads up, it is positive, which in our example means that the family is actually traveling away from home. And if the slope leads down, they are going towards home. The triangle in the right upper corner of Fig. 3.10 provides the slope of the curve $f(x)$ at 6 p.m., and its value is - 45 miles per hour.

As I mentioned, the lower curve shows how the speed changes over time which, more generally, means how the slope of the upper curve changes with respect to the horizontal axis x. Thus, at 6 p.m., the lower curve provides the speed value of - 45 miles per hour which we found as the slope of the upper curve at 6 p.m. Where the slope of the upper curve is zero, i.e., at points x where the tangent to the upper curve is horizontal, the function $s(x)$ must have the value zero, i.e., the corresponding points of the lower curve must intersect the horizontal x-axis. In Fig. 3.10, this is the case for the interval $-1<x<0$ and for the point x = 4 hours. In the intervals where the slope of the upper curve is constant, the lower curve is horizontal. In our example, this is the case in the intervals $-2<x<-1$ and $-1< x<0$. The grey triangles are used to determine the slopes in the upper diagram, and the grey areas in the lower diagram correspond to the distances $f(x_{right})-f(x_{left})$. The grey rectangle in the interval $-2<x<-1$ has the area (50 miles/hour)*(1 hour)= 50 miles, and this corresponds to the difference f(-1)-f(-2) =(50–0) miles. While it is

Differentiation and Integration 67

easy to get the area of a rectangle, it is difficult to get the areas of the shapes in the intervals 0<x<4 or 4<x<7 – unless we have specific knowledge about the upper curve. However, from the upper curve, we get f(4)-f(0)=(90-50) miles and f(6.852)-f(4)=(0-90) miles. An area which lies below the x-axis of the lower diagram corresponds to a negative value of the difference f($x_{right}$)-f($x_{left}$), meaning that the direction of motion is toward home.

As long as we actually draw the graphs of these curves and get our results by analyzing these curves, we may say we are solving problems of geometry. But now we assume that either the function f(x) or the function s(x) is given as a formula which tells us how we can compute the value of the function when the value of x is given. In our example in Fig. 3.10, the formulas for the two curves in the interval 0≤x≤7 are f(x)=50+7.5*$x^2$-1.25*$x^3$ and s(x)=15*x-3.75*$x^2$, respectively. Wouldn't it be great if we had methods for deriving the formula of s(x) from the formula of f(x) and vice versa? Such methods actually exist, and I shall give you at least an idea about what they are. Once the method for deriving the formula of s(x) from the formula of f(x) has been found, the method for deriving the formula of f(x) from the formula of s(x) could be obtained by just reversing the first method.

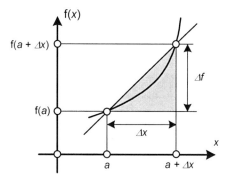

**Fig. 3.11**   Getting the slope of a tangent by computing a limit

The slope Δf/Δx which is defined by the grey triangle in Fig. 3.11 is not the slope of a tangent but of a straight line which intersects the curve at two points. But by making Δx smaller and smaller, this slope will become a better and better approximation of the slope of the tangent at the point [a, f(a)]. Clearly, we will get the exact slope of this tangent by computing the limit of the fraction Δf/Δx = ( f(a+Δx) – f(a) )/Δx for Δx becoming infinitely small. Do you remember that I used the strange phrase "the fraction zero divided by zero" in the title of this section? I used this as a hint about the computation of the limit of the fraction Δf/Δx by making Δx smaller and smaller, with both the numerator and the denominator approaching zero. In Fig. 3.12, this computation of a limit is performed for the special case where $f(x)=x^2$, and then the result is generalized for functions $f(x)=x^n$. The pair of formulas $(f(x), s(x)) = (x^n, n*x^{n-1})$ which was deduced in Fig. 3.12 can be applied to all polynomials, since any polynomial can be written as a weighted sum of powers of x.

---

If the function f(x) is given as a formula, the formula of the slope function s(x) can be obtained by computing a limit:
$$s(x) = \lim_{\Delta x \to 0} \frac{f(x + \Delta x) - f(x)}{\Delta x}$$

---

For the case $f(x) = x^2$ we get:

$$s(x) = \lim_{\Delta x \to 0} \frac{(x + \Delta x)^2 - x^2}{\Delta x} = \lim_{\Delta x \to 0} \frac{(x^2 + 2*x*\Delta x + \Delta x^2) - x^2}{\Delta x}$$

$$= \lim_{\Delta x \to 0} \frac{2*x*\Delta x + \Delta x^2}{\Delta x} = \lim_{\Delta x \to 0} (2*x + \Delta x) = 2*x$$

---

For the case $f(x) = x^n$ we get: $\quad s(x) = \lim_{\Delta x \to 0} \frac{(x + \Delta x)^n - x^n}{\Delta x} = n * x^{n-1}$

---

If we apply this to the function f(x) in Fig. 3.10, we get:

$$f(x) = 50 + 7.5*x^2 - 1.25*x^3$$

$$s(x) = \quad 15*x - 3.75*x^2$$

**Fig. 3.12** Derivation of the slope formula for the function $f(x)=x^n$

Once we know that the partner of the function $f(x)=x^n$ is the slope function $s(x)=n*x^{n-1}$, we can easily conclude that the partner of the function $s(x)=x^n$ will be the function $f(x)= x^{n+1}/(n+1)$. That's what I had in mind when I said that the method which provides the formula of s(x) for a given formula of f(x) can be reversed. But there is a problem you might not yet be aware of. Assume that now the family whose journey was described by the functions in Fig. 3.10 does not start from home, but from the lake which is 50 miles from home. And assume further that the information about who is driving and at which speed stays as it was. Then, the function s(x), which describes how the speed changes over time, will stay exactly as it was, but the function $f_{new}(x)$ which gives the position of the family for any point in time during their journey will no longer be as shown in Fig. 3.10. Since, at any point in time, the family's distance from home now is 50 miles more than the function $f_{old}(x)$ says, we get the new position function by adding 50 to the old function, i.e., $f_{new}(x)=50+f_{old}(x)$. The curve for this new

# Differentiation and Integration

function will look similar to the one shown in Fig. 3.10, but will be shifted vertically by a distance of 50 miles. Such a shift will not change the slope of the curve at any point, and therefore the shifted curve will have the same slope function s(x) as the un-shifted curve.

Since there is no difference between the new and the old speed functions, the areas under these curves will not be affected by changing the starting position of the family, i.e., by the vertical shift of the position curve. This is a consequence of the fact that such an area does not correspond to one value of the position function, but to the difference $f(x_{right})-f(x_{left})$ of two values. If the curve of f(x) is shifted vertically by a distance of $\Delta f$, the difference will stay the same: $[\Delta f+f(x_{right})]-[\Delta f+f(x_{left})]=f(x_{right})-f(x_{left})$. From these considerations it follows that, while a given function f(x) has exactly one partner s(x), a given function s(x) has infinitely many partners f(x) where any two of these differ only by a constant, $\Delta f$.

This unsymmetrical relationship between f(x) and s(x) can also be illustrated by a completely different approach. We now consider the problem of constructing an f-curve for a given s-curve on the basis of the areas under the s-curve. The area between an s-curve and the x-axis is well-defined only if an interval $x_{left} \leq x \leq x_{right}$ is given. This area is equal to the difference $f(x_{right})-f(x_{left})$. Thus we have a conflict: if we leave both ends of the interval open, we cannot have a well-defined area, and if we choose a well-defined interval, we cannot get a function f(x). This problem can be solved by choosing a particular value for $x_{left}$ and leaving the value of $x_{right}$ open. In order to express the fact that $x_{left}$ is still constant, but $x_{right}$ is now a variable, we use $x_0$ instead of $x_{left}$ and x instead of $x_{right}$. However, this creates a new problem: we now have lost the ability to use the letter *x* within the limits of the interval. It would be nonsense to write $x_0 \leq x \leq x$. Again, we can overcome this problem by using a substitution. We can use a different letter within the limits. I chose the letter v, and with this the interval becomes as $x_0 \leq v \leq x$.

The choice of the constant value $x_0$ determines a position where the f-curve intersects the *x*-axis because the area will be zero if the length of the interval is zero, and this is the case for $x=x_0$. Clearly, our choice of a value for $x_0$ determines what function f(x) we get.

Fig. 3.13 shows how an area between an s-curve and the *x*-axis can be computed as the limit of a sum. The interval between $x_0$ and *x* is divided into *n* subintervals of equal length $\Delta v=(x-x_0)/n$. Each of the shaded rectangles has the width $\Delta v$, and its height is chosen so that the s-curve intersects the upper edge at its center. In the formula which describes the sum of the *n* shaded rectangles, the upper-case Greek letter sigma, $\Sigma$, is used. This corresponds to the first letter of the word sum. The values of the summands depend on the value of the variable *j* which is used to enumerate the summands. The lowest value of *j* is 1 and it is written below the $\Sigma$. The highest value of *j* is *n* and it is written above the $\Sigma$. While the sum of *n* rectangles provides only an approximation of the area between the s-curve and the *x*-axis, the exact area is obtained by computing the limit of the sum. The width Δv of the rectangles is made smaller and smaller, and at the same time the number *n* of rectangles is made greater and greater. It was this process of coming closer and closer to the limit which I had in mind when I included the phrase 'the product "infinity times zero"' in the title of this section. The transition from a finite number of summands to an infinite number is specified by replacing the symbol Σ by the symbol ∫ which looks like a capital letter S which has been compressed horizontally. This symbol is called the *integration symbol*. While the boundaries of the interval $1 \leq j \leq n$ are written at the bottom and the top of the Σ, the boundaries of the interval $x_0 \leq v \leq x$ are written at the bottom and top of the integration symbol.

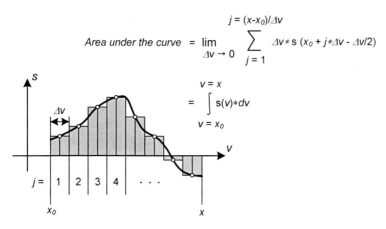

**Fig. 3.13** Obtaining the area under a curve by computing a limit

In Fig. 3.14 you can see how the slope of a function f(x) and the area between a curve s(x) and the *x*-axis in the interval $x_0 \leq v \leq x$ are expressed in formulas. Here the symbols Δf, Δ*x* and Δv, which stand for small but finite differences, are replaced by df, dx and dv which indicate the fact that limits are computed by making Δf, Δ*x* and Δv smaller and smaller.

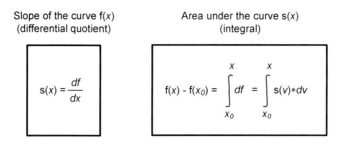

**Fig. 3.14** Notations for derivatives and areas in formulas

The process which produces the slope function s(x) for a given function f(x) is called *differentiation*, and the slope function s(x) is called the *derivative* of f(x). The process which produces the function f(x) from its derivative s(x) is called *integration*, and the resulting function f(x) is called the *integral* of s(x).

Differentiation, i.e., providing the derivative s(x) for a given function f(x), is of great practical importance when values $x_{extreme}$ must be found which determine the positions of extreme points of the f-curve. Such a point is either a mountain top or a valley bottom, i.e., a maximum or a minimum. At these points the tangent is horizontal and the slope $s(x_{extreme})$ is zero. In the example in Fig. 3.10, the f-curve has a mountain top at $x_{max} = 4$ and a valley bottom at $x_{min} = 0$. Finding the positions $x_{extreme}$ of a given function f(x) is no problem when the f-curve is drawn, since then we can see these extremes just by looking at the curve. But in most cases, drawing the f-curve can be avoided and the positions of its extreme points can be found just by computation. This requires that the function f(x) be described by a formula.

We now consider an example which illustrates the practical relevance of being able to compute extreme points of a given function f(x). Fig. 3.15 shows a section of a floor plan with a corner where a hall turns by 90 degrees and changes its width. We assume that some construction is going on in this building and that the construction workers must carry a ladder through the hall and around the corner. If the length of this ladder exceeds a certain maximum value, it will be impossible to move the ladder around the corner. How do we get this maximum length? We define a function f(x) which gives the length of the ladder for any value of the variable distance x shown in Fig. 3.15. The formula for this function can be easily

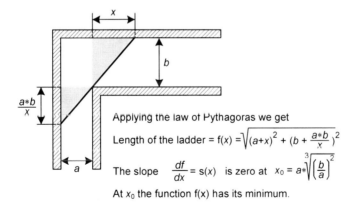

Applying the law of Pythagoras we get

Length of the ladder = $f(x) = \sqrt{(a+x)^2 + (b + \frac{a*b}{x})^2}$

The slope $\frac{df}{dx} = s(x)$ is zero at $x_0 = a*\sqrt[3]{\left(\frac{b}{a}\right)^2}$

At $x_0$ the function f(x) has its minimum.

Example with numbers: a=2 m, b=3 m, Length of the ladder = 7.02 m

**Fig. 3.15** Example of a problem solved by computing an extreme value

obtained by applying the law of Pythagoras. We also use the fact that the two shaded triangles are similar, with the same shapes but with different sizes. The details of the computation are not shown since they are not relevant for understanding this example. In the resulting formula, the widths a and b of the hall are left open. If we assume that these widths are 2 meters and 3 meters, the valley bottom (minimum) of the f-curve lies at $x_0$=2.621 meters, and therefore the maximum length of the ladder is $f(x_0)$=7.02 meters.

So far, we looked only at the relationship between a function $f(x)$ and its derivative $s(x)$. But the function $s(x)$ also has a derivative, which also has a derivative, and so on. Obviously, there is the need to find a way to symbolize the position of a derivative in a sequence of derivatives which begins with a function $f(x)$. The following representation is used for a fourth derivative:

$$\frac{d^4 f}{dx^4}$$

This formula identifies a function which is the fourth derivative in a sequence of derivatives, the first being the derivative of the function $f(x)$. This form of writing derivatives was introduced by the philosopher and mathematician, Gottfried Wilhelm Leibniz (1646-1716), over 300 years ago. It is important for you to notice the difference of the positions of the 4 in the numerator and the denominator of the fraction. In the denominator, the 4 is really an exponent of a power, namely $(dx)^4$. In the numerator, the 4 is used only to identify the position of the derivative in the sequence of derivatives. This derivative is the fourth in the sequence.

The example in Fig. 3.10 may help you to understand why there is an exponent of a power only in the denominator of the fraction, but not in the numerator. In Fig. 3.10, the values of x are points in time, measured in hours; the function $f(x)$ describes the position of the vehicle, i.e., its distance from a reference point, measured in miles, and the function $s(x)$ describes the speed of the vehicle in miles per hour. Since here f represents a distance, its physical unit, miles, will be unchanged in all derivatives in the sequence. In contrast to this, the physical unit of x which is a time unit, hours, will appear in a power with an increasing exponent in the sequence of derivatives of $f(x)$. The first derivative is the speed which has the physical unit distance per time. The derivative of the speed function describes how the speed changes with time (acceleration); its unit is ((distance per time) per time) which has the square of the time unit in the denominator.

Now we ask what happens when we continue computing derivatives of given functions. Besides the polynomials, I introduced two other functions which are of great importance in the world of applied mathematics, namely the sine function $\sin(x)$ and the exponential function $e^x$ (see Figs. 2.15 and 2.20).

# Differentiation and Integration

These two functions together with the simple polynomial $x^4$ appear in the leftmost column of the table in Fig. 3.16. To the right of these functions, you see the corresponding sequences of their derivatives. In the case of the function $x^4$, each step which leads to the next derivative reduces the exponent by one. Consequently, after 4 steps the exponent will be zero with the corresponding curve being a horizontal straight line whose slope is zero for all values of x. This slope of zero will remain throughout the infinite remaining sequence. In the case of the sine function and the exponential function, however, the situation is quite different. In both cases, there are positions (shaded cells) in the sequence which are equal to the original function f(x). This means that the whole sequence of derivatives is a periodic repetition of a short section of finite length.

| $f(x)$ | $\dfrac{df}{dx}$ | $\dfrac{d^2f}{dx^2}$ | $\dfrac{d^3f}{dx^3}$ | $\dfrac{d^4f}{dx^4}$ | $\dfrac{d^5f}{dx^5}$ |
|---|---|---|---|---|---|
| $x^4$ | $4*x^3$ | $4*3*x^2$ | $4*3*2*x$ | $4*3*2*1$ | 0 |
| $\sin(x)$ | $\cos(x)$ | $-\sin(x)$ | $-\cos(x)$ | $\sin(x)$ | $\cos(x)$ |
| $e^x$ | $e^x$ | $e^x$ | $e^x$ | $e^x$ | $e^x$ |

**Fig. 3.16**  Examples of sequences of derivatives

With the results shown in Fig. 3.16, I could easily compose the equations which are given in Fig. 3.17. The equations in the top row are called *differential equations*, since each of them describes a relationship between the elements of a sequence of derivatives of a function f(x). Solving a differential equation means finding the formula of the function f(x) for which the equation holds. I guess that most of you haven't learned how to solve differential equations, and that is quite ok. Mathematicians, physicists and engineering professionals, however, are very familiar with such equations, and they know how to deal with them. For you, it is quite sufficient that you know what a differential equation is. And on the basis of the information given in Fig. 3.16, you should be able to verify that the solutions given in Fig. 3.17 are correct.

| Differential equation | $x^2 * \dfrac{d^3f}{dx^3} - 6 * \dfrac{df}{dx} = 0$ | $\dfrac{d^2f}{dx^2} + f(x) = 0$ | $\dfrac{df}{dx} - f(x) = 0$ |
|---|---|---|---|
| Solution | $f(x) = x^4$ | $f(x) = \sin(x)$ | $f(x) = e^x$ |

**Fig. 3.17**  Examples of differential equations

## Relations Which We Can Deduce, but Not Really Understand

Do you still remember that, at the end of my discussion about the process for creating numbers, I said that the power term $2^i$ has the value $0.769 + 0.639i$ ? At that time, I could not yet show you how this result was obtained, and I had to postpone this until later sections. But now, everything has been introduced which I need to deduce this result. Hopefully, reading my deduction in this section will make you realize that in mathematics almost everything is connected to everything else. Now I am ready to "prepare a soup" using as ingredients the concepts of complex numbers, polynomials, limits, derivatives and the transcendental functions $e^x$, $sin(x)$ and $cos(x)$.

| | | | | | | | | | | |
|---|---|---|---|---|---|---|---|---|---|---|
| $p(x)$ | = | $c_0$ | + | $c_1 * x$ | + | $c_2 * x^2$ | + | $c_3 * x^3$ | + | $c_4 * x^4$ |
| $\dfrac{dp}{dx}$ | = | 0 | + | $c_1$ | + | $2 * c_2 * x$ | + | $3 * c_3 * x^2$ | + | $4 * c_4 * x^3$ |
| $\dfrac{d^2p}{dx^2}$ | = | 0 | + | 0 | + | $2 * c_2$ | + | $3 * 2 * c_3 * x$ | + | $4 * 3 * c_4 * x^2$ |
| $\dfrac{d^3p}{dx^3}$ | = | 0 | + | 0 | + | 0 | + | $3 * 2 * c_3$ | + | $4 * 3 * 2 * c_4 * x$ |
| $\dfrac{d^4p}{dx^4}$ | = | 0 | + | 0 | + | 0 | + | 0 | + | $4 * 3 * 2 * c_4$ |
| $\dfrac{d^5p}{dx^5}$ | = | 0 | + | 0 | + | 0 | + | 0 | + | 0 |

**Fig. 3.18** Derivatives of a polynomial

The preparation of this soup begins with Fig. 3.18. The first row of the table in this figure contains a polynomial where the coefficients are not specified numbers, but the variable coefficients $c_0$, $c_1$, $c_2$, etc. Although a polynomial of fourth degree is shown, you should assume that there are more columns of the table to the right, i.e., that the polynomial could be of higher degree. Finally, we shall consider polynomials which don't have any degree at all, but have an infinite number of summands. Starting with the polynomial in the top row, I moved down step by step by computing the derivatives, i.e., the polynomial in any row which is not the top row is the derivative of the polynomial in the row just above it. Thus, in each column, the exponent of x is decremented by one on the way down from row to

Euler's Relation

row. You will notice that the cells of the right-most column would have the same contents as the second row in Fig. 3.16 if $c_4$ had the value 1. Now we move on to Fig. 3.19 whose structure should remind you of Fig. 3.16. In both figures, sequences of derivatives are shown for the different original functions f(x). While in Fig. 3.16 the value of the variable x has been left open, x is set to zero in Fig. 3.19. The correspondence between Fig. 3.18 and 3.19 has been made clearer by shading those cells which have the same contents in both figures.

|  | f(x) at x=0 | $\frac{df}{dx}$ at x=0 | $\frac{d^2f}{dx^2}$ at x=0 | $\frac{d^3f}{dx^3}$ at x=0 | $\frac{d^4f}{dx^4}$ at x=0 | $\frac{d^5f}{dx^5}$ at x=0 |
|---|---|---|---|---|---|---|
| p(x) | $c_0$ | $c_1$ | $2*c_2$ | $3*2*c_3$ | $4*3*2*c_4$ | $5*4*3*2*c_5$ |
| $e^x$ | 1 | 1 | 1 | 1 | 1 | 1 |
| sin(x) | 0 | 1 | 0 | -1 | 0 | 1 |
| cos(x) | 1 | 0 | -1 | 0 | 1 | 0 |

**Fig. 3.19**   Values of different functions and their derivatives at x=0

The fact that the sequence of derivatives of $e^x$ is just a sequence of ones in Fig. 3.19 is a consequence of the last row of Fig. 3.16 which says that all derivates of $e^x$ are equal to $e^x$ itself, and $e^0=1$ (see Fig. 2.20). The zeros and ones in the rows of sin(x) and cos(x) are a consequence of the definitions of these functions which were given in Fig. 2.15. In this figure, the curve of the function sin(x) is shown, and as I mentioned previously, the curve of cos(x) is obtained by just shifting the sine-curve to the left by a distance of $\pi/2$. From this it follows that the value of cos(0) is equal to the value of sin($\pi/2$) which is 1. Looking at the sine-curve in Fig. 2.15, you can see that the curve of its first derivative, i.e., its slope, must be the same as the cosine-curve. This knowledge was used previously to fill the cells of the third row of Fig. 3.16, and from there it is only a short step to the fill the cells in the last two rows of Fig. 3.19.

I hope you could follow me as I deduced the contents of the tables in Figs. 3.18 and 3.19. However, in following me, you might have had a feeling of uneasiness caused by the fact that you didn't know my goal. Therefore, it is high time to give you some information about this goal. Long ago, some mathematicians asked themselves whether it could be possible for any curve to be approximated by a polynomial of adequate complexity. Once this idea was born, the problem to be solved became how to find the coefficients of the approximating polynomial for a given function. The functions we want to approximate by polynomials are the exponential function, the sine function and the cosine function. With the results in the table in Fig. 3.19, we already are very close to finding the coefficients of these polynomials.

But before we go to the last step which leads us to the goal, we have a brief look at a function that cannot be approximated by a single polynomial. The graph of this function is shown in the upper part of Fig. 3.20. It is composed of three sections, each defined as a polynomial, and therefore its slope curve is also composed of three sections. The graph of this slope curve is shown in the lower part of the figure, and you can see that this curve has two corners where no single tangent can be drawn. This means that for x=-1 and x=+1 the derivative of this function is not defined. As you soon shall see, the method which provides the coefficients of an approximating polynomial requires that the function to be approximated has an infinite sequence of well-defined derivatives. Since this is not the case for the f-function in Fig. 3.20, it cannot be approximated entirely by a single polynomial.

From Fig. 3.16, we know that the exponential function and the sine function have infinite sequences of well-defined derivatives. The same is true for the cosine

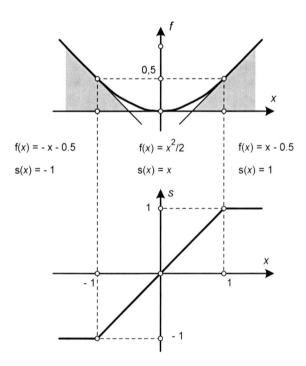

**Fig. 3.20** Example of a three-section curve

# Euler's Relation

function. Now we go the last step which leads us to the coefficients of the approximating polynomials for the three non-polynomials in Fig. 3.19. These coefficients are shown in Fig. 3.21. The step which leads from Fig. 3.19 to Fig. 3.21 is rather simple: we just ask what the values of the coefficients $c_0$, $c_1$, $c_2$, etc. must be in order to obtain the values of the derivatives in Fig. 3.19. If, for example, we want all the values in the row of p(x) in Fig. 3.19 to be 1, we must set the variable coefficients $c_i$ to the values given in the row of $e^x$ in Fig. 3.21. In the third column of Fig. 3.19 involving $c_2$ for $e^x$, for example, $2*c_2$ must equal 1, so $c_2$ must equal 1/2. Although the table in Fig. 3.21 ends with the column for $c_6$, one can easily deduce the general rule for getting from any coefficient to the next higher coefficient. In order to get an exact approximation of a non-polynomial function by a polynomial, we need an infinite number of coefficients. Since the term "polynomial" is restricted to cases where the number of coefficients is finite, mathematicians use the term "*series*" for the cases where the number of coefficients is infinite.

|        | $c_0$ | $c_1$ | $c_2$ | $c_3$ | $c_4$ | $c_5$ | $c_6$ |
|--------|-------|-------|-------|-------|-------|-------|-------|
| $e^x$  | 1 | 1 | $\frac{1}{2}$ | $\frac{1}{3*2}$ | $\frac{1}{4*3*2}$ | $\frac{1}{5*4*3*2}$ | $\frac{1}{6*5*4*3*2}$ |
| sin(x) | 0 | 1 | 0 | $\frac{-1}{3*2}$ | 0 | $\frac{1}{5*4*3*2}$ | 0 |
| cos(x) | 1 | 0 | $\frac{-1}{2}$ | 0 | $\frac{1}{4*3*2}$ | 0 | $\frac{-1}{6*5*4*3*2}$ |

**Fig. 3.21** Values for the coefficients in Fig. 3.19

At the top of Fig. 3.22, the series for the three functions from Fig. 3.21 are arranged in such a way that you can easily see the rather simple relationship between corresponding series. In contrast to Fig. 3.21, I now use the abbreviation n! for the product chains $1*2*3*4*5* \ldots *n$. This abbreviation is common in mathematical formulas, and is spoken "*n factorial*". The summands from the $e^x$ series appear alternately in the series of sin(x) and cos(x). But since their signs also alternate, the series of $e^x$ cannot be obtained simply by adding the series for sin(x) and cos(x). Once mathematicians had gotten that far, they began to look for a way to express the relationships among these three series by a single formula. Once again, they dipped into their bag of tricks and remembered that they had created the imaginary number i by defining a number whose square is -1. From this follows the sequence of powers of i shown in the shaded rows in the middle of Fig. 3.22. The information in these shaded rows gives a basis to the assumption that the relationship between the three series could be expressed by one formula using $e^{i*x}$ instead of $e^x$. Actually, this assumption proved to be correct. Below the shaded rows, you find the series of $e^{i*x}$ and its decomposition into two summands, one being the series for $\cos(x)$ and the other being the series for $i*\sin(x)$, with the final formula on the bottom line of Fig. 3.22.

$$e^x = 1 + \frac{x^1}{1!} + \frac{x^2}{2!} + \frac{x^3}{3!} + \frac{x^4}{4!} + \frac{x^5}{5!} + \frac{x^6}{6!} + \frac{x^7}{7!} + \frac{x^8}{8!} + \cdots$$

$$\sin(x) = \frac{x^1}{1!} - \frac{x^3}{3!} + \frac{x^5}{5!} - \frac{x^7}{7!} + \cdots$$

$$\cos(x) = 1 - \frac{x^2}{2!} + \frac{x^4}{4!} - \frac{x^6}{6!} + \frac{x^8}{8!} \cdots$$

| $i^1$ | $i^2 = i*i^1$ | $i^3 = i*i^2$ | $i^4 = i*i^3$ | $i^5 = i*i^4$ | $i^6 = i*i^5$ | $i^7 = i*i^6$ | $i^8 = i*i^7$ |
|---|---|---|---|---|---|---|---|
| i | -1 | -i | 1 | i | -1 | -i | 1 |

$$e^{i*x} = 1 + \frac{i*x}{1!} - \frac{x^2}{2!} - \frac{i*x^3}{3!} + \frac{x^4}{4!} + \frac{i*x^5}{5!} - \frac{x^6}{6!} - \frac{i*x^7}{7!} + \frac{x^8}{8!} + \cdots$$

$$= 1 - \frac{x^2}{2!} + \frac{x^4}{4!} - \frac{x^6}{6!} + \frac{x^8}{8!} + \cdots$$

$$+ i*\left( \frac{x^1}{1!} - \frac{x^3}{3!} + \frac{x^5}{5!} - \frac{x^7}{7!} + \cdots \right)$$

$$e^{i*x} = \cos(x) + i * \sin(x)$$

**Fig. 3.22** Derivation of Euler's formula via polynomial approximation of functions

This formula was discovered 250 years ago by the genius mathematician Leonhard Euler, and therefore it is called Euler's formula. You should also notice that Euler is honored twice with this formula: it not only refers to him by name, but it also contains the limit number e which refers to the first letter of his name.

The definitions of the functions $\sin(x)$ and $\cos(x)$ were referred to in Figs. 2.14 and 2.15 where a complex number having radius r and angle $\alpha$ was written $r*[\cos(\alpha)+i*\sin(\alpha)]$. Comparing this to Euler's formula, we see that the variable x corresponds to the angle $\alpha$. Thus, once we have the radius r and the angle $\alpha$ of a complex number, this number can be expressed by the product $r*e^{i*\alpha}$.

Now, please don't say that you cannot understand Euler's formula because it combines transcendental functions and imaginary numbers in a way which strikes you as odd. Don't you realize that I have never referred to your experiences from everyday life? The important feature of my deduction lies in the fact that, when

Euler's Relation

writing the formulas, I arranged their elements in such a way that you could easily see where similar patterns are neighbors in their row or their column. Throughout, the order of the natural numbers played a central role. This is true in the case of the powers of x and i: $x$, $x^2$, $x^3$, $x^4$ etc., or $i$, $i^2$, $i^3$, $i^4$ etc., and in the case of the derivatives: $df/dx$, $d^2f/dx^2$, $d^3f/dx^3$, $d^4f/dx^4$ etc., and also in the case of the factorials: 1!, 2!, 3!, 4!, etc. Be assured that for me, and also for any mathematician, this formula is just a result which can be helpful in solving different kinds of mathematical problems, and which was deduced formally step by step. You should not believe that there is any hidden meaning behind this formula which only some expert geniuses can understand.

We now use this formula to answer our original question about the value of $2^i$, the question posed at the beginning of this section. If we had asked for the value of $e^i$, we would have obtained the result directly from Euler's formula (see the second row in Fig. 3.23). But since we want the value of $2^i$, we first have to look up the value of x where $e^x$ is 2 (see Fig. 2.20). This is a problem where we need the logarithm, i.e., the function which provides the exponent *exp* to a given base *b* when the power $p=b^{exp}$ is given. This is expressed as $exp = \log_b p$. In our case, the base *b* is e and the power *p* is 2. When the base is Euler's number e, the logarithm is called the *natural logarithm* and the function symbol is "ln." So in this case, $2 = e^{\ln 2}$. You are likely to find the LN-key on your pocket calculator. Then you may check the result yourself: $\ln 2 = 0.693147$. Using this, the result of $2^i$ can be obtained as shown in the third row of Fig. 3.23, $2^i = 0.7693 + 0.6389 * i$.

Euler's formula can even provide a result for the power $i^i$. First we have to find a way of writing i as a power of e. Looking at Euler's formula, we notice that the

| x | cos(x) | sin(x) | $e^{i*x}$ |
|---|---|---|---|
| 1 | 0.5403 | 0.8415 | $e^i = 0.5403 + 0.8415 * i$ |
| $\log_e 2 = \ln 2 = 0.6931$ | 0.7693 | 0.6389 | $e^{(\ln 2)*i} = 2^i = 0.7693 + 0.6389 * i$ |
| $\frac{\pi}{2}$ | 0 | 1 | $e^{\left(i*\frac{\pi}{2}\right)} = i$ |
| $i * \frac{\pi}{2}$ | | | $i^i = \left(e^{\left(i*\frac{\pi}{2}\right)}\right)^i = e^{-\frac{\pi}{2}} = 0.2079$ |

**Fig. 3.23** Examples using Euler's formula

result will be $e^x = i$ if $\cos(x)$ is 0 and $\sin(x)$ is 1. This is the case for $x=\pi/2$. Therefore, we will obtain the result of $i^i$ by setting $x=i*\pi/2$ in Euler's formula (see the last row of Fig. 3.23). Isn't this amazing? By jauntily defining a number whose square is -1, we created a number i and called it an imaginary number. And now we are confronted by the fact that a power term which is a simple abbreviation in the world of natural numbers, e.g., $5^5=5*5*5*5*5$, still has a meaning in the case of the formal term $i^i$ where the exponent is an imaginary number. In this case, we cannot obtain the result by writing a product where all factors are equal and appear "i times" in a chain. A factor may appear five times, but not i times; the term "i times" is always complete nonsense.

Now our flight in a space ship over the mathematics continent has come to its end. I wanted to show you two things which, hopefully, you shall never forget. One thing is the structure of this continent, i.e., which countries are present and how they are related to each other. The other thing is the fact that all journeys on this continent start with simple structures which can be interpreted by referring to our experiences in the real world, but soon lead us into regions which are purely formal and cannot be interpreted as abstractions from reality. You will never get lost, however, if you stay on the paths of logical consistency.

# Chapter 4
# When It Helps to Ignore Any Meaning

In this chapter, you may find many concepts which you think belong to the two preceding chapters on mathematics. But it is also possible that a computer scientist, when he reads the following sections, will claim that the subjects covered here belong to the area of computer science. Both views have certain justifications. Therefore, I decided to assign my description of the world of formalisms neither to mathematics nor to computer science, but to grant it its own chapter.

## Where Discretionary Powers Are Not Allowed

When a structural engineer has analyzed the structure of a building with respect to its static equilibrium, another engineer must check whether the computations of the first engineer are correct. While a judge in court may use his discretionary powers in making his judgements, the engineer has no discretionary powers at all on the way to his judgements. When the computations of the engineer are correct, no one has any reason to object to his judgement. But in court, it is possible that two different judges may come up with different judgements without one of them being wrong. In general, a judge has to weigh the different arguments against each other, and the result may depend on his personal preferences. In the world of formalisms, however, powers of discretion do not exist. We may even say that it was the goal of getting rid of all powers of discretion which led to the creation of the formal world.

You almost certainly will remember statements which contain the word "formal." For example, consider someone who makes a written application for something and finds that his application is rejected. Then he claims that the person responsible for the rejection had quite formally decided about it based upon a set routine, without considering the deeper meaning of the text of the application. Or think of an article in a newspaper saying that a higher court has repealed the judgment of a lower court on the basis of a formal irregularity in the process. For most people who are not graduates of a law school, decisions based purely on formal reasons have a negative taste. The reason for this lies in the assumption that the decider refused to use his powers of discretion, since otherwise he would have come up with a more favorable decision. Or consider the people who believe that formal rules, whose sense and purpose cannot be understood at all, have been applied. I am not a judge or a lawyer and therefore I am allowed to come up with a rather silly example. Let's suppose that there was a rule in the legal process regulations requiring that the accused person must stand when the judge reads the verdict. And let's suppose further that a verdict is negated on the grounds of a video showing that the defendant stayed seated during the reading of the verdict. Would you think this was a good decision? I certainly would not. Fortunately, the formal rules presented in the following sections are not of the kind where we could argue about their sense or nonsense.

The criterion for detecting whether a rule is completely *formal* or not is as follows: a rule is formal if, just by observing the process, anyone can deduce whether the rule was obeyed or violated. It is not difficult to invent rules which at the first glance look like formal rules but which, nevertheless, are not completely formal. For example, consider the rule, "A college student must write his name and address on the first page of his exam paper in such a way that it can be read clearly." In this case, it can happen that the text which has been written is accepted by one professor, but rejected by another one. For our following discussions, we use this definition: A formal rule is characterized by the fact that it can be obeyed in such a way that no one is required to refer to his discretionary powers in order to verify that the rule has been obeyed.

It would not be reasonable to try to write all laws and regulations as formal rules in order to eliminate the need for the discretionary powers of judges, thus enabling machines to provide the judgements. But there are some domains where it is absolutely necessary for decisions to be made without any reference to discretionary powers. For example, the question of whether a proof of a mathematical theorem is correct must have an answer that does not depend on the personal preferences of the examiners.

When you were in high school, did you ever write a composition which the teacher returned to you with the remark "subject not addressed"? In such a case, there was a question about how the subject assigned by the teacher was to be interpreted. Obviously, the teacher and the student had answered this question differently, and one of three possible cases had occurred. In case (1), the assigned subject had been worded so vaguely that it could be interpreted in different ways; in this case, the interpretation depended on the discretionary powers of the reader. In cases (2) and (3), the wording of the subject permitted only one interpretation, and the interpretation of either the student (case 2) or the teacher (case 3) was incorrect while the interpretation of the other was correct.

There are many areas where we must require that a given text be interpreted in only one way. In these cases, it must be guaranteed that the interpretation of a text does not depend on the discretionary powers of the reader. One of these areas is the area of computer programming. A computer program is a text which tells the computer what to do, and obviously it would not be acceptable if such a text has more than one valid interpretation. Therefore, so-called formal languages are used in the area of computer programming.

## Games Which Can Be Played without Thinking

The games of the kind which shall be considered in subsequent sections are played using finite sets of elements which can be recognized and classified by merely looking at them. The rules of such a game define which combinations of the elements are allowed as initial states of the game, and what kind of steps are allowed to transfer from an initial state via intermediate states to a final state. A simple example of such a finite set of elements is the set of the 32 chessmen with 16 being white and the other 16 being black. In this example, there is only one combination of the elements which is allowed as an initial state: the chessmen have to be placed in a well-defined way on the board. The rules specify that the two players alternate in changing the situation on the board by making moves. In each situation, the set of allowable moves, one of which must be selected, is well-defined. Here you might object by saying, "You said that we shall consider areas where no discretionary powers are allowed. But playing chess requires lots of discretionary powers. There are no rules telling us which moves to select. If we select the proper moves, we may win, or else we shall lose." But you should realize that I never said that all decisions in our game must be determined completely by the rules. I specified only that the answer to the question about whether the rules of the game have been obeyed should not depend on any discretionary powers. There is a rule saying that the player who has to make a move, must select his move from a formally well-defined set of moves. But which move he selects from the actual set is not determined by a rule.

For the most part, people play chess or card games or other interesting games as a pleasant pastime and, less often, to win money. You may recall that I mentioned playing games in the chapters on mathematics. The purpose of those games was not passing time or winning money, but gaining insight. The games we are considering in the present chapter are played for getting answers to questions. Before I tell you what kind of questions these are, I shall first present a simple game, and afterwards I shall show that this game really provides answers to certain questions.

In Fig. 4.1 you see a table on which nine rectangular white and black pieces are placed in a certain configuration. Next to the table, on its right hand side, there is a box. It contains eight pieces of the same kind as of those on the table, three being black and five being white. On the table, there are two shaded areas on which pieces will be placed in the course of playing the game. The situation shown on the table is an initial state according to the rules. These rules require that in the initial state, nine pieces must be sitting on the table on the locations filled in Fig. 4.1. The color of eight of these nine pieces may be chosen arbitrarily; only the piece on the lower right position must be white. Since for eight positions there is the binary choice of the color, the number of possible initial states is $2^8=256$.

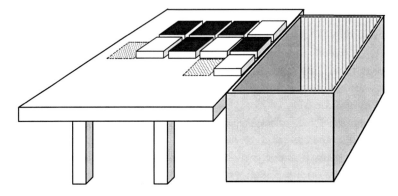

**Fig. 4.1** An initial state for our game

Once the initial state is given, the whole sequence of steps leading to the final state is completely determined, i.e., in the course of playing the game, there are no choices to decide about. The steps are determined by the rules presented in Fig. 4.2.

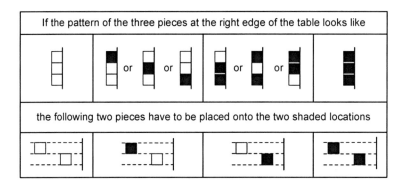

**Fig. 4.2** Rules for moves in our game

In each possible situation, the three pieces at the right edge of the table form one of the eight possible patterns given at the top of the figure. The actual pattern determines the colors of the two pieces, shown in the bottom of the figure, which must be taken out of the box and placed on the two marked locations. Shifting all pieces on the table one position to the right causes three of these pieces to fall into the box, and then a new pattern appears at the right edge of the table. After having

Playing without Thinking 85

applied the rules of Fig. 4.2 four times where each time three pieces fall into the box, the four pieces which initially were sitting in the upper row now have all disappeared into the box. Then there are only two pieces left at the right edge of the table, and the rules of Fig. 4.2 may no longer be applied. This situation tells us that we have reached the final state.

Surely, you had no problem understanding the rules of this game and now you could play it yourself. But the game is much too trivial to be played as a pastime. So now it is time to tell you what the purpose of this game is and how I invented the rules. Look at Fig. 4.3.

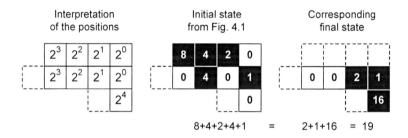

**Fig. 4.3** Interpretation of our game

The left side of this figure shows the eleven positions which, in the course of playing the game, can have pieces sitting on them. The nine positions for the initial state are inscribed with powers of two which means that a black piece sitting on that position has a weight equal to its corresponding number. The initial state from Fig. 4.1 is shown again in the middle of Fig. 4.3. In this state, the sum of the weights of all black pieces is 19. The final state is represented on the right side of the figure where you can see that the sum of the weights of the black pieces is again 19. Obviously, the purpose of this game is the implementation of some arithmetic operation.

Fig. 4.4 helps you to see how I found the rules of the game. In the upper half of the figure, I illustrated the well-known procedure for adding decimal numbers. The two numbers to be added are 3,926 and 6,348. We write these summands one below the other such that corresponding digits are located in the same columns. The process of adding begins at the rightmost column. 6+8 is 14 where the 4 is placed in the row reserved for the sum and the 1 is copied as a carry into the left neighboring column. We continue moving stepwise to the left: "2+4+1 is 7, write 7, transfer 0; 9+3+0 is 12, write 2, transfer 1; 3+6+1 is 10, write 10." As the final result we obtain the number 10,274.

Now look at the structure in the lower half of the figure. At first glance, it looks exactly like the structure above. Only the inscriptions differ. While the numbers in the upper half are *decimal numbers*, the numbers in the lower half are the

86                                                                                                           4. Formalisms

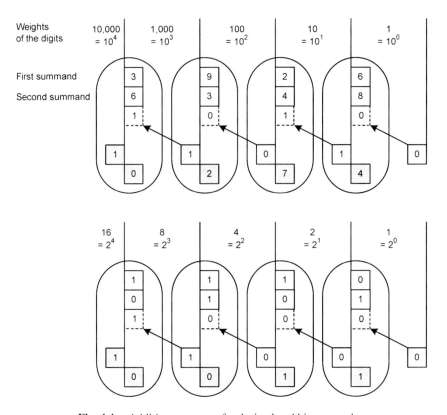

**Fig. 4.4**   Addition structures for decimal and binary numbers

so-called *binary numbers*. The term "decimal" means that the weights of the digits in a number representation are powers of 10: at the rightmost position, the weight is 1; in the next position to the left, it is 10; then 100, then 1,000 and so on. The base of these powers is called the base of the number system. While the base is 10 in the case of the decimal number system, the base is 2 in the case of the binary number system. The exponents of the powers are the same in all number systems. The rightmost weight is always 1 since here the exponent is 0. Moving stepwise to the left, the weight increases by a constant factor which is the base of the number system. While in the case of the decimal numbers we need a set of ten different digits {0, 1, 2, 3, 4, 5, 6, 7, 8, 9}, a set of two different digits {0, 1} is sufficient in the case of the binary numbers. In each number system, the number of different digits needed is equal to the basis B, and the weights of these digits are {0, 1, 2, 3, . . . , (B-1)}. In the lower half of Fig. 4.4, I have illustrated the addition of two binary numbers: $1110 + 0101 = 10011$, corresponding to the addition of the decimal numbers: $14 + 5 = 19$.

You may already have realized that the patterns within the oval shapes correspond to the game rules in Fig. 4.2. That was exactly how I discovered these rules: I looked at an example of adding two multiple digit binary numbers and asked myself how many different patterns could occur in a column. I found the eight patterns in Fig. 4.2. Since I didn't want you to know from the very beginning that we were doing arithmetic with binary numbers, I didn't inscribe the pieces with the digits 0 and 1, but I chose the white color for the 0 and the black color for the 1. When the game is implemented by technical means using a computer, it is up to us to decide what kind of physical properties to use for representation of the two binary digits. In the era of data processing with punched cards, the 1 was assigned to positions with a punched hole, and the 0 to positions which had not been punched.

Looking at Fig. 4.4, you can see that our game must not necessarily be restricted to four columns; the structure may be easily extended to the left. Applying the formal rules in Fig. 4.2, we can add binary numbers of any length.

The games we are considering here can always be interpreted as procedures for computing the result of a function for given input information. I could now present the games which provide the results for subtraction, multiplication or division. But from this, you would not gain any fundamentally new insight. Mathematicians and computer scientists use the word *algorithm* for such games which are played to get the results of functions. It is assumed that the root of the word *algorithm* is the Arabian name Al-Hwarizmi, the name of the mathematician whom I already mentioned in connection with the term algebra. Although the two words *algorithm* and *logarithm* differ only in the order of their first four letters, their meanings have nothing in common.

## How Logical Thinking Can Be Replaced by Pattern Recognition

Instead of saying that the purpose of a formal game is getting the result of a function for given inputs, I could say that the purpose is getting the answer to a given question. These two definitions of the purpose of the game are equivalent. Instead of saying that the game from Fig. 4.1 provides the result of adding two given numbers, we could say that the game answers the question about what the sum of the two numbers is. In the following section, we consider only questions whose answers are restricted to yes or no. Such questions are called *binary questions*, since there are only two possible answers. Some important binary questions are, in particular, whether a given proof of a mathematical theorem is correct, or whether a hypothesis can be proven, or whether a linguistic term obeys the corresponding grammar.

The binary question to be answered must be completely specified in the initial state of the game, and the sequence of moves must lead to a final state which can be interpreted to mean either yes or no. If the initial state contains a hypothetical proof of a mathematical theorem or a linguistic term, the pieces of the game must be such that we can use them for writing text and formulas. The use of only two kinds of pieces, for example black and white pieces, doesn't seem to be sufficient for these cases. In order to find out what kind of pieces we need, we must analyze the structure of texts and formulas.

It is well known that the Greek philosopher Aristotle (around 340 BC) had the idea that the components of a text are always either factual or logical. Therefore, Aristotle is often called the father of formal logic. Fig. 4.5 illustrates the difference between factual and logical components. This figure shows a logical equation, a sequence of rectangles, with some of them being shaded while the others are white. The inscriptions $a$ and $b$ in the shaded fields are variables for propositions which can be only true or false. For example, as proposition $a$ we could choose "Abraham Lincoln died in Washington D.C.", and as proposition $b$ we could choose "Thomas Jefferson was the second president of the United States." Whether these propositions are true or false – the first is true and the second is false – cannot be decided by logical thinking since they say something about the real world where, at particular locations and at particular points in time, certain events occur which someone may observe. Factual components of texts are those that refer to such events or situations in the real world. According to this criterion, words such as tree, dog, red, wet, drink or die are factual words. In Fig. 4.5 factual words appear only in the shaded fields. All the components in the white fields are logical. I hope that you still remember what I told you about equations: the two terms on the left and the right side of the equals sign refer to the same thing. In the case of the logical equation in Fig. 4.5, the identified thing is the meaning of what is stated on either side. Someone who says the phrase which stands left of the

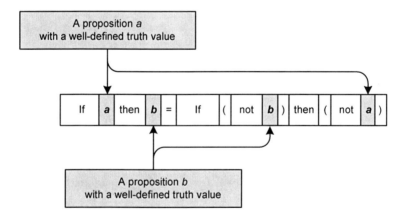

**Fig. 4.5** Example of a formula in propositional logic

equals sign could equally well have said the phrase which stands on the right side of the equals sign, since the meaning of these two phrases is the same.

While in school, you certainly had to write many compositions, and one reason for this was to teach you that there are many ways to express a particular meaning. Therefore, you should not be surprised that we can put an equal sign between two phrases. If we now insert the propositions which we chose as examples for $a$ and $b$, the phrase on the left side is "If Abraham Lincoln died in Washington D.C., then Thomas Jefferson was the second president of the United States." The logical equation says that instead of saying this, we could say what stands on the right side of the equal sign, "If Thomas Jefferson was not the second president of the United States, then Abraham Lincoln did not die in Washington D.C." I am sure that now you will severely object by pointing out that the location where Lincoln died had nothing at all to do with where Jefferson stands in the sequence of United States presidents. But your objection misses the point, since nowhere did I say that these two facts are causally related. What I said was that the phrase on the left side of the equals sign has the same meaning as the phrase on the right side. If you think that the phrase on the left side is nonsense – with which I perfectly agree – then the equation says that the same nonsense is stated with other words on the right side.

So far, I have pretended that the equation in Fig. 4.5 is correct, but I have not proved it yet. The proof will now be presented. Equations of the type considered here can contain only a finite number of variables for propositions. Whether the two phrases on the two sides of the equal sign are true or false depends on whether the actual propositions which are chosen to replace the variables are true or false. Logicians state this by saying, "The truth value of the phrases on the two sides of the equations depends on the truth values of the propositions replacing the variables." Since the number of the variables is finite and since each variable can provide only one of two truth values, the number of possible different combinations of the truth values is also finite. In the case of Fig. 4.5, there are the two variables, $a$ and $b$, and the possible combinations of truth values for the pair $(a, b)$ are {(false, false), (false, true), (true, false) and (true, true)}. These four combinations are listed on the left side of the table in Fig. 4.6. In order to find out whether the equation in Fig. 4.5 is correct, it is sufficient to check whether the truth values for the two sides of the equation are really identical for all four possible input combinations. First, we consider the left half of the table in Fig. 4.6 which corresponds to the left side of the equation in Fig. 4.5. You see that only one of the four input combinations results in the truth value false for the left side of the equation, namely when proposition $a$ is true and proposition $b$ is false.

An example may help you to understand this result. Suppose I say to my son, "If I win $20,000 in the lottery next Sunday, I will buy you the car you have been wanting for so long." There is only one situation when what I said will be a lie, namely when I win and don't keep my promise. The first two rows in Fig. 4.6

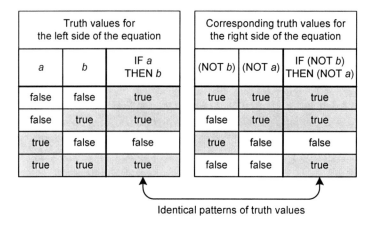

**Fig. 4.6** Proof of the correctness of the equation in Fig. 4.5

correspond to the situations where I don't win. In the first row, I don't buy my son a car, but in the second row I do although I didn't win. Nevertheless, I didn't tell a lie in this case, since I promised to buy him a car in the case of a win. I didn't say anything about what I would do if I didn't win.

Now look at the right half of the table in Fig. 4.6 which corresponds to the right side of the equation in Fig. 4.5. Here you see the truth value combinations for the pair (not *b*, not *a*). These combinations are obtained by simply inverting the corresponding values of *a* and *b* in the same row (true → false and false → true). For each combination on the right side, the result of the if-then proposition can be looked up on the left side. By comparing the patterns in the two columns which contain the results for the two sides of the equation, you can see that these patterns are identical, and this means that the equation is correct. Obviously, we could prove the correctness of the equation without considering any actual propositions instead of using the variables *a* and *b*.

By using white and shaded fields for the truth values false and true, I built a bridge to our game in Fig. 4.1. There we had only two kinds of pieces, white and black, and now again we have two kinds of pieces, white and shaded. In the case of the game for adding binary numbers, there was no need for the player to know how to interpret the colors white and black. And now, in the case of propositional logic, again there is no need for the player to know how to interpret the white and shaded fields. Comparing the patterns can be done by anyone who is not blind. A formal game can always be played without knowing the purpose which led to the rules of the game.

In Fig. 4.7, five so-called logical functions are shown, the first of which is the "If-then function" we considered in Fig. 4.6. Each of these functions is defined by its characteristic pattern of white and shaded fields. In Chapter 14, we shall refer to these functions again since they are the basis for all kinds of computing machines.

Logic by Pattern Recognition

| Truth values for the input variables || Results of the corresponding logical function f(a, b) ||||  |
|---|---|---|---|---|---|---|
| a | b | IF a THEN b | a AND b | EITHER a OR b | a AND/OR b | NEITHER a NOR b |
| false | false | true | false | false | false | true |
| false | true | true | false | true | true | false |
| true | false | false | false | true | true | false |
| true | true | true | true | false | true | false |

Fig. 4.7 Truth values for different logical functions

The concepts of propositional logic as we know them today were developed by the English mathematician George Boole (1815-1864). Not much later, it became clear that propositional logic does not cover all the possibilities for formalizing logical thinking. Around the year 1900, an important step forward occurred when a major contribution was made by the German mathematician Gottlob Frege (1848-1925). In propositional logic, the internal structure of the propositions is not considered; it is relevant only whether the propositions are true or false. Going further, logicians looked inside of propositions hoping to find structures which could be formalized.

From your English class, you know the terms subject, predicate and object for particular parts of a sentence.. Subject and object are connected by a certain relationship called the predicate. A very unpleasant relationship is expressed by the sentence "Cain killed his brother, Abel." Sentences are represented in the so-called *predicate logic* where constants and variables are used to represent the subjects, the objects, and the predicates. The constants and variables for subjects and objects are called "individuals." Using the variables $x$ and $y$ for the individuals and "killed" for the predicate, the sentence now reads "$x$ killed $y$." In the next step, we replace the actual predicate "killed" by another variable. This requires that we introduce a criterion for distinguishing between variables for individuals and those for predicates. It is common to use capital letters for the predicate variables and lower-case letters for the variables associated with the individuals. Now our sentence reads "xPy". This so-called infix form is possible only in cases where the predicate relates two individuals to each other. There are also predicates which relate more than two individuals to each other, e.g., "John owes Ruth $100." Here the individuals are John, Ruth and $100. You may wonder why I call $100 an individual, but in predicate logic, "individual" is the common name for anything that may become a subject or an object. While the infix form is a special case, the most general form for a sentence in predicate logic is $P(x_1, x_2, \ldots, x_n)$. The predicate variable is placed in front of the brackets and the individual variables are listed within the brackets. In the definition of such a predicate, it must be determined how the different positions within the bracket are to be interpreted, since there is more than one possible order for listing individuals. For example, we could use owing(John, Ruth, $100), owing(Ruth, John, $100), owing(John, $100, Ruth), owing(Ruth, $100, John), owing($100, John, Ruth) or, owing($100, Ruth, John).

The introduction of new types of variables is a prerequisite for the concept of predicate logic, but it is not the key point. The key point is the concept of so-called quantifiers. The word quantifier is derived from the Latin word *quantus* (how much). In reference to these quantifiers, predicate logic is alternatively called quantificational logic. We need only two different kinds of quantifiers to express all logically relevant relations between a universe (containing all possible elements of a particular type) and its subsets. On the one hand, we want to say, "For all elements of the actual universe, it is true that … ", and on the other hand we want to say, "There is at least one element in the actual universe for which it is true that … " The first of these is called *universal quantification* and the second is called *existential quantification*. The first letters of the words *all* and *existing* have been chosen as symbols for these quantifiers, and they are used in their capital letter form and inverted: ∀ and ∃.

Fig. 4.8 represents an equation from predicate logic. What I said in my comment about Fig. 4.5 is also valid here. The equal sign says that the two phrases on its two sides have the same meaning. In natural language, the phrase on the left side reads: "Not all elements in the universe U have the property P." On the right side, this fact is expressed again in a different form: "There is at least one element in the universe U which does not have the property P."

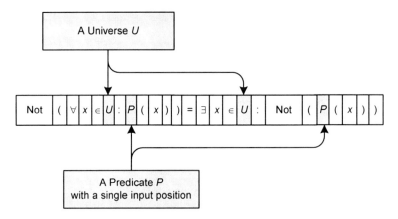

**Fig. 4.8** Example of a formula in predicate logic

Considering the fact that there are books with about 300 pages dealing exclusively with formal logic, you certainly shall forgive me for not presenting the rules of formal games, the states of which are formulas of predicate logic. But I

# Logic by Pattern Recognition

shall show you how any formula of predicate logic can be transformed into a structure containing only white and black pieces. Consider Fig. 4.9. In the upper part of this figure, you find a formula saying that all elements of the universe U have the property P. Underneath, there is a sequence of pieces where each is a square made of 3×3 elemental squares. An elemental square is either white or black, and with 3•3=9 such elemental squares we can create $2^9$=512 different patterns. Since the rules of a formal game must be defined exclusively with reference to the looks of the pieces and without any reference to their meaning, it is possible to play the game with pieces having the black and white patterns of those in Fig. 4.9. A human player will undoubtedly not be able to memorize the rules of the game only by referring to the looks of the pieces, and a computing machine cannot assign meanings to patterns. Therefore, letting a computer play the game requires only that the pieces can be distinguished merely by their looks.

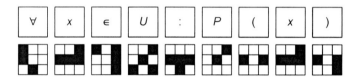

**Fig. 4.9** Human-oriented versus machine-oriented patterns

Without predicate logic it would not be possible to represent proofs of mathematical theorems in such a form that the question about whether a proof is correct can be answered by playing a formal game. When the correctness of a proof is checked formally, no interpretation of the states is allowed. However, human testers are not able to turn off their capability of interpreting texts and formulas. When looking at a proof written in common mathematical language, we cannot look at it as if it were just a structure of distinguishable patterns, but we automatically think about what this structure means to us. Therefore, when a proof is written for human testers, it does not contain those assumptions which are self evident to all of us. In particular, we take it for granted that the tester knows how to count using natural numbers, and therefore there is no need to include the concept of numbers and counting in a proof to be tested by humans. But when the tester is a computer which knows nothing about numbers and counting, it needs a formal text specifying what these concepts are. The first person who successfully formalized the concept of counting was the Italian mathematician Giuseppe Peano (1858-1932). Around the year 1900, he published the five formal expressions which are shown in Fig. 4.10. Of course, it was not Peano's problem to create the natural numbers since for him as for everybody else, Kronecker's statement was true: "The natural numbers have been created by the dear God and everything else is the work of humans." The problem Mr. Peano had to solve was taking a concept which every educated human individual is familiar with and formalizing it in such a way that the form could be manipulated by a machine which cannot assign any meaning to the formal structures. Our idea of counting is illustrated in the shaded field in the lower right corner of Fig. 4.10. We start with a first element, then go to the next, and so on, and we know that the chain continues indefinitely on the right side.

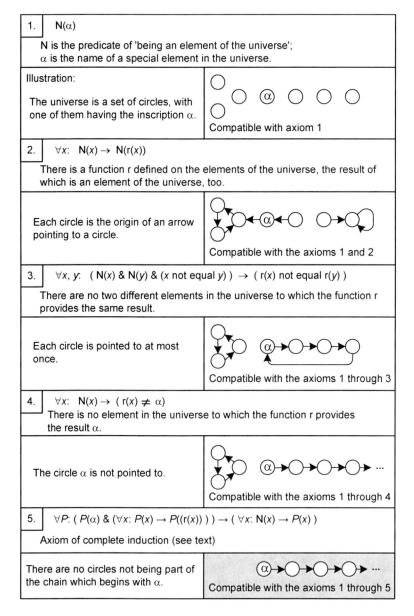

**Fig. 4.10** Peano's axioms

The five expressions in Fig. 4.10 are called axioms. An *axiom* is a mathematical proposition of a special kind. Its purpose is to define a structured universe, and therefore it would be nonsense to require that such a proposition be proven. It is true simply because the person who writes it down defines it to be true. By the first axiom in Fig. 4.10, the first element in the chain with the name $\alpha$ is created. By the second axiom, the arrows are created which are needed for building the chain. By the axioms 3, 4 and 5, certain combinations of elements and arrows are enforced and certain others are prohibited. While the axioms 1 through 4 can be easily understood, the fifth axiom cannot be understood at all, at least not at first glance. The great achievement of Mr. Peano was finding this fifth axiom.

While I was able to comment on the first four axioms in Fig. 4.10 in natural language, a corresponding comment about the fifth axiom is missing. The first four axioms are sufficient for the creation of the chain which begins with $\alpha$ and continues indefinitely on one side. Therefore, the fifth axiom is needed only to exclude all other structures except the one chain. The fifth axiom is more complex than the other four since it is the only one containing not only quantifiers for individuals, but also quantifiers for predicates. The formal expression $\forall P$ expressed in natural language is: "For any property P an element of the universe may have, it is true that ...". Compared to this, the expression $\forall x$ is much simpler: "For any element x of the universe, it is true that ...". You shall not be surprised to see that the complexity of the formula in Fig. 4.10 also shows up in the natural language form of the axiom:

For the universe considered and for any property P an element may have, the following statement is true:
>If the two conditions are met,
>>that the element $\alpha$ has the property P
>>and that each element *x* having the property P
>>transmits it to the element r(*x*),
>
>then all elements of the universe have the property P.

The fifth axiom of Peano is often called the "axiom of complete induction." The principle of complete induction allows the conclusion that, under certain conditions, an infinite number of elements have a certain property. Let's assume that $\alpha$ is the natural number 9 and that the function r(*x*) is defined as $2x+3$. The property P we are looking at is natural divisibility by 3, i.e., the fact that the result of the division *x*/3 is a natural number. The element $\alpha$ has this property since 9/3=3, and, if *x* has this property, then r(*x*) also has this property, since r(*x*)/3=2•(*x*/3)+1. Therefore, all elements in the chain which begins with 9 and is continued by r(*x*)=$2x+3$ (i.e. 9, 21, 45, 93...) have natural divisibility by 3.

Formal games having a set of axioms as their initial state are called *calculus*. The purpose of the axioms is the creation of a universe of structures which are defined by the individuals, functions and predicates introduced by the axioms. These individuals, functions and predicates are introduced merely by giving them names. In the case of Peano's axioms, the individual is α, the function is r(x) and the predicate is N(x). Sometimes it is said that Peano introduced the axioms defining the natural numbers, but Fig. 4.11 shows that the universe of the natural numbers is obtained only by a special interpretation of the axioms. Different interpretations result in different universes. These interpretations all have in common that an infinite set of elements, which have a well-defined position in an enumeration, is created. It is correct to say that Peano formalized the concept of enumeration while leaving open what the elements are. I deliberately included an example where the elements are not numbers but geometric shapes (see the third column in Fig. 4.11).

| N | Natural numbers | Positive odd numbers | Shapes which are either squares or rectangles where the edge relation is horizontal : vertical = $(n+1):n$ where $n$ is a natural number. |
|---|---|---|---|
| α | 1 | 1 | □ |
|  | 2 | 3 | 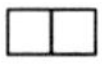 ▭ |
|  | 3 | 5 | 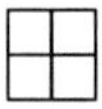  |
|  | 4 | 7 | 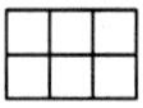  |
|  | 5 | 9 | 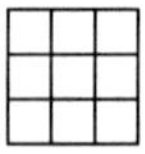  |
|  | 6 | 11 | 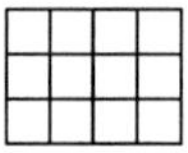  |
| r(x) | x + 1 | x + 2 | If $x$ is a square with the edge $n$: the rectangle obtained by adding $n$ elementary squares to the right. If $x$ is a rectangle with the width $n$: the sqare obtained by adding $n$ elementary squares at the top |

**Fig. 4.11**  Alternative interpretations of Peano's axioms

A structured universe which is defined by providing a set of axioms is called a *theory* in the professional language of the mathematicians. Each interpretation of the names of individuals, functions and predicates which occurs in the axioms is called a *model* of the theory. Using these terms we may say that Fig. 4.11 shows three models of Peano's theory in its three columns. Two models belonging to the same theory are said to be *isomorphic* to each other. The Greek term *morph* means shape or structure. Correspondingly, the English word amorphous means "without structure". The prefix *iso* comes from the Greek language, too, and means equal. Perhaps you know the word "isobar" for locations having the same atmospheric pressure, or the word "isotherm" for locations having the same temperature. Thus, isomorphic means having the same structure. Isomorphism is of great importance in the formal world, and mathematicians wandering through the world of formalisms are always looking for isomorphisms.

In a way similar to the way Peano formalized the concept of enumeration, other mathematicians were successful in formalizing the basic concepts of geometry. In particular, the German mathematician David Hilbert (1862-1943) made the major contributions [HIL]. He characterized his axioms of geometry by saying [SCR], "Instead of using the words point, straight line and plane you could as well say table, chair and beer mug." By this statement, Hilbert emphasized the fact that the names for individuals, functions and predicates introduced in the axioms have no meaning of their own and therefore they could be chosen without any reference to geometry. Such a name could even be "qq13" or "Z28" or "dog food". The only condition required is that different things must have different names.

Fig. 4.12 shows a small excerpt from axiomatic geometry. In the shaded fields you find the names of the abstract concepts introduced by the axioms, namely the predicates $U(x)$, $P(x)$ and $G(x)$ and the function $g(x, y)$. Fig. 4.12 shows two different interpretations of axiomatic geometry. That one represented in the middle column corresponds to Euclidean geometry which is called geometry in a plane, while the one represented in the right column corresponds to spherical geometry.

| The predicate $U(x)$ | $x$ is an element of the universe | |
|---|---|---|
| The universe | All points of a plane | All diameters of a sphere |
| The predicate $G(x)$ | $x$ is a straight line on the plane | $x$ is a great circle on the surface of the sphere |
| The result of the function $g(x, y)$ in the axiom $\forall x, y: U(x) \& U(y) \& (x \neq y) \rightarrow G(g(x, y))$ | The straight line containing the two given points $x$ and $y$ | The great circle having the two given diameters $x$ and $y$ |

**Fig. 4.12** Alternative interpretations of axiomatic geometry

In Chapter 3 in the section on geometry, I pointed out that it is difficult or maybe even impossible to give a precise definition of the concepts of a straight line or a plane. There, I quoted the following "definition" from Euclid's book: "A plane is an object which lies in between its straight lines in a homogeneous way." In axiomatic geometry, definitions of the elementary geometric objects like points, straight lines or planes are not provided at all. It is enough to see that our freedom of interpretation allows us to interpret the axioms in such a way that the formal objects denominate points, straight lines or planes.

## Detours Which Are Shorter Than the Direct Route

When someone comes late to a meeting, he often excuses his lateness by saying that the direct route was blocked and that he had to take a detour. In everyday language, we use the term detour only for routes which are longer than a direct route. But in the formal world of mathematics, it is possible that we reach our goal faster than via the direct route by taking a detour. In these cases, the detours are alternate routes found by using what are called *isomorphisms*. This is illustrated by the example in Fig. 4.13 where the starting point is the number 17, and we are looking for the best way to reach our goal, finding the third root of 17. We reach our goal when we find the number 2.5713, since $2.5713^3=17$. You probably have a pocket calculator which finds the third root of 17 very quickly, and therefore you might think that this is the shortest way to your goal. But actually, by using a pocket calculator, you are not going your way, but the way of the electronic process which occurs inside the calculator. That way is not necessarily the direct way.

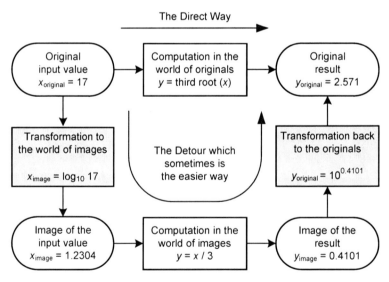

**Fig. 4.13** Operations in two worlds

Convenient Detours

When I was in high school, electronic pocket calculators were not yet available, but nevertheless we were expected to be able to compute the third root of 17. The direct way we used was first to guess a solution $y_1$, and then to check and see whether $y_1^3 = y_1 \cdot y_1 \cdot y_1$ is close enough to 17. If not, a next guess was made which again was checked, and this procedure was continued until an acceptable solution was found. But our mathematics teacher showed us a very elegant detour. Each student had to buy a booklet which contained the so-called "tables of logarithms". In these tables, we could look up the logarithm to the base 10 of any given positive rational number. In this booklet, we found the number 1.2304 which is $\log_{10} 17$, since $10^{1.2304} = 17$. In the language of mathematicians, obtaining the number 1.2304 that corresponds to the given number 17 is called a transformation which leads from the so-called "original" 17 to its "image," the logarithm whose value is 1.2304. Correspondingly, moving from 1.2304 to 17 is called the reverse transformation. The logarithm of 17 will be used to find the third root of 17.

The concept of detours using transformations can be explained by referring to the world of axiomatically defined theories and their isomorphic models which are obtained by different interpretations of the axioms (see Figs. 4.11 and 4.12). While the direct way lies within a single model, the detour consists of three sections: the first step leads from the original model to a different model within which the computation is performed (the second step), and the final step leads back to the original model. In Fig. 4.13, the direct way leads from the starting point in the upper model horizontally to the goal, while the detour has its computational section in the isomorphic model underneath. The advantage of taking the detour comes from the fact that instead of computing the third root in the original model, it is only necessary to divide the logarithm by 3 in the isomorphic model. However, we would not take the detour if the two transformational steps leading back and forth between the two models were too difficult or extensive. Using our tables of logarithms made it very easy to perform these transformational steps, although someone had to put a lot of effort into computing these tables. All kinds of transformations, not only the logarithmic transformation, are provided as look-up tables which make it easy to take the detours, although computing these tables surely took a lot of effort.

The general structure shown in Fig. 4.13 does not depend on which axiomatic world we are dealing with. The logarithmic transformation was chosen as a simple example. However, there are transformations which have nothing to do with simple arithmetic, although the relation between an original road and a detour is still as shown in Fig. 4.13. Perhaps you will occasionally encounter the name of one of the transformations which are of great importance in engineering. Therefore, I mention some of their names without trying to explain what they are about: *Fourier-Transformation, Laplace-Transformation* and *Z-Transformation*.

## How We Can Enter into Four- or Higher-Dimensional Spaces Using Simple Steps

In Chapter 8 we shall be confronted with a statement of Albert Einstein saying that the continuum of space and time is a four-dimensional space. Everybody who hears this for the first time is rather irritated since he has no idea what could be meant by four- or higher-dimensional spaces. He only knows the space where he is moving around and where he spends his life, where satellites circle the earth, where planets move around the sun and where all the other stars are located somewhere. We call this space our three-dimensional space, and we think it is quite obvious. But the idea of a three-dimensional space is not even five-hundred years old, and is a consequence of the bridge between the world of numbers and the world of geometry.

Fig. 3.2 shows a so-called two-dimensional coordinate system where each point in the plane can be represented by a pair of numbers (x, y). In order to describe a point in the three-dimensional space, three numbers are required. The interpretation of such a triplet of numbers is in reference to a three-dimensional coordinate system. In most cases, a so-called Cartesian coordinate system is chosen which has three axes perpendicular to each other as shown in Fig. 4.14 where there are arrows on the axes. In this figure, a half-sphere is represented, and a point on the surface of this sphere is selected which is the upper corner of the grey-shaded triangle. The distance between this point and the center of the sphere is the radius r. In the given coordinate system, this point can be identified by the three numbers $(x_1, x_2, x_3)$. That's the reason such a space is called a three-dimensional space.

In Fig. 3.2, the plane was introduced as being two-dimensional, and the curve in this plane was described by an arithmetic relationship between the two numbers in the pair (x, y). Correspondingly, we can describe the surface of our sphere by an arithmetic relationship between the numbers of the triplet $(x_1, x_2, x_3)$ which holds only for the points on this surface. This relationship can be found by applying the law of Pythagoras to the two triangles in the upper part of Fig. 4.14. This formula holds not only for all points on the surface of the half sphere, but for the whole sphere, and now can be taken as the starting point for a very simple formal generalization. We just increase the number i of $x_i$ in the sum and pretend quite boldly that all the new formulas are also descriptions of surfaces of spheres, although they are no longer three-dimensional, but of higher dimension. This is shown in the lower part of Fig. 4.14.

By doing this, we disconnect the term "sphere" from its everyday meaning and use it as the name of a special formal object. Nobody would normally call a circle a "two-dimensional sphere" or call a straight line of finite length a "one-dimensional sphere". But by this generalization, we made it possible to consider n-dimensional spheres, with n being any natural number. Nobody can have a mental

# Higher-Dimensional Spaces

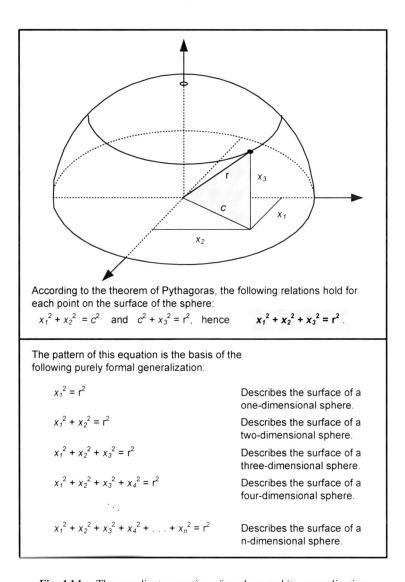

**Fig. 4.14** The coordinate equation of a sphere and its generalization

image of these formal objects, but only of the formulas describing them. Making the step from three to higher dimensions means crossing a border which should be emphasized explicitly. In the years shortly after the Second World War, Berlin, the old German capital city, was divided into four sectors, and I still remember the signs with the following warning:

> **ATTENTION!**
> 100 METERS FROM HERE ENDS THE AMERICAN SECTOR!

Think of a similar sign whenever you cross the boarder between the world of perceivable objects and the formal world:

> **ATTENTION!**
> HERE ENDS THE WORLD OF PERCEIVABLE OBJECTS!
> YOU ARE NOW ENTERING THE FORMAL WORLD!

In order to help you to overcome your fear of n-dimensional spaces, I shall show you how we even can deduce formulas for the surface area and the volume of n-dimensional spheres. As an example of a three-dimensional sphere, we may think of our earth and its surface area and volume. We know that the area is measured in square meters $m^2$ and the volume in cubic meters $m^3$. Thus, we may conclude quite formally that the surface area of a four-dimensional sphere can be measured in $m^3$ and its volume in $m^4$.

We found the formulas describing higher-dimensional spheres by starting with the formula for the three-dimensional sphere and formally increasing the number of coordinates $x_i$. We now can proceed correspondingly in order to find the formulas for the surface area and the volume of higher-dimensional spheres. We first must analyze the process which leads to the formulas for the surface area and the volume of a three-dimensional sphere. The half-sphere from Fig. 4.14 is shown again in Fig. 4.15, but now in a different view. The grey-shaded triangle in this figure has the same position as the corresponding triangle in Fig. 2.15 which was used to introduce the two functions, sine and cosine. There, $\cos(\alpha)$ was the length of the horizontal edge of the triangle, and this was equal to the real part of a complex number being determined by its angle $\alpha$ and its radius one. In Fig. 4.15, the radius r of the circle is not set to 1, but left open, and therefore the length of the horizontal edge of the triangle in Fig. 4.15 is r•$\cos(\varphi)$. The section of the circumference which belongs to the angle $\varphi$ is r•$\varphi$ since we do not measure the angle in degrees but as the corresponding section of a circle with radius 1. Thus, a right angle encompassing one-half of a semi-circle has the value $\pi/2$.

We get the surface area and the volume of the sphere by integration, i.e., by computing the limit of a sum of an infinite number of infinitely small summands. We start with small, but not infinitely small summands and let them become smaller and smaller. Imagine that the half-sphere in Fig. 4.15 were half of an apple. Its surface area is the area that can be peeled. We now assume that we peel this half apple in such a way that we get a lot of thin rings where each ring has the same angle $\Delta\varphi$. The area of such a ring is given as the product of its length and its width. Its length is equal to the length of the circumference of the circle having the

Higher-Dimensional Spaces 103

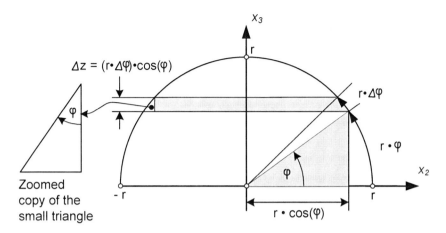

**Fig. 4.15** Quantities needed to get the surface and volume formulas for n-dimensional spheres

radius r•cos(φ), and this length is 2π•r•cos(φ), since the circumference of a circle is 2π times its radius. The width of the ring is r•Δφ, and thus the area of the ring is 2π•r²•cos(φ)•Δφ. If we now add up the areas of all of these rings in the interval 0 ≤ φ ≤ π/2, we get an approximate value for the surface area of the half-sphere. The exact value for the surface area of the whole sphere is obtained by computing the limit of the sum for Δφ → 0 and then doubling the result.

In order to move forward into higher-dimensional spaces, I now introduce the abbreviation $S_n(r)$ which means "formula for the surface area of an n-dimensional sphere having the radius r." Now look at the formula in the top-most oval in Fig. 4.16. This formula describes exactly what I said before about adding up the areas of the rings of peeling and then doubling the result. In front of the integration symbol ∫ you find the factor 2. At the right-most end of the formula, you find the width r•dφ of the actual ring. Here a "d" replaces the former Δ which is a consequence of computing the limit for Δφ → 0. The limits 0 and π/2 of the integral ∫ express the fact that we add up the areas of the rings at all heights which are determined by the angle φ between 0 and π/2. Finally, the function $S_{n-1}$ provides the length of the circumference of the actual ring. Not all rings have the same length since their radius depends on φ according to the product r•cos(φ). For φ=0 we get cos(0)=1, and for φ=π/2 we get cos(π/2)=0. This means that the ring at the bottom has the radius r and therefore is the longest, whereas the ring on the top has the radius 0 and is the shortest.

In a way similar to the way we obtained the formula for the surface area of the sphere, we can obtain the formula for its volume. The only difference is that we no longer must add up areas of rings of peeling, but now we must add up volumes of

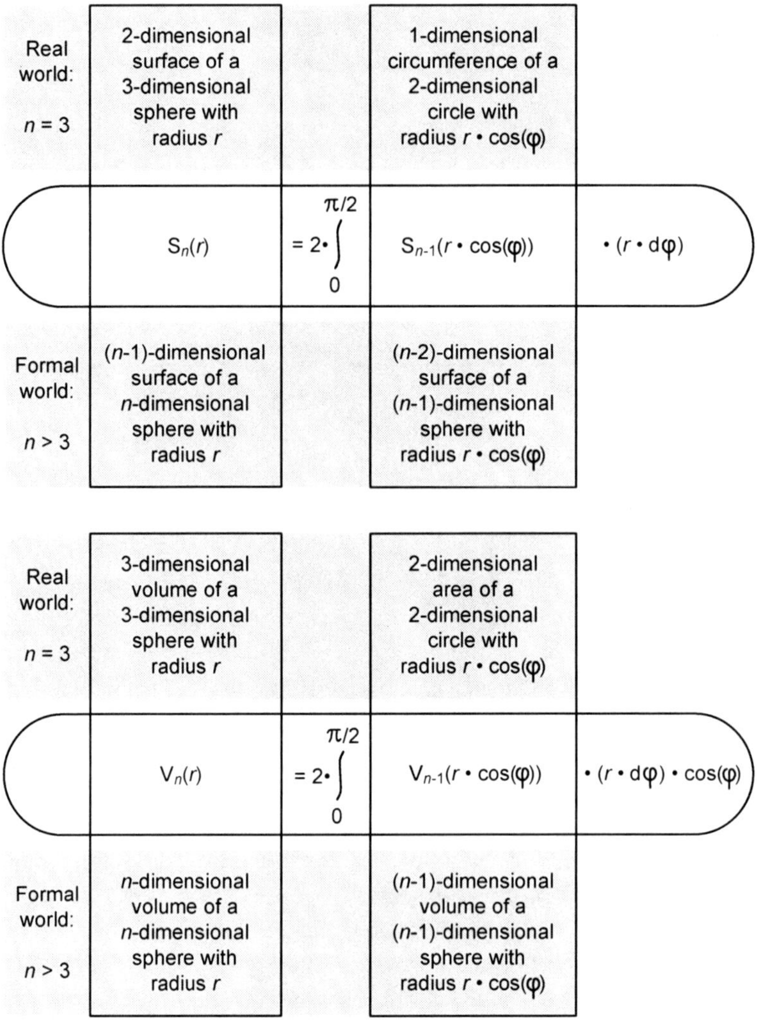

104                                                                                    4. Formalisms

| Real world: $n = 3$ | 2-dimensional surface of a 3-dimensional sphere with radius $r$ | 1-dimensional circumference of a 2-dimensional circle with radius $r \cdot \cos(\varphi)$ |
|---|---|---|
| | $S_n(r) = 2 \cdot \displaystyle\int_0^{\pi/2} S_{n-1}(r \cdot \cos(\varphi)) \cdot (r \cdot d\varphi)$ | |
| Formal world: $n > 3$ | $(n-1)$-dimensional surface of a $n$-dimensional sphere with radius $r$ | $(n-2)$-dimensional surface of a $(n-1)$-dimensional sphere with radius $r \cdot \cos(\varphi)$ |
| Real world: $n = 3$ | 3-dimensional volume of a 3-dimensional sphere with radius $r$ | 2-dimensional area of a 2-dimensional circle with radius $r \cdot \cos(\varphi)$ |
| | $V_n(r) = 2 \cdot \displaystyle\int_0^{\pi/2} V_{n-1}(r \cdot \cos(\varphi)) \cdot (r \cdot d\varphi) \cdot \cos(\varphi)$ | |
| Formal world: $n > 3$ | $n$-dimensional volume of a $n$-dimensional sphere with radius $r$ | $(n-1)$-dimensional volume of a $(n-1)$-dimensional sphere with radius $r \cdot \cos(\varphi)$ |

**Fig. 4.16** Integral formulas for the surface and the volume of n-dimensional spheres

thin disks. These disks are exactly the same which we had to peel previously in order to get our rings. Such a disk is shaded in grey in Fig. 4.15. All the rings have the same width $r \cdot \Delta\varphi$, but the disks do not all have the same thickness $\Delta z$. The expression describing this thickness as a function of the angle $\varphi$ is $r \cdot \Delta\varphi \cdot \cos(\varphi)$. This is a consequence of the fact that the very small white triangle, which you see at the left side of the grey disk, has the same angles as the big grey triangle. The volume of the disk is equal to the product of its base area and its thickness where the base area is a circle with radius $r \cdot \cos(\varphi)$. Thus, the volume of the disk is $\pi \cdot$

$(r \cdot \cos(\varphi))^2 \cdot r \cdot \Delta\varphi \cdot \cos(\varphi) = \pi \cdot r^3 \cdot \Delta\varphi \cdot \cos^3(\varphi)$. The volumes of all of these disks must be added up while $\Delta\varphi$ is made smaller and smaller. This is indicated by replacing $\Delta\varphi$ by $d\varphi$. This procedure results in the formula which you find in the lower oval in Fig. 4.16. In analogy to $S_n(r)$, the abbreviation $V_n(r)$ means "formula for the volume of an n-dimensional sphere having the radius r."

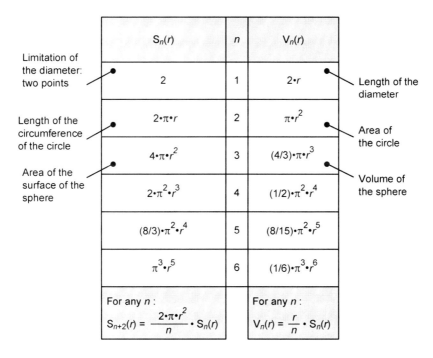

**Fig. 4.17** Surface and volume of spheres for dimensions 1 through 6

Starting from the one-dimensional sphere, i.e., from the straight line having the length 2r, I now have successively computed the formulas in Fig. 4.17. Once I had the formula for a certain n, I could compute the formula for n+1. However, I avoided the bother of deriving the formula for the integral of the expression $\cos^n(\varphi)$ (see Fig. 4.18); I looked it up in my book of mathematical formulas [HUE].

$$\int_0^{\pi/2} \cos^n(\varphi) \cdot d\varphi = \begin{cases} \dfrac{\pi}{2} \cdot \dfrac{1 \cdot 3 \cdot 5 \cdot 7 \cdot \ldots \cdot (n-1)}{2 \cdot 4 \cdot 6 \cdot 8 \cdot \ldots \cdot n} & \text{if } n \text{ is positive and even} \\[2ex] \dfrac{2 \cdot 4 \cdot 6 \cdot 8 \cdot \ldots \cdot (n-1)}{3 \cdot 5 \cdot 7 \cdot 9 \cdot \ldots \cdot n} & \text{if } n \text{ is odd and greater than 1} \end{cases}$$

**Fig. 4.18** Formula for integration of powers of the cosine function

Once I had computed the formulas in the white fields of the table in Fig. 4.17, I started to search for patterns in the relationships between ($S_n$ and $V_n$), ($S_n$ and $S_{n+1}$) and ($V_n$ and $V_{n+1}$). It was easy to find the simple relationship between ($S_n$ and $V_n$) which is represented in the grey field in the lower right corner. A much longer search finally let me find the relationship between ($S_n$ and $S_{n+2}$) which is represented in the grey field in the lower left corner. The relationships represented in the grey fields make it possible to obtain the formulas for $S_n(r)$ and $V_n(r)$ for any natural number n without performing any integration. We need only the simple formulas $S_1(r)=2$ and $S_2(r)=2\pi r$ as starting points.

In the many years I worked with students in engineering and computer science, I found that the experience of dealing with formalisms which must not be interpreted can become a passion. Actually, the formalism then becomes a game which attracts and holds on to its players. Although I am not really an addict, I nevertheless sometimes enjoy playing with formalisms, and therefore I continued playing with the formulas for n-dimensional spheres. I wanted to check the correctness of these formulas numerically, and I thought the following procedure would be very suitable. Each (three-dimensional) sphere with radius r can be placed inside a (three-dimensional) cube with edges of length 2r. When the sphere is n-dimensional, the cube must also be n-dimensional. But what is an n-dimensional cube? We start from a "real" three-dimensional cube which sits in a Cartesian coordinate system where each point is determined by three numbers ($x_1$, $x_2$, $x_3$). We now ask for the two-dimensional cube. It will be what is left from the three-dimensional cube when we omit the $x_3$-axis; therefore it is the square with edges of length 2r sitting in the ($x_1$, $x_2$)-plane. The one-dimensional cube is what is left from this square when we omit the $x_2$ axis; it is the straight line with length 2r. The "volumes" of these cubes are $(2r)^1$ for the straight line, $(2r)^2$ for the square and $(2r)^3$ for the "real" cube. Therefore, we may assume that the volume of an n-dimensional cube with edges of length 2r is $(2r)^n$. The table in Fig. 4.19 shows the ratios of the volumes $V_n$ from Fig. 4.17 to the volumes $(2r)^n$ of the corresponding cubes. Would you have expected that higher dimensional spheres fitted into such cubes take smaller and smaller fractions of the cube's volume as the dimension increases? I really was surprised when I first saw the numbers in Fig. 4.19.

| n | 1 | 2 | 3 | 4 | 5 | 6 |
|---|---|---|---|---|---|---|
| $\dfrac{V_n(r)}{(2r)^n}$ | 1 | $\dfrac{\pi}{4}$ | $\dfrac{\pi}{6}$ | $\dfrac{\pi^2}{32}$ | $\dfrac{\pi^2}{60}$ | $\dfrac{\pi^3}{384}$ |
|  | 1 | 0.785 | 0.524 | 0.308 | 0.164 | 0.081 |

**Fig. 4.19** Ratio of the volume of a sphere to its outside cube

Higher-Dimensional Spaces

I then checked these numbers using a procedure which is illustrated in Fig. 4.20 for the two-dimensional case. The cube is filled with a regular pattern of equally-spaced points, and then those points which lie inside the sphere are counted. The counting can be restricted to the points with positive coordinate values only, i.e., to the points in the grey square, since the symmetric structure guarantees that we get the same relationship between the results of the counting as we would if we counted in the whole cube. Since the density of points in Fig. 4.20 is not very high, the relationship we get here is only a rough approximation of the exact value in Fig. 4.19. By increasing the density of points, this procedure provides values which get closer and closer to the exact value.

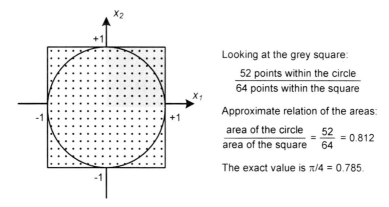

**Fig. 4.20** Checking the relations in Fig. 4.19 by computer

Though the concept of computer programming will not be introduced until Chapter 14, I now must mention a computer program. I wrote a computer program to compute the volume relationships for dimensions 1 through 6 according to the procedure illustrated in Fig. 4.20, but with higher densities of points. In the program, each point is represented by its tuple of coordinates $(x_1, x_2, x_3, \ldots x_n)$ where each coordinate $x_i$ has the same domain of possible values, namely $\{1/(2p), 3/(2p), 5/(2p), \ldots, (2p-1)/2p\}$. The letter p stands for the number of points in one row or column of the grey subcube. In Fig. 4.20, p has the value 8. In this figure, we not only see all the points, but we can also see which of the points lie within the circle and which don't. The computer executing my program does not see anything at all. There, the decision about whether a point lies within the cube or not, is made by computing the value of $x_1^2+x_2^2+x_3^2+ \ldots +x_n^2$: if this value is less than 1, the point lies within the cube; otherwise it lies outside. The domain of possible values for the coordinates was chosen such that no point lies exactly on the surface of the cube. The values obtained with p=20 were very close to the exact values in Fig. 4.19; the relative difference was less than one in one thousand.

You should not think that the availability of the formulas for the surface area and the volume of a five-dimensional or 25-dimensional cube is of great practical relevance. There are only a small number of problems in physics and computer science where n-dimensional spaces are useful. I showed you all of this primarily because I want you to see that it is not very difficult to move around accurately in these spaces, although we do not have any idea what they look like. Don't forget that we still are in the chapter which introduces formal worlds. It is the purpose of this chapter to make you aware of the fact that, besides our world of perceivable objects, there exists a formal world which we may enter and leave as we like, but which we must enter if we want to do information processing using computers. Even the computation of surface areas and volumes of n-dimensional cubes, which I demonstrated here, is nothing but a formal game which can be played by agents who have no idea about the considerations that led us to the rules of this game.

# Chapter 5
# About the Methods for Computing the Future

When discussing the consequences of the theory of quantum mechanics, Albert Einstein is said to have maintained that, "God doesn't throw dice." Others were convinced that the only acceptable interpretation of that theory required acceptance that there are elementary events which cannot be predicted accurately. Of course, Albert Einstein shared the conviction that mankind will never be in a position to predict the future. However, he was convinced that the course of the world is determined completely by natural laws, although no one will ever be able to grasp all of these laws. In the terms of philosophy, the question is whether or not the world is causally determined. In simple language, we may ask, "Does the state of the world at a particular point in time imply that all future events are already completely determined?" If this were true, the actual configuration of all elementary particles in the universe would not only determine which football team will win the championship next year, but it would also determine that a murder which will happen twenty years from now is completely unavoidable and could not be prevented. In this case, the murderer could not be held responsible for what he did since he was only a victim of the laws of physics. I shall not discuss this issue any further, since it is enough for us to accept the fact that nobody can entirely predict the future, and that we must live with surprising events.

The reason the problem of predicting the future is a subject of this book is that calculating probabilities is actually part of the job of scientists and engineers. For example, an engineer who plans a telephone network must make assumptions about the future behavior of the people using this network. And no physicist would ever have come up with the idea of building a laser (**L**ight **A**mplification by **S**timulated **E**mission of **R**adiation) unless he had understood the theory of the interactions between atoms and photons which is based heavily on the concept of probability.

Did you groan because we have entered the field of mathematics again? Let me remind you that at the beginning of Chapter 2, when mathematics was discussed in geographical terms, I said, "We had no reason to study the continent of mathematics if it was not fundamental for physics and technology." By doing math with probabilities, we do not introduce new mathematical structures. We only apply the old structures to a special problem domain. There is no reason to be frightened, since the rock wall is not very high, and your mountain guide will lead you along the easiest possible path.

## Attempts to Reduce Expectations to Numbers

When people use the word probability in ordinary conversation, they do it for the purpose of telling us something about their expectations about certain future events. They might say, "The probability is not very high that ...", and sometimes they might even use explicit numbers such as in, "The probability that we shall spend this year's vacation in Canada is over 90 percent." How can we interpret this statement? Where did they get this 90 percent? Why didn't they choose 60 or 80 percent? If they had said 100 percent, we would have known that they were certain to spend their vacation in Canada, and if they had said 0 percent, this would have been equivalent to the statement that under no circumstances would they spend their vacation in Canada. Thus, we must now find out how an exact interpretation of probability values between 0 and 100 can be defined. However, instead of using percentages, we now shall use the interval between 0 and 1. The value 1 for a probability corresponds to 100 percent. The reason for this new scale is rather simple: we desire that the product of two probability values is also a probability value, and this would not be the case if we used the percentage scale. For example, the product of 50 % and 40 % is 2,000 $(\%)^2$ which cannot be interpreted as a probability unless we reduce it to 20 %. If, however, we restrict ourselves to the interval between 0 and 1, the product in our example is 0.5*0.4=0.2. Here, both the two factors and the result can easily be interpreted as probability values, and no additional reduction of the result is necessary.

The considerations which lead us to an appropriate definition of numerical probabilities should begin with the study of devices which have been designed for generating random processes. An example of such a device, a so-called *board of Galton*, is shown in Fig. 5.1. Francis Galton was an Englishman who lived from 1822 until 1911. He was a relative of Charles Darwin who introduced the theory of evolution which, among other things, says that men and monkeys have common ancestors. Mr. Galton did research concerning the inheritance of human properties – a field which nowadays is called human genetics. Readers of detective stories may find it interesting to know that Francis Galton was the first to come up with the idea that fingerprints might be helpful in solving crimes.

Now we look at the device which he invented and which is shown in Fig. 5.1. The drawing on the right side of this figure shows what can be seen when the board is cut in two along the dashed line in its middle, and its edge is viewed. On the right side of the figure, three layers can be seen. They are, from left to right: a base board (usually wooden) which appears in the left view as the white area; the profile boards which appear in the left view as the grey areas; and the covering

Expectations as Numbers 111

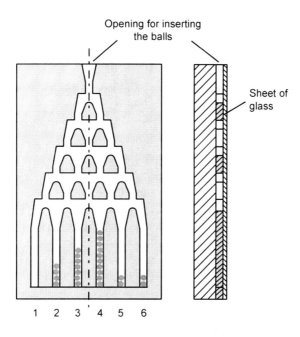

**Fig. 5.1** Board of Galton

sheet of glass which is transparent in the left view. Galton's board is placed vertically on a table such that the opening for the insertion of the balls is on the top. Each ball inserted will move down along a path which contains five binary forks. At each fork a random decision determines whether the ball continues its way downwards on the left side or the right side of the fork. Both sides have the same probability, namely 0.5. The path of a ball ends in one of the six compartments, here numbered 1 through 6. Fig. 5.1 shows the result of an experiment where I inserted 25 balls, one after the other.

Of course, besides Galton's board, many other devices have been designed for generating random processes. Some of them are known much more commonly than Galton's board. For example, think of using dice to generate random sequences of the natural numbers 1 through 6 using one die, or a roulette wheel with its 37 compartments which are numbered 0 through 36. All of these devices are used to generate random events from a given finite set of numbers where all events have the same probability. The corresponding probability of such an event is the reciprocal of the number of elements in the set. In the example of throwing a die, the probability that a predicted number will occur, such as the number 5, is 1/6, and the probability that the ball of a roulette wheel will fall into a predicted compartment is 1/37. But in the case of our Galton's board in Fig. 5.1, which has 6 possible compartments for a ball, computing the probability of a ball landing in a certain compartment is not nearly as simple as in the cases of the dice or roulette. The field of mathematics which we must enter in order to find these probabilities is called *combinatorics* or *combinatorical analysis*.

## How We Can Calculate the Number of Possible Cases

The name combinatorics suggests that we shall now do some kind of combining. The things to be combined are elementary sequential random events. Their combination is called a structured random event. The concept of structured random events is best illustrated by Fig. 5.1: the distribution of 25 balls among the six compartments of the board is a structured random event which results from combining 125 elementary sequential random events. Each of the 25 balls, which are inserted one after the other, falls down along a path which is determined by five elementary sequential random events, namely by the decisions at the five binary forks where the ball can go either left or right. Another simple example of a sequence of elementary random events is the selection of the 6 numbers in a game called "Lotto." In a particular version of this game, there is a machine which randomly picks 6 balls out of a container originally containing 49 balls. The balls are labeled with the numbers 1 through 49, and the gamblers try to predict the set of 6 numbers which the machine will pick next. For the selection of the first number, the probability is 1/49, since it is picked from the original set of 49 numbers. The probability for the second number is no longer 1/49, but 1/48 since there are only 48 balls left in the container after the first ball has been removed. Whenever we consider a sequence of elementary random events, we must specify whether all these events occur with the same probability, or how the probability depends on the position of the event in the sequence. It may well be that past events have an effect on the next event.

The main operation in combinatorics is filling a finite sequence of positions by random selection from given finite sets. This is illustrated in the top part of Fig. 5.2. The set from which an element $x_i$ for position j is selected at random is called $X_j$, and the number of elements it contains before an element is picked is $m_j$ = $|X_j|$. Thus, the probability of picking a certain element from this set is $1/m_j$. If the gamblers sitting at the roulette table in a casino would think of the structure shown in Fig. 5.2, they would immediately stop writing down the sequence of numbers from the past, because it is irrelevant. Once a year, I have the pleasure of spending a week or two at Baden-Baden which is not only a famous German spa, but which is also said to have the most beautiful casino in the world outside of Monte Carlo. During these stays, there is always one night which I reserve for going into this casino – not to gamble, but to watch the gamblers. On one of these occasions, I happened to see that the ball in roulette landed in red compartments twelve times in a row. Of the 37 compartments, 18 are red, 18 are black and one,

# Number of Possible Cases

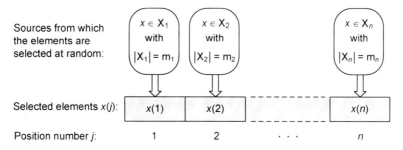

**Fig. 5.2** Filling a sequence of positions at random

the compartment for 0, is green. The gamblers who had written down the sequence of numbers got quite excited when they realized that the color red had occurred twelve times in a row, since the probability of this structured event is extremely small, namely $(18/37)^{12} = 0.0001757$. From the fact that an event whose probability is so small had occurred, they concluded that the probability of the next number again being a red one was also extremely small. Therefore, they put large amounts of money on the prediction that the next ball will land in a black compartment. But the fact that a sequence of red numbers had occurred did not have any effect at all, neither on the mechanism of the roulette wheel nor on the way the croupier would throw the ball. Therefore, the probability for the next number being red or black had not changed at all, it was still 18/37, and would remain so forever.

Now we must consider the difference between the random selection of a sequence versus the random selection of an unordered subset. In the bottom line of Fig. 5.2, I gave you the number of different sequences the Lotto machine may select. But the rules of the Lotto game do not require that the gamblers predict the exact sequence in which the 6 numbers are picked by the machine, they only have to predict these 6 numbers in an unordered set. Therefore, the probability of becoming a Lotto winner is higher than the reciprocal of the number given in Fig. 5.2. The factor by which this probability is higher is given by the number of different possible orderings of six numbers. This again is obtained by applying the scheme in Fig. 5.2. For filling the first position, one out of 6 numbers has to be selected. For filling the second position, one out of 5 numbers has to be selected since the number selected for the first position is no longer available for the second position. After the fifth position has been filled by selecting one out of the two remaining numbers, there is only one number left which must be taken to fill the sixth position. Thus, there are 6*5*4*3*2*1=720 different possible orderings of a set of six elements.

What has been shown for the case of the Lotto game can, of course, be generalized to the question of how many different subsets with s elements can be selected from a universe with u elements. The result of this generalization is represented in Fig. 5.3. In the formulas shown there, you see a vertical symbolic structure or notation which looks like a fraction in brackets where the fraction line is missing. This symbolic structure is an abbreviation of the expression "number of different subsets with s elements which can be selected from a universe with u elements," and I think this is a very helpful notation. In the case of the Lotto, each subset with 6 elements selected from the universe with 49 numbers has the same probability, and the probability of becoming a Lotto-winner is 1/13,983,816.

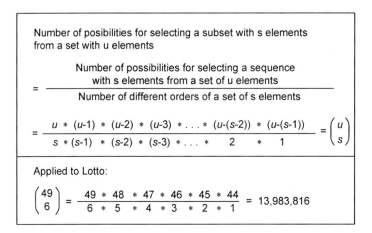

**Fig. 5.3**  Getting unordered sets from sequences

The French mathematician Blaise Pascal (1623-1662), as with many other mathematicians, liked playing around with numbers, and one day he happened to find a rather interesting structure which is related to the problem of selecting subsets from universes. This structure, which is called Pascal's triangle, is represented in Fig. 5.4, and the principle for its construction is shown at the top of this figure.

Number of Possible Cases 115

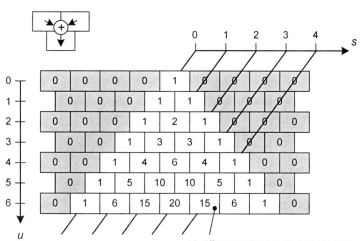

**Fig. 5.4** Pascal's triangle and its interpretation

We now return to the board of Galton shown in Fig. 5.1. An interesting question is how many different distributions are possible if we have b balls and c containers. In the example in Fig. 5.1, we have b=25 and c=6. It is not at all obvious that the number of different distributions is equal to the number of different subsets with b elements which can be selected from a universe with (c+b-1) elements. This surprising equivalence of the two problems – selecting a subset on the one hand and distributing balls among containers on the other – cannot be understood unless the origin of the strange mathematical expression (c+b-1) is explained. This explanation is given in Fig. 5.5. By selecting as many numbers from the universe as we have balls, each ball is associated with some kind of information. But this information cannot always be the number of the container where the ball must be placed. In a subset, a single element of the universe can be contained at most once, although it must be possible to assign the same container number to different balls. The procedure shown in Fig. 5.5 makes it possible to assign the same container number to different balls although in the subset selected from the universe, no number can occur more than once. The case where all b balls are placed in the container with the highest number c requires only that the smallest number in the subset is equal to c. The remaining numbers must be (c+1), (c+2), ... until (c+b-1).

We now apply the results that are represented in the figures 5.2, 5.3 and 5.5 to Galton's board in Fig. 5.1. How can we compute the probability of a certain distribution, for example that in Fig. 5.1? Fig. 5.5 tells us how to get the number of different distributions for the case where b=25 and c=6. This number, 142,506, is

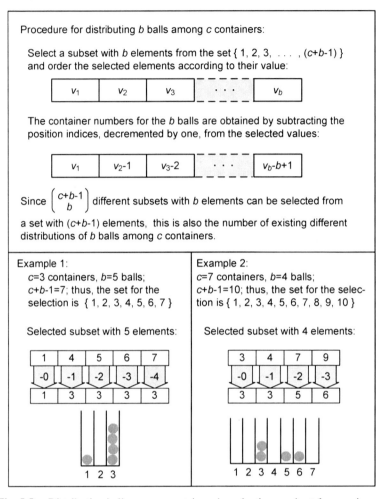

**Fig. 5.5** Distributing balls among containers by selecting a subset from a given set

given in the top section of Fig. 5.6. However, these different distributions don't have the same probability. The probability of a ball reaching a certain compartment depends on the number of different paths which lead to that compartment. The number n of a compartment is always one more than the number of decisions to move to the right at the forks along the path of the ball. In order to obtain the number of different paths leading to compartment n, we must compute the number of different subsets with (n-1) elements from the universe of five fork levels, since this is the number of different fork selections for the (n-1) right-moving decisions. The probability of a specific distribution of the balls among the compartments is then obtained as a fraction where the numerator is the number of different

Number of Possible Cases 117

> Inserting 25 balls into the board of Galton in Fig. 5.1 leads to
> one of the $\binom{6 + 25 - 1}{25} = \binom{30}{25} = 142{,}506$ possible distributions.

> There are $2^{25*5} = 2^{125}$ different courses of the experiment,
> since for each of the 25 balls, 5 arbitrary binary decisions
> with respect to their path - left or right - are made.
> $2^{125}$ is approximately $42*(1\text{ billion})^4$.

> The number of courses of the experiment which result in the distribution shown in Fig. 5.1, is
>
> $$\binom{25}{4}*\binom{5}{1}^4 * \binom{21}{7}*\binom{5}{2}^7 * \binom{14}{10}*\binom{5}{3}^{10} * \binom{4}{2}*\binom{5}{4}^2 * \binom{2}{2}*\binom{5}{5}^2$$
>
> For each non-empty container, this product has a pair of factors
>
> $\binom{r}{b} * \binom{5}{n-1}^b$ with
> 
> $n$ = index of the compartment;
> $b$ = number of balls in the compartment n;
> $r$ = remaining balls, which have not yet been assigned to a compartment with a lower index;
> $5$ = number of the fork levels;
>
> The first factor determines the number of different subsets with $b$ balls which can be selected from the set of $r$ balls.
> The second factor determines the number of different path combinations by which the $b$ balls will reach their compartment $n$.

**Fig. 5.6** Numbers related to Galton's board in Fig. 5.1

path combinations for all balls, resulting in the specific distribution, and where the denominator is the number of all possible path combinations for all balls.

The formula which provides the number of all path combinations for the 25 balls resulting in the distribution shown in Fig. 5.1 is given in the lower section of Fig. 5.6. In this long product expression, there is a pair of two factors for each of the 5 non-empty compartments. The first factor provides the number of possibilities for selecting the s balls for the actual compartment n from the set of those balls which have not yet been assigned to a compartment with an index smaller than n. The second factor is a power, the base of which is the number of different paths via which a ball can reach compartment n. The exponent of this power is the number of balls which lie in the compartment in the actual distribution. I did not go through the trouble of actually computing the values of these ten factors and multiplying them to obtain the overall product, since I think it is sufficient that you understand how this long product expression is obtained.

Now, at the end of this section on combinatorics, I present a problem which has caused numerous controversial discussions, and the solution of which seems paradoxical to many people. There was a quiz show on television where a contestant had been told that a new car was hidden behind one of three closed doors and that, if he selected that particular door, the car would be his. Only consolation prizes were hidden behind the other two doors. After the contestant had pointed to the door he had decided to select, the quiz master opened one of the two other doors where a consolation prize became visible. Then, the quiz master offered the contestant a chance to change his first decision and select the other of the two still-closed doors. The question is whether the probability of finding the car is the same for both doors, namely 0.5, or whether the probabilities for the two doors differ. Many people argue that the two probabilities cannot differ since there was no information available for giving a higher probability to either of these two doors. However, such information was actually provided by the quiz master when he decided which of the two non-selected doors he would open. He certainly would not open the door with the car behind it. At the beginning, all three doors were closed, and the probability of the car being hidden behind the door selected by the candidate was 1/3. Correspondingly, the probability of the car not standing behind the selected door was 2/3. This could not be changed by the quiz master's opening of one of the non-selected doors, and therefore, the contestant should change his decision. If he stays with his first decision he will win the car with a probability of 1/3, whereas if he changes his decision, he will win with a probability of 2/3.

## What You Can Do If You Don't Want to Know All the Details

Politicians and top level managers don't have much time for individual subjects because they must multitask, i.e., switch from one task to another very often. Therefore, they expect that all information material they get has been prepared in such a way that all details have been omitted and only the essentials are included. Those who prepare this material have noted that their superiors look first at the length of a text and then ignore it if it is longer than one page. Very often such a superior expects his information providers to explain a complex problem and its solution within ten minutes, although it is quite clear that even an hour would not be enough.

I made these introductory remarks in order to characterize the subject which we shall now consider. We are looking for ways to reduce information about a distribution of probabilities to not more than a couple of numbers. This really can be done, but only under the following restriction: the random events must be such that each event has a number assigned to it, and, if there are more than two events, the assignment should not be arbitrary, i.e. the numbers assigned should correspond to some meaningful property of the events. In order to illustrate this restriction, let's consider the following example. We assume that a young man will certainly marry

Expected Value and Standard Deviation

one of the three sisters living next door - Anne, Bess or Carol, but he has not yet decided which one. The probabilities may be p(A)=0.1, p(B)=0.6 and p(C)=0.3. Surely, we can assign numbers arbitrarily to the three girls, e. g., 1 to Ann, 2 to Bess and 3 to Carol, but the differences between these numbers have absolutely no meaning. One way to get meaningful numbers would be to consider the ages of the girls which certainly will differ because they are sisters.

From here on we assume that the restriction specified above is satisfied, i.e., we assume that the random process is generating meaningful numbers. It seems rather obvious that a so-called "average number" provides some information about the distribution of the probabilities. But the example specified in Fig. 5.7 may help you see that the concept of an average number is not as clear as it may seem.

| Possible number $x$, which is selected at random | 0 | 7 | 9 | 11 | 13 | 15 |
|---|---|---|---|---|---|---|
| Probability of selecting $x$ | 0.1 | 0.3 | 0.05 | 0.05 | 0.3 | 0.2 |

**Fig. 5.7** Example of a discrete probability distribution

In Fig. 5.8 the values of the probabilities for the numbers in Fig. 5.7 are interpreted as weights of bodies sitting on a see-saw, and the positions of these

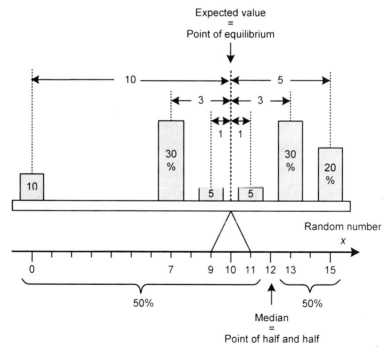

**Fig. 5.8** Expected value versus median for the example in Fig. 5.7

bodies are determined by the random numbers to which the probabilities belong. In this figure, two numbers are indicated which might be meant when the term "average" is used. One number is given by the point where the see-saw must be supported in order to bring it into equilibrium, and the other number, which is called the *median*, determines the point where the distribution is split into halves.

There are distributions where these two numbers are the same, but these are only special cases as you can see in Fig. 5.8. Here the see-saw is in equilibrium although the sums of the weights on either side of the support are not the same. This is a consequence of the distances of the weights from the support. Probably you have seen two children sitting on a see-saw where one child was heavier than the other one. In such a situation, the heavier child must sit closer to the support than the other one in order to produce the equilibrium. The equilibrium point is determined by the fact that the sum of the products of weight and distance are the same on both sides of the support. In Fig. 5.8, these product sums are

$$10*10 + 30*3 + 5*1 = 5*1 + 30*3 + 20*5$$

In probability theory, this point of equilibrium is called the *expected value*, although in certain cases this term is misleading. If you had to predict which one of the six random numbers given in Fig. 5.7 will be selected next, you certainly would not choose the number 10, the number for equilibrium, because 10 does not belong to the set of the six possible numbers and therefore cannot possibly occur.

Once, an average value of a distribution – the expected value or the median - is known, we may question how the distances from this average point vary. We could consider these distances with their corresponding probabilities as a new distribution of random numbers, and we also can compute an expected value and a median for this. Clearly, we must decide whether we should take the distances measured from the expected value or from the median. The German mathematician Carl Friedrich Gauss (1777-1855) spent much time searching for interesting properties of probability distributions, and he found out that it is more useful to take the distances as measured from the expected value than from the median. He also suggested that, instead of considering the distribution of the distances themselves, the distribution of the squares of these distances should be considered. The expected value of such a distribution is called the *variance* of the original distribution, and the square root of the variance is called the *standard deviation*. For our example in Fig. 5.8, the distribution of the squares of the distances is shown in Fig. 5.9. The expected value of this distribution is 20.5 and its square root is 4.5277. In other words, the variance of the distribution in Fig. 5.8 is 20.5 and its standard deviation is 4.5277.

Expected Value and Standard Deviation 121

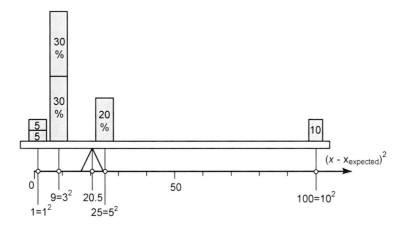

**Fig. 5.9** Computing the variance for the distribution in Fig. 5.7

The reason for using distributions of squares of distances lies in the simplicity of the calculations which must be performed when a probability distribution of structured random events is derived by combining the probability distributions of their elementary events. However, I shall not present an example to illustrate this since, in actual context, it is not really important. I think you can see the big picture without knowing those details.

While the median, the expected value and the standard deviation are numbers which characterize a single distribution of probabilities, the so-called *correlation factor* says something about the relationship between two distributions. The concept of the correlation factor goes back to the same Francis Galton who invented the board which is shown in Fig. 5.1. As you may recall, Mr. Galton did research in the field of inheritance of human properties. Thus, it was quite natural that he asked how he could combine the distributions of two human properties in such a way that the result would tell him whether or not these properties are independent of one other. The example I use to explain the concept of correlation is introduced in Fig. 5.10. Here, the question is whether there is a probability connection between a man's being short-sighted and his being bald. In order to keep the example as simple as possible, I neglect the existence of different degrees of short-sightedness and baldness, i.e., we assume these properties are binary – yes or no. Since the only purpose of this example is illustrating the concept of correlation, I did not go through the trouble of finding the actual probabilities which some expert might have found by counting many thousands of real cases. My numbers given in Fig. 5.10 most likely differ greatly from the real numbers.

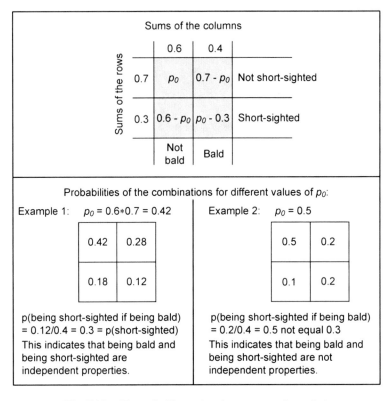

**Fig. 5.10** Example illustrating the concept of correlation

With respect to the two binary properties considered, there are four categories of men, and we can ask what the probability is that a man who is selected at random belongs to a particular one of these four categories. The four probabilities which are written in the four fields of the grey square are called *combined probabilities* since they belong to specific combinations of the two properties. The elementary probabilities for being short-sighted or being bald do not completely determine the values of the combined probabilities. Therefore, I introduced the probability variable $p_0$ for the combination of a man being neither short-sighted nor bald. The arithmetic expressions in the remaining three fields provide the corresponding probabilities once a value for $p_0$ has been chosen. Note that the sum of the probabilities in the four grey squares is 1.0 because every man must be included in one of the four categories. The value of $p_0$ must not be smaller than 0.3 and not greater than 0.6 in order to avoid negative values for all of these probabilities.

Instead of saying that two properties are independent of one another, the experts say that they are uncorrelated. In the case of uncorrelatedness of two properties, the distribution of either of the properties does not depend upon the actual value of the other property. In our example, this means that if short-sightedness

Expected Value and Standard Deviation

and baldness are uncorrelated, both 30 percent of the bald men and 30 percent of the non-bald men are short-sighted. In this case, the combined probability $p_0$ of not being bald and not being short-sighted will be 0.42 which is the product of the two probabilities, 0.6 for not being bald and 0.7 for not being short-sighted. The more the two properties are correlated, the more the value of $p_0$ deviates from 0.42. In Example 2 which is represented in the lower right corner of Fig. 5.10, the value for $p_0$ has been chosen to be 0.5. In this case, the probability of being short-sighted depends on the question of whether a particular man is bald or not. This shows that a bald man will be short-sighted with a probability of 50 percent, while a man who isn't bald will be short-sighted with a probability of only 16.7 percent.

Now we shall consider how to compute the *correlation factor*. This number is a measure of how strongly two distributions are interdependent. In the case of total uncorrelatedness, this number is zero, and in the case of total correlatedness, this number is either -1 or +1. A positive correlation factor indicates that the events with both properties being on the same side of the see-saw have higher probabilities than they would have in the case of uncorrelatedness. Correspondingly, the negative sign indicates a preference of the events where the two properties are on different sides of the see-saw. From these definitions, it follows that the computation of a correlation factor requires that the elementary probabilities can be placed on a see-saw. This means that each property must have a number assigned to it. The additional requirement that the distances between these numbers must be meaningful does not apply here because our example is restricted to binary properties. When a distribution has only two events, the "distance" between these events may be chosen arbitrarily since it cannot be compared with other distances on the same see-saw – there are no such other distances.

In the top section of Fig. 5.11, the distributions of baldness and short-sightedness are shown on their see-saws, and you can see the numbers which I assigned arbitrarily. In order to simplify the following computation, I chose the numbers for the properties in such a way that in both distributions the equilibrium point sits at the zero position. Of course, my choices have an effect on the standard deviations. The next see-saw which we have to consider is the one on which we place the four combination probabilities. Fig. 5.11 shows the corresponding see-saw for both of the two examples in the lower part of Fig. 5.10. As you can see, the positions of the probabilities are the same on both see-saws. These positions are obtained by multiplying the positions from the two elementary distributions: -22.4 = (-16)*1.4, -14.4 = 24*(-0.6), 9.6 = (-16)*(-0.6) and 33.6 = 24*1.4. The probabilities on the see-saw for Example 1 are also obtained by multiplication where the factors are the probabilities from the elementary distributions. When the support of this see-saw is at position zero, equilibrium is obtained. This situation is the result of total uncorrelatedness.

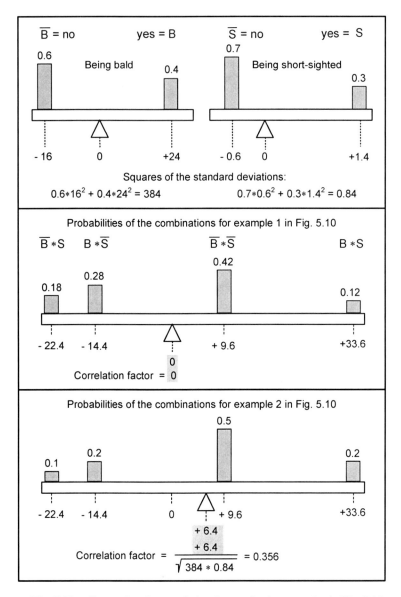

**Fig. 5.11** Computing the correlation factors for the examples in Fig. 5.10

The see-saw for Example 2 is obtained by exchanging the probabilities from Example 1 for those of Example 2, and then moving the support to the position which brings the see-saw to its equilibrium again. The fact that the position of the support had to be shifted away from its original zero position, indicates that now the properties are no longer uncorrelated. But the distance of this shift cannot yet be the correlation factor we are looking for, since it is influenced by the arbitrary choice of the distances in the elementary distributions. This influence can be eliminated by dividing the distance of the shift by the product of the elementary standard deviations. As a final result, we get the number 0.356. This is a positive correlation factor which indicates that the probability of being both bald and short-sighted on the one hand and the probability of being neither bald nor short-sighted on the other hand are higher than they would be in the case of uncorrelatedness. You can check this by comparing the inscriptions in the squares for the two examples in Fig. 5.10. Again, I emphasize the fact that the probabilities I chose do not correspond at all to reality, and therefore the real correlation factor between baldness and short-sightedness might well be totally different from 0.356.

Computing correlation factors for human properties does not necessarily correspond to political correctness, since correlation factors may be used to discriminate against large groups of people. In the past, correlation factors have been used to "prove" that there is a correlation between intelligence and color of the skin, or between crime rate and nationality.

## How to Handle the Situation When the Cases Are No Longer Countable

In the previous examples, we always had a finite set of numbers from which the random numbers could be selected. Think of Galton's board, where we had the finite set of compartments for the balls. Now we shall discuss situations where the random numbers are no longer selected from a finite set, but from a continuous interval of real numbers. Think of a wheel which a motor spins rather fast, and then the motor is switched off. After some time, the wheel will come to a stop, and the position of this stop can be at any angle φ in the interval $0° \leq φ < 360°$. Without giving it much thought, we might say that each angle has the same probability – but which one? As long as the random numbers are picked from finite sets, each of these numbers has a specific probability assigned to it, and the sum of the probabilities of all the numbers in the set is 1. But when the random numbers come from a continuous interval of real numbers, we have to consider an infinite set of numbers and, in this case, the random numbers cannot have non-zero probabilities assigned to them, because otherwise the sum of these probabilities would not be 1, but infinity. This problem should remind you of the process of integration where we obtained the area under a curve by adding an infinite number of infinitely small rectangles. The same procedure can be applied to the case of random numbers being selected from a continuous interval. We now consider curves where the total area under the curve has the value 1. These curves represent so-called *probability density functions*, with the notation dp/dx. An example of such a curve is shown in Fig. 5.12. Here, it is no longer meaningful to ask for the probability of a certain number x, since this probability is infinitely small. But non-zero probabilities are obtained for intervals between two points on the x-axis. The area above such an interval corresponds to the probability that the random number x will lie within the limits of the interval $x_1 \leq x < x_2$. In the example of the spinning wheel coming to a stop, there is no non-zero probability that it will stop exactly at the angle $\varphi=2°$, but there is a non-zero probability that it will come to a stop within the interval $1.5° \leq \varphi < 2.5°$; this probability is $1°/360° = 0.00278$.

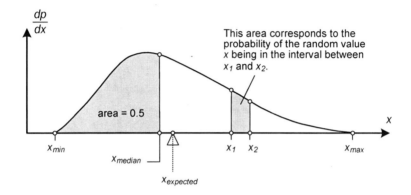

**Fig. 5.12** Example of a probability density function

Fig. 5.12 shows that the concepts of the expected value and the median which have been defined in Fig. 5.8 also can be applied in the case of distributions of probability densities. If we would take a thick wooden board and cut it into a form which exactly corresponds to the area between the given probability density curve and the x-axis, then we could experimentally find the point (the expected value of the distribution) where the board must be supported in order to stay in horizontal equilibrium and rotates neither to the left nor to the right. And certainly we can find the point (the median of the distribution) where a cut along a vertical line will split the two halves of the area under the curve.

There is one type of distribution of probability densities which is called the "*normal distribution.*" This does not mean that any other distribution should be thought of being abnormal. It only means that many distributions which can be found in nature or every day life come close to this mathematically defined distribution, for example, the weight of babies at the time of their birth.

The concept of the normal distribution goes back to Carl Friedrich Gauss who was mentioned previously. The curve of this distribution and the formula for the probability density function describing it are shown in Fig. 5.13.

Continuous Domains for Random Selections 127

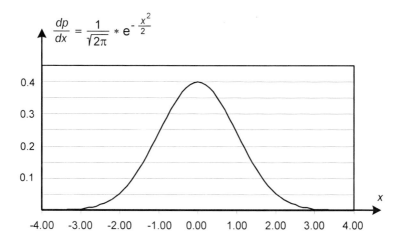

**Fig. 5.13** Gaussian probability density function

I think it is very interesting that this formula contains the two most important transcendental numbers, namely $\pi$ and e. I don't know how Mr. Gauss found this formula, but at least I can show you how the curve is obtained. I shall get to this distribution by starting with a distribution of a finite set of structured random events. If we throw one die eight times in a row and add up the eight random numbers, the sum will be a number from the finite set $\{8, 9, \ldots, 47, 48\}$. While there is only one sequence of eight thrown dice numbers having the sums 8 or 48, the other sums in the interval $9 \leq \text{sum} \leq 47$ can be obtained by many sequences. For example, the sum 9 results from eight different sequences. The sum having the highest number of different sequences is 28; there are 135,954 different sequences having the sum 28. The number of all possible sequences is $6^8 = 1,679,616$. Thus, the probability of throwing the sum 28 is $135,954/1,679,616 = 0.08094$ which is slightly more than eight percent. The distribution of all possible sequences over their sums is shown in Fig. 5.14.

Obviously, there is a certain resemblance between this distribution and the normal distribution in Fig. 5.13, although there are some noticeable differences. The expected value of the normal distribution is 0 and its standard deviation is 1 while in Fig. 5.14 these two values are 28 and 4.83, respectively. But it is possible to modify the experiment from Fig. 5.14 in such a way that the resulting distribution has an expected value of 0 and a standard deviation of 1. In this modified experiment, we no longer add up eight natural numbers which we select at random from the set $\{1, 2, 3, 4, 5, 6\}$, but now we select the eight summands from the set $\{-0.5, -0.25, 0, +0.25, +0.5\}$. Then we actually get a distribution which has an expected value of 0 and a standard deviation of 1.

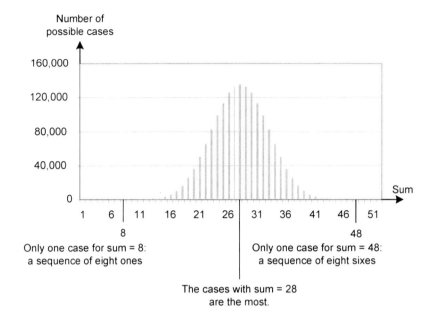

**Fig. 5.14** Distribution of sums of all possible sequences of 8 numbers from the set {1, 2, 3, 4, 5, 6}

We must find a trick which leads us from a distribution of discrete probabilities to a continuous function which can be interpreted as a distribution of probability densities. If we already had found the function we were looking for, the probabilities p(x) of the discrete distribution would correspond to areas under the density curve d(x). The intervals of these areas have equal lengths since the discrete probabilities have equal distances – see Fig. 5.14. This distance is the minimal difference between two sums, Δx, which is equal to the minimal difference between two summands. The area equal to p(x) will then be equal to the area of a rectangle having width Δx and height d(x). This area is given by the product Δx*d(x). The fraction p(x)/Δx will be an approximate value of the probability density for a given sum. Why is this approximate and not exact? Because we are still not considering a continuous function, but a sequence of discrete values, since the sums considered have the distance Δx. Such an approximation of the area under a curve was used in Fig. 3.13. There, the approximation had been made better and better by making the width of the rectangles smaller and smaller. If the summands are selected from the set {-0.5, -0.25, 0, +0.25, +0.5}, Δx is 0.25. If we make Δx smaller than 0.25 and take the eight summands from the set {-2*Δx, -Δx, 0, Δx, 2*Δx}, the standard deviation of this new distribution will no longer have the value 1, but will be less than 1. But there is a way to keep the standard deviation constant at 1: when Δx is made smaller than 0.25, the number n of the summands must be made greater than eight. For the standard deviation to be 1, it is required that Δx and n are related by the equation $2*n*\Delta x^2 = 1$. Examples of pairs (Δx, n) which satisfy this condition are (0.25, 8), (0.1, 50) or (0.02, 1250). Now we really can make Δx smaller and smaller by making n greater and greater according to the given condition. In this process, the extreme values of the sums, which are –2*n*Δx and +2*n*Δx will grow beyond all limits, since they are -1/Δx and 1/Δx because of the relation between n and Δx.

When the number of summands grows beyond all limits, the number of different sums grows beyond all limits, too. Therefore, the probability of a sum, p(x), must become infinitely small, because otherwise the sum of the probabilities of all different sums could not be 1. But the fraction p(x)/Δx of two infinitely small numbers has a well-defined limit which is the probability density d(x) of the normal distribution in Fig. 5.13.

## Statistics Are More Than Just Listing the Results of Counts

Sir Winston Churchill (1874-1965), who was Prime Minister of Great Britain during the Second World War, once said, "I don't believe any results from statistics unless I myself falsified them." In this statement, the term statistics is used in its restricted meaning – being a report about the results of counts. Sometimes newspapers or broadcast media publish more or less interesting results of counts which were performed by some governmental agency. For example, such information could be that 19 percent of all couples who were married within the last five years have now been divorced.

But the term statistics does not have only this restricted meaning of just doing counts. When it occurs in the titles of university textbooks, e.g., "Probability Theory and Statistics," it means the area of relating probability distributions to the results of counts. There are mainly two questions which are of interest in this area: (1) if the results of a count are given, how can a corresponding probability distribution be assigned to these results? and (2) if one assumes a certain probability distribution, how many events must be observed in order to verify the justification of this assumption?

In general, we might say that the subject of statistics as a field of mathematics is the probability of probability distributions. For example, let's assume someone suspects that the machine which selects the numbers of the state Lotto has been tampered with in such a way that the probability of even numbers is higher than that of odd numbers. If this person has no direct access to the machine, he or she can never be sure that the probability distribution really is as suspected. But using the results produced by the machine, the probability which says how likely it is that the suspected distribution really is implemented by the machine can be computed. The question is how many selections of the machine must be taken into account in order to reach a certain probability level - say 95 percent - to justify the suspicion. I shall not answer this question, since the only reason I introduced it, was to give you an idea of the kind of problems a statistician must solve.

# Chapter 6
# What Talking and Writing Have in Common

The two words 'communication' and 'communism' both begin with *communi* which is a consequence of the fact that they have something in common with respect to their meaning. We must use the word *common* or *shared* in both cases when we describe what these words mean. In the case of the word communism, we must talk about common or shared property, and in the case of communication, we must talk about common or shared information. Think of Saint Martin who shared his overcoat with a beggar, and also think of someone who is the only one with certain information and communicates it with others in order to share it with them. There are two fundamental differences between these two types of sharing. After Saint Martin shared his overcoat with the beggar, only half of the overcoat was left for himself, while a person who shares information doesn't lose a thing by sharing. The second difference has to do with observing the process of sharing. People observed how Saint Martin cut his overcoat into two halves, one of which he gave to the beggar. These people, without any doubt, could be sure that a sharing had occurred. But now assume you were listening to a conversation between two persons who use a language unfamiliar to you. How could you know for sure that they were really sharing information? You could only assume that such a sharing had occurred. The only thing you could know for sure is the fact that these two persons were together and producing strange sounds. Therefore, when we consider communication processes, we always have to consider both the observable process of producing physical signals and the interpretation of these signals by assigning meaning to them.

We think primarily of people communicating when the word communication is used, but animals can communicate, too. Long ago, someone discovered how a bee communicates its knowledge about the location of a rich source of food to its fellow bees. This method of communication has even been called the "language of the bees." We shall use the term *language* only in cases where

- a set of symbols has been defined,
- a set of rules has been defined which describe how structures of symbols can be composed, and
- the interpretation of both the symbols and the structures composed of symbols is known to all communication partners.

A *symbol* is an observable pattern which can be easily reproduced and has a meaning assigned to it. This definition applies not only to written or spoken letters or numbers, but also to nodding one's head or to pawing at the ground with the right hind leg.

According to the definition given above, what the bees use to communicate really is a language. The use of the term "defined" in connection with symbols and rules does not mean that this defining must necessarily be done by a human action. It only means that the symbols and rules are determined in such a way that they become common knowledge of all individuals in the actual community. In the case of the bees, this determination is the result of evolution and has been incorporated into their genes.

## How Speech and Writing Are Interrelated

We now consider a process of two people communicating without using a language. Assume that a young man and a young girl have fallen in love with each other, and they want to talk on the phone as often as possible. But the girl's father doesn't like this, and will not allow his daughter to use the phone as long as he is at home. As a result, the young man needs information about when the father is at home and when he is away. In this case, the two young people came up with the idea of using the flower pot in the living room window as a symbol for their communication: if this pot, as seen from the street, stands on the right side of the window, the father is at home. Otherwise the young man may call without risk. Certainly, in this case, the two possible positions of the pot have been defined as two symbols. They can be observed and reproduced easily, and they have meanings assigned to them. But they are not elements of a language because there are no rules describing how these symbols could be used to build meaningful structures.

From now on, I shall restrict my considerations to written or spoken texts. These texts are structures built by putting elementary symbols in a linear order. How the evolution of speech occurred is undoubtedly very interesting, but this book is not the place to discuss this subject. Our ability to speak and write not only enables us to communicate with a much higher degree of efficiency than other creatures, but it also makes it possible for a single person to try to find answers to questions which could not even be thought of before. The use of texts is not restricted to inter-human communication, but also applies to the basis of individual thinking. Assume that a philosopher is sitting at his desk, meditating over a difficult problem, and someone comes in and takes away all his papers and writing tools. This certainly would prevent the philosopher from trying different approaches for solving his problem, comparing the advantages and disadvantages of these approaches, and finally finding the best, or maybe the only, solution to his problem. When we watch what such a person is scribbling down while he is thinking, we see that he uses not only letters from the common alphabet, but also digits (numbers 0 through 9) and other symbols. It is important to realize the fundamental difference between the letters on the one hand and all other symbols on the other hand. Try reading the following sentence aloud:

*The symbol + must be read as "plus".*

You probably got stuck when you reached the symbol having the shape of a cross. The letters from the common alphabet are symbols for transforming a written text into a spoken text and, therefore, we don't have any problem reading aloud a written text which contains only letters. Also, we often use symbols which have been created primarily not for written speech. The best known symbols of this kind are the numerical digits. Consider the following two sentences:

*In the year 1998, our son Andrew was born.*

*Last year, 199 students graduated from our college.*

Both sentences contain the sequence 199, but this sequence is transformed to different acoustic patterns when the sentences are read aloud. In the first case we read "nineteenhundredandninetyeight", and in the second case we read "onehundredandninetynine." In a word which is written as a sequence of letters, the interpretation of a letter does not depend on its position in the sequence. Consider, for example, the word "position." Here, the first i doesn't have a meaning different from the second i, and the same is true for the two letters o. But in a sequence of digits, the position of a digit determines its weight: e.g., the right-most digit has the weight 1. Then the next digit to the left has the weight 10, followed by the weight 100, and so on. While the meaning of a word which is written as a sequence of letters cannot be derived from the meanings of its letters, the meaning of a sequence of digits is completely determined by the meanings of its individual digits and their positions in the sequence. Therefore, a written or a spoken word is a composite symbol which got its meaning merely by an arbitrary assignment. In contrast to this, a sequence of digits is not a composite symbol, but a structure of symbols. It didn't get its meaning by an arbitrary assignment, but by a well-defined derivation from its digits and their positions. According to the definition of the concept of a language given above, digits can be considered elements of a language: they are symbols where each has a meaning, and the linear structures composed of such digits have meanings which can be derived according to a given rule.

## What Grammar Has to Do with the Meaning of Texts

There are two different types of professionals dealing with language, the philologists and the linguists. I found the following definition in an encyclopedia:

*Philology*: Academic discipline with the focus on texts, analyzing cultures on the basis of their specific ways of using language and of their literary texts.

While linguists are interested in the means and methods which can be used for producing texts, the philologists look at the texts produced. The fact that the means and the products are subjects of interest to different professional groups can be found not only in the field of texts, but in other areas, too. For example, think of paintings. With paintings, the means are paints which are a subject for chemists, light which is a subject for physicists and human eyes which are a subject for physiologists. Members of these professional groups may have no idea about what the people who view the products, namely the paintings, might be interested in.

It will not surprise you that, in the following section, texts are not looked at in the way philologists look at them. We ask the questions which are asked by linguists who are mainly interested in the rules for the composition of structures using given elementary symbols, and for the interpretation of such structures. Linguists strictly separate the so-called syntax and the so-called semantics of the structures. *Syntax* means the set of rules that determine which structures can be built using the elementary symbols as components, and *semantics* means the set of rules which determine the interpretation of these structures. If, for example, the elementary symbols are the words of a given natural language, the syntax is the grammar that determines which sequences of words are allowed. Such sequences are the sentences. You know that the meaning of a sentence is not completely determined by the meanings of the individual words occurring in the sentence, but depends on the position of the words in the sequence. Thus, the two sentences "Cain killed his brother Abel." and "Abel killed his brother Cain." are not equivalent.

Describing the syntax of a language is a much simpler problem than describing its semantics, since the objects considered in syntax are exclusively perceptible patterns while semantics deals with meaning.

The example I chose for explaining the concepts of syntax and semantics is rather simple. Here I won't consider sentences in a natural language, but rather arithmetic expressions for addition and multiplication. The elementary components in these expressions are numbers, the two arithmetic operators + and * and the opening and closing parentheses. In linguistics, the elementary components of expressions or sentences are called *terminals* (from Latin terminus for boarder, barrier, and end). Thus, our set of terminals is the union of the set of numbers with the set $\{+, *, (, )\}$.

The syntactical rules cannot be expressed using terminals exclusively. In addition, the so-called *non-terminals* are required. If you think back to your English lessons in high school, you will remember the terms subject, predicate and object which were introduced as means to show the general structure of sentences. Probably at that time, you were not told that these terms were non-terminals - although, of course, they are. The set of non-terminals needed for expressing the grammar of our arithmetic expressions is {expression, sum, component, product, factor, number}. Using these non-terminals, I could express the grammar of our arithmetic expressions in rules as shown in Fig. 6.1. These ten rules are required for capturing the principles for building the simple arithmetic expressions considered. Each of these rules says that the non-terminal standing on left side of the arrow can be replaced by the sequence of terminals and non-terminals standing on the right side of the arrow. In our case, seven of the rules say that a certain non-terminal can be replaced by a certain other non-terminal. Only rules 3, 7 and 10 have a sequence of more than one element on their right sides.

| The non-terminal for the root is E (E stands for *expression*.) <br> (1) E → S   An *expression* is either a *sum* <br> (2) E → P   or a *product*. ||
|---|---|
| A *sum* is a sequence of two *components* separated by the plus-symbol. <br> (3) S → C + C | A *product* is a sequence of two *factors* separated by the multiplication-symbol. <br> (7) P → F * F |
| A *component* is either <br> a *number*          (4) C → N <br> or a *product*    (5) C → P <br> or a *sum*.        (6) C → S | A *factor* is either <br> a *number*              (8) F → N <br> or a *product*        (9) F → P <br> or a *sum* enclosed  (10) F → (S) <br> in brackets. |

Fig. 6.1   Grammar for sum and product expressions

At the top of Fig. 6.2 is an example of an arithmetic expression, and underneath is a graphical structure which has been derived by applying the rules from Fig. 6.1. Perhaps, when you first looked at Fig. 6.1, you wondered what the term "for the root" in its top row means. In Fig. 6.2 you now can see what kind of root was meant. The graphical structure can be viewed as a tree which has the root E at the bottom. (Obviously, in the given context, the term "root" has nothing to do with the arithmetic operation of computing a root, for example the square root).

The leaves of the tree in Fig. 6.2 are the terminals at the very top. Going from the root to the leaves, you encounter many round nodes labeled with the non-terminal symbols (E, S, C, P, F, and N). Each step upwards from one such a node corresponds exactly to one rule from the grammar in Fig. 6.1. Beginning at the root, rule 2 takes us to the node P. Since there is only one rule having P on its left side, namely rule 7, this rule must be applied in the next step, resulting in the left F, the multiplication symbol * and the right F. The left F now corresponds to the left side of rule 10 which provides a sequence of three elements, the opening parenthesis, the node S and the closing parenthesis. Continuing in this way, you can generate the entire tree from the root to the leaves.

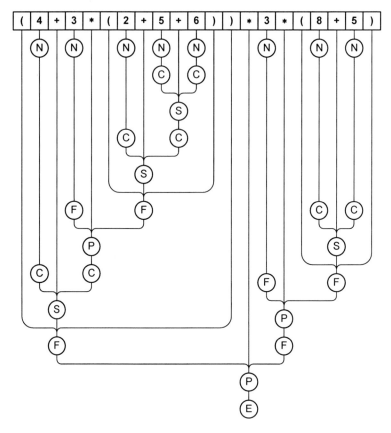

**Fig. 6.2** Tree showing the syntactical structure of an arithmetic expression using the grammar in Fig. 6.1

Our grammar has a specific property which captures the fact that both addition and multiplication are associative operations. This means that in a chain of three operands, either a+b+c or a*b*c, the order in which the operations are performed is irrelevant because (a+b)+c=a+(b+c) and (a*b)*c=a*(b*c). Therefore, if such chains do occur in a given arithmetic expression – as is the case in Fig. 6.2, - the grammar will not completely determine the tree structure. The person who generates the tree by applying the rules of the grammar has the freedom to make these decisions. For example, look at the node P which is reached directly from the root. To this P, I assigned the * which is sitting at the left side of the number 3. Thus, I decided that first the number 3 is multiplied by the factor to the right of it, and then this product is multiplied by the factor left of the number 3. But certainly, I could have selected the * sitting at the right side of the number 3 and assigned this to the node P sitting directly above the root E. This would have generated a different tree.

# Grammar

Since we know the meaning of the words and symbols in Fig. 6.1, we automatically associate rules of interpretation to the syntactical rules although, formally, the syntactical rules do not provide any information concerning the interpretation of the terminal sequences they describe. When we look at the arithmetic expression at the top of Fig. 6.2, we automatically begin to add and to multiply until we reach the final result which is 1,677. By observing what you actually do when you evaluate the given expression, you come up with a rather simple insight: going from the top to the bottom, you assign numbers to the round nodes, and when you find the number which belongs to the root, you have found the total result for the expression. Before you reach the root, you must have found the numbers 43 and 39 belonging to the two F-nodes sitting closest to the root. Not only in the given example, but in general, the nodes of a syntactical tree are containers for meaning, and the contents of such a container are derived from the containers sitting above it. While the rules of the grammar are read from left to right when the tree is generated, these rules are read from right to left in the interpretation process.

In the case of the grammar given in Fig. 6.1, it is rather easy to see what it means to read a rule from right to left. We always have to ask what the information contents of the left non-terminal is if the contents of the non-terminals on the right side of the rule are known. Seven of the ten rules have only a single non-terminal standing on the right side, and therefore the contents of the right side non-terminal is just copied to the left side non-terminal. For example, let's consider rule 2 which was the first rule applied when the tree was generated. This rule will be the last one to be applied when the expression is interpreted. When this rule is applied in the interpretation process, all nodes except node E will already have a number assigned to them, and the contents of the node P above node E is the number 1,677. This number now has to be copied into the root node according to rule 2. Only the two rules 3 and 7 are such that an arithmetic operation must be performed when they are read from right to left. In these cases, there are two non-terminals on the right side of the rule, and the numbers assigned to them are the operands of the corresponding operation, either addition or multiplication.

In the case of Fig. 6.2, the type of information which is assigned to the nodes of the tree in the interpretation process is the same for all nodes, namely numbers. But this is only a very special case. The fact that normally the non-terminal nodes are containers for different types of information is illustrated in Fig. 6.3. There, you see two trees having the same structure and being mirrored at a horizontal line. These trees belong to two expressions, each determining a certain object. The object for the lower expression is the number 48, and the object for the upper expression is a person from the Bible, namely the young man Cain who, as I mentioned earlier, killed his brother Abel. The texts in the non-terminal nodes in Fig. 6.3 indicate that these nodes, at least in part, are containers for different types of information. For example, look at the lower tree. In the interpretation process, the node "List of operands" will be filled with an ordered set of two numbers, (3; 8), the node "Function name," with the meaning of the word `product` and the node "Functional expression" with the number 24.

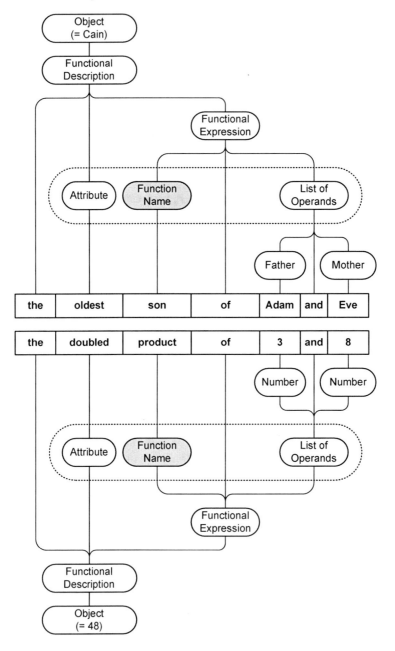

**Fig. 6.3**  Example illustrating the concept of context sensitivity of a grammar

By representing two trees having the same structure, I wanted to illustrate the concept of so-called *context-sensitivity*. In both trees, you find a dashed line enclosing three nodes, one of which, namely the node "Function name," is shaded grey. Using these graphical elements, I tried to indicate that the three nodes within the dashed line belong together, and that the choice made for the shaded node has a restricting effect on the two other nodes. While in the lower tree, the terminal assigned to the shaded node is the word `product`, the corresponding terminal in the upper tree is the word `son`. Once the terminal for the shaded node has been chosen, we obviously are no longer free to choose any adjective as an "Attribute" or any pair of objects as a "List of operands." For example, it would be nonsense to speak of "the oldest product of 3 and Eve" or of "the doubled son of Adam and 8." Such nonsense expressions are avoided by context-sensitivity. When the tree is generated bottom up from its root, the actual context, i.e., the choices made for the left and right neighbors of an actual node, may restrict the freedom of choice for this actual node.

The term context-sensitivity sometimes is used without respect to grammars. Assume I stand in front of a grocer's table in the food market, point to a melon and say, "Could you please give me this tomato?" Undoubtedly, the grocer would be confused, since in the actual context, my behavior was inconsistent. But context-sensitivity sometimes helps to interpret ambiguous symbols unambiguously. If a word is used which has two or more absolutely different interpretations, its actual meaning in most cases is quite clear because of the context which includes the information about where and when the word is used and who used it. For example, consider the three uses of the word *present* in the following sentences:

> Everyone needs to be *present* for a special meeting.
> I need to buy my sister a *present* for her birthday.
> The company executive will *present* his ideas to the Board of Directors tomorrow.

Both the arithmetic expression in Fig. 6.2 and the expressions in Fig. 6.3 are texts which can be read and understood by human beings who have learned the corresponding languages. However, the concept of a grammar is useful not only in connection with texts which can be read and understood by people, but also in the field of information technology. There, sequences of symbols are used which contain information that we cannot extract directly by interpreting the perceived patterns. These patterns are not produced as input for human interpretation, but as input for machines such as computers. Therefore, if we want to have access to the information that is contained, these patterns must be decoded. By writing or speaking, we produce perceptible patterns which contain the information we have in our mind and which we want to make accessible. This process of bringing information to a form which is accessible to human beings is called formulating or expressing, but not encoding. Encoding is performed on perceptible patterns, not on information sitting in one's mind. Both the input and the output of a process of encoding are perceptible patterns. But such a process is called encoding only if the rules applied do not refer to the interpretation of the input patterns. As a result, the translation of a given text from the original language into a different language is not an encoding process because the output text cannot be produced without interpreting the input text. If you look back at Fig. 4.9 which shows how the symbols for writing formulas in predicate logic could be encoded by patterns, you see that each is a specific combination of nine small black and white squares arranged in a 3x3 square. Here the black and white patterns are assigned without referring to the meaning of the formula or the meaning of the symbols that occur.

One of the first codes defined for transmitting information by technical means was the so-called Morse code. It was named for the American Samuel Morse who, in 1830, began to build electromagnetic systems such as the telegraph for transmitting texts. Although the final code which was standardized in 1865 was designed by others, its original name was retained. The Morse code is based on only five perceptible patterns which are produced by turning on and off a source of flowing energy, e.g., a source of light, sound or electrical energy. The five patterns are *on short, on long, off short, off long* and *off very long*. The "off long" is used to separate the letters, while the "off very long" separates the words. Combinations of the other three patterns are used to represent the letters. The relative lengths of the patterns are, according to the standard, 1:3:1:3:7. In tables which represent the assignment of Morse combinations to letters (see Fig. 6.4), the "on short" is represented by a dot, the "on long" by a horizontal line or dash, and the "off short" by the distance between two of these visible patterns. Both the "off long" and the "off very long" do not appear in such tables. While the Morse code was used heavily in earlier times, especially for communication between ships and between ships and their companies on shore, it is no longer needed for modern systems of information technology. However, amateurs sometimes still use it: think of girl scouts and boy scouts who communicate at night using their flashlights.

| e | t | i | m | a | n | s | o |
|---|---|---|---|---|---|---|---|
| . | — | .. | — — | .— | —. | ... | ——— |

**Fig. 6.4** A partial table defining the Morse code

There are two different reasons for encoding information. One reason is that technical systems for information transmission and processing can handle information much more efficiently if it is adequately encoded. In Chapter 14, this subject is discussed in more detail. The other reason is that some information must be kept secret, i.e., nobody outside of a group of insiders should have access to this information, although the patterns used must be public information. In this case, only insiders should be able to decode the text. It is amazing what ideas have been developed by the experts for encoding texts in such a way that outsiders have an extremely difficult time trying to interpret the resulting sequences of patterns.

## How to Control Conversations in Order to Make Sure All Participants Get a Fair Chance to Say Something

The systems considered up to now were such that a sender formulates and possibly encodes information, and the receiver gets this information by decoding and interpreting the transmitted patterns. Now we shall consider systems with two or more communicating partners who, in the course of the communication process, switch back and forth between the roles of sender and receiver. Think of a meeting of the board of directors of a company. Such a meeting is moderated by a president who has special rights compared to the other members. The set of rules which the communicating group members must obey in order to get a well-organized communication process is called a *protocol*. You probably know that every government has a so-called "Chief of Protocol" whose job it is to make sure that diplomatic communication with the governments of other countries is done according to certain rules. The ceremonies you might watch on television when a foreign politician comes for a state visit are strictly defined by the rules of such a protocol.

Protocols which are of special interest in this book are those which have been defined for communication processes where at least one partner is a machine, such as a telephone system. In these cases, it is impossible to have efficient communication without well-defined protocols. When two people communicate via a telephone system, they do not communicate with each other all the time. For part of the time, they communicate with the technical parts of the telephone system. Only after they have been connected and can talk to each other can they forget about the system. But when they dial or hear the dial tone or the busy signal, their communication partner is a machine. Fig. 6.5 shows a graphical representation of the protocol for a telephone communication which was applied when an older, traditional type of telephone system was used. The protocols for using modern telephone systems and cell-phone systems look slightly different.

The elementary activities which must be executed as steps of the process are written in the rectangles. The time for executing such a step is when the circle, from which an arrow leads to the rectangle, is filled with a so-called token. You might think of the circles as being saucers and the tokens being coins. In Fig. 6.5, the only circle having a token, drawn as a solid circle, is the one in the upper left

142                                                                 6. Structures of Languages

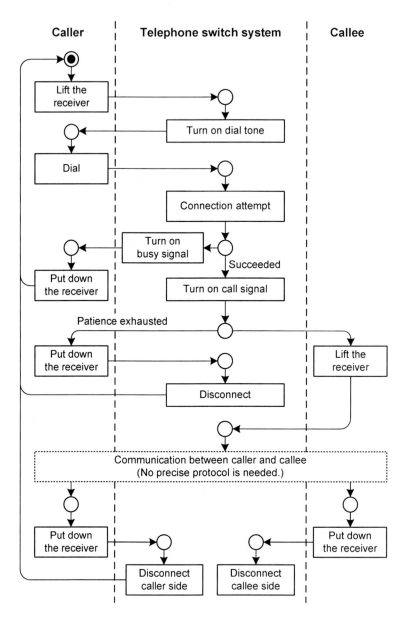

**Fig. 6.5**   Protocol for using a traditional telephone system

corner of the diagram. In this situation, lifting the receiver is the only activity which must be performed next. After the activity which is specified in the rectangle, has been executed, the token is taken away from the circle in front of the rectangle and a token is placed inside each circle at the end of an arrow coming from the rectangle. All rectangles except the one with the dashed outline have only one circle which must be filled after the execution of the corresponding activity. Filling two circles at the same time means that there are two steps which can be executed concurrently. The only concurrency shown in Fig. 6.5 comes from the fact that both the caller and the callee can terminate the communication independently by putting down their receiver. Some circles have two arrows leading away from them. In these cases, a condition must be given which determines which of the two steps is to be executed next. When there are two alternative steps to be selected for a condition, and these belong to the same agent – either the caller, the telephone-system or the callee – it is this agent who decides what will be done next. But there is one case where the two alternative steps belong to two different agents; in such cases, it is unpredictable what's going to happen. Look at the circle which is filled after the call-signal has been turned on. Then, the telephone is ringing on the callee's side and the caller is waiting for the callee to lift his receiver. Either the callee's receiver is lifted before the caller's patience is exhausted, or the caller puts down his own receiver. In this case, there is no single agent having the power to decide which of the two alternatives will actually happen.

By presenting the protocol in Fig. 6.5, I wanted to show you how precisely one can define protocols. Such precision is needed whenever technical systems are involved in the communication process.

# Part II
# Fundamentals of Natural Sciences

# Chapter 7
# What the Moon Has to Do with Mechanical Engineering

You are probably familiar with situations where a child doesn't stop asking its why-questions. My brother-in-law once told me the story of his little daughter, Barbara, who went on and on asking why things are as they are, until he finally had to say, "Now look, my dear girl. I really don't know the answer myself." However, the child didn't accept this and said, "That doesn't matter. Answer anyway!" Obviously, our urge to know and understand the principles behind our experiences seems to be determined deeply in our genes. Most likely, the subjects of these questions have been the same for thousands of years; only the answers have changed over time. In order to get an idea what these subjects are, we only must listen to the questions of the children: "Why doesn't coffee smell like perfume? Why did the stone I threw into the pool sink to the bottom? Why is the sky blue? Where do clouds come from? Why do these metal pins stick to this piece of iron? Why is there thunder and lightning? How is it possible that a big plant grows out of such a small seed? Why can't grandma who died yesterday be brought back to life again? Why are the stars so small and the moon so big?"

The old Greeks – and many other people before them – answered such questions quite differently than we would today. To us, many of these old answers sound funny and, since we know better answers, we are tempted to look down at the former scientists and philosophers. But we should try not to be arrogant, since we have no reason whatsoever to believe that we would have come up with better answers if we had been in the situations of those people. For example, the Greek philosopher Aristotle spent much time thinking about the motion of stones that are thrown or falling. His conclusion was that all things on earth have their natural location, a kind of home, and if, for some reason, such a thing was forced to leave its home, it would attempt to return. Thus, it was quite clear to him that the natural home of the stones was the earth, since this was the location they always attempted to return to. About a hundred years before the time of Aristotle, the idea was born that anything is composed of only the four elements: fire, earth, water and air. I could present to you many more examples of outdated answers, but this book is not a report on the history of science. It is my goal to bring you up to the plateaus of the actual answers without taking any detours.

In many cases, there is more than one way leading to the top of a mountain. There is, of course, exactly one specific way mankind has taken for its journey to the plateaus of its present knowledge and understanding, but in the meantime, other ways have been found which are shorter and easier to travel. The way I have chosen for you leads us first to the plateau of mechanics. From there, we continue our journey to the plateau of electrodynamics. Then we reach the plateau of the elementary components of which all things around us are composed. This is the plateau of chemistry and quantum theory. Finally, we climb to the plateau of genetics which combines chemistry and biology.

## What Galileo Galilei Could Teach Us without Upsetting the Pope

When they hear or read the name Galileo Galilei (1564-1642), most people immediately think of the quarrel he had with the pope. Galilei pretended that, by looking through his telescope, he had discovered facts which could be explained only by assuming that Nicolaus Kopernikus (1473 -1543) was correct. Kopernikus had come up with the hypothesis that all planets, including the earth, orbit the sun and that the earth rotates around its axis with one complete rotation per day. The pope objected by presenting texts from the bible which said that the earth sits in the center of the universe. You certainly can imagine how upset the pope was when Galilei bluntly answered that the bible could be partially wrong.

If this quarrel with the pope were the only reason for remembering Galilei, his name would not have been mentioned in this book. But he left us something extremely valuable which has nothing at all to do with the question about where the earth is located in the universe. Galilei can be called the "father of physics," since he used mathematical formulas to express relationships arising from the results of measurements. Such formulas are considered laws which not only describe experiences made in the past, but make it possible to predict the results of future experiments. It was a rather revolutionary idea that laws exist which "force the universe to behave" according to certain rules which had been valid in the past and which will stay valid as long as the universe exists. Most of the laws Galilei found by experimenting belong to the field of gravity. He found that the speed of a falling body does not depend on its mass, i.e., that a light stone falls with the same speed as a much heavier one. The formula for the time a body needs to fall a certain distance when it begins its fall with zero speed is

$$\frac{duration}{second} = 0.4515 * \sqrt{\frac{distance}{meter}}$$

The notation used here on the left side of the equation, for example, means that the time duration has the units of seconds. Similarly, distance has the units of meters. Time, distance and weight were the first quantities of physics which were combined in formulas. It is the essence of all formulas of physics that quantities such as time durations, distances, weights, temperatures, electrical voltages, etc., together with natural constants are operands of the arithmetic operations addition, subtraction, multiplication and division. Sometimes even the square root must be computed. Arithmetic operations have been introduced with numbers being both the operands and the results. But what could it mean if a temperature is multiplied by a time duration or a weight is divided by a distance? Maybe you haven't yet realized that, in your daily life, you very often encounter such products or fractions of physical quantities. For example, you measure the speed of your car in miles per hour which is a fraction having a distance as the numerator and a time duration as the denominator. And the amount on your bill for the electrical energy consumption of your house is calculated in kilowatt-hours (kWh) which is a product where one factor is an electrical power (kilowatt) and the other factor is a time duration (hour). While the operands of products and fractions can be of any type of physical quantity, the operands of sums and differences must be of the same type. It doesn't make sense to add a length and a temperature.

Whenever you apply a physics formula to compute a result, it is very important that you always introduce the complete information about the physical quantities involved. This information is represented by a number and a physical unit. Since the physical units can be chosen arbitrarily – for example inches, feet, yards, miles, meters or kilometers for a distance – the same physical quantity can be expressed in many different equivalent ways. Thus, it is correct to write 0.5 miles = 880 yards = 2,640 feet = 31,680 inches = 80,470 cm = 804.7 m = 0.8047 km, since the equal signs in this expression do not relate the numbers, but relate the distances, and these really are all the same.

In physics formulas, letters are used as variables for physical quantities, as symbols of natural constants and as abbreviations of units. Since there are not enough different letters, the assignment of different meanings to the same letter cannot be avoided. For examples, we consider the letters m, g and s. When interpreted as abbreviations of units, they mean meter, gram and second, respectively. But the two letters m and s are also used as variables in formulas describing relationships between a mass $m$ and a distance $s$. The letter g has a second meaning, since it is used to symbolize the constant acceleration of falling objects, 9.81 m/s$^2$. In printed texts, it is possible to indicate the actual meaning of these letters by using different font styles. Wherever there might be the danger of confusing the meanings of the letters, variables are printed in italic style. For example, s means seconds and $s$ means a variable distance. Of course, it requires very attentive reading to notice this difference. Look at the formula $s/t$ = 15 m/s which says that a speed is obtained by dividing a distance $s$ by a time $t$ and that the actual speed is 15 meters per second. Now we want to express this speed in miles per hour, since this is the speed unit we are used to and which enables us to get a feeling of whether the actual speed is slow or fast. The calculation which transforms the speed from the given unit (meters per second) to the requested unit (miles per hour) is as follows:

$$\text{speed} = \frac{15 \text{ m}}{\text{s}} * \overbrace{\frac{60 \text{ s}}{\text{min}}}^{1} * \overbrace{\frac{60 \text{ min}}{\text{h}}}^{1} * \overbrace{\frac{\text{mile}}{1609.4 \text{ m}}}^{1}$$

$$= \frac{15 \cancel{m}}{\cancel{s}} * \frac{60 \cancel{s}}{\cancel{min}} * \frac{60 \cancel{min}}{\text{h}} * \frac{\text{mile}}{1609.4 \cancel{m}} = \frac{15*60*60}{1609.4} * \frac{\text{mile}}{\text{h}} = 33.55 \frac{\text{miles}}{\text{h}}$$

The fraction which describes the speed using the given units is multiplied by a chain of three factors, each with the value 1. These 1's are given in the form of fractions whose numerators equal their corresponding denominators. The first fraction has the value 1 since 60 seconds is the same time duration as one minute. The second fraction has the value 1 since 60 minutes is the same duration as one hour. And the third fraction has the value 1 since one mile is the same distance as 1609.4 meters. The multiplication of the speed by these three factors of 1 leaves the speed unchanged and changes only its representation in units. If the same physical unit appears as a factor in both the numerator and the denominator of a fraction, this factor can be eliminated (or cancelled) leaving the value of the fraction unchanged. In our formula, such eliminations were possible with respect to the time units s and min, and the distance unit m. The final result says that 15 meters per second is the same speed as 33.55 miles per hour.

In the next example of transforming an original physical unit into a desired one, the calculation not only requires multiplications and divisions, but also the computation of a square root. Assume that an American is visiting Europe and gets the information from the TV news that a fire in Portugal has destroyed a forest with an area of 20,000 hectares. Since he is not familiar with the area unit hectare, he cannot imagine the size of the area involved. He might assume that a European who is familiar with hectares would be able to imagine such an area, but this assumption is wrong. Both the American and the European can only imagine areas for which they know the lengths of the sides. The simplest shape of an area is a square, and so we can imagine such a square if we know the length of its edge. Therefore, the American tries to transform the information 20,000 hectares into the form $(x \text{ miles})^2$. His calculation is as follows:

$$20{,}000 \text{ hectares} = 20{,}000 \, \cancel{\text{hectares}} * \overbrace{\frac{100 \cancel{m} * 100 \cancel{m}}{\cancel{\text{hectare}}}}^{1} * \overbrace{\frac{\cancel{\text{yard}} * \cancel{\text{yard}}}{0.9144 \cancel{m} * 0.9144 \cancel{m}}}^{1} * \overbrace{\frac{\text{mile} * \text{mile}}{1760 \cancel{\text{yards}} * 1760 \cancel{\text{yards}}}}^{1}$$

$$20{,}000 \text{ hectares} = \frac{2{,}000{,}000 * \text{mile}^2}{(0.9144 * 176)^2} = \left(\frac{\sqrt{2{,}000{,}000} * \text{mile}}{0.9144 * 176}\right)^2 = (8.79 \text{ miles})^2$$

Looking it up on the internet or in a library or asking a friend what a hectare is, he will get the information that one hectare corresponds to the area of a square whose sides have a length of 100 meters. Since he doesn't want to imagine lengths measured in meters or kilometers, but in yards or miles, he needs to know what a meter corresponds to in American units. He has learned that one yard is a length which is a little less than one meter. The exact relationship is 0.9144 meters per yard. Finally he applies his knowledge that one mile corresponds to 1,760 yards. Now, no additional information about relations between physical units is needed and the remaining arithmetic operations can be performed. The result tells him that the area of the forest destroyed by the fire corresponds to a square with sides having a length of 8.79 miles. This description is what he was aiming at since he has no problem imagining a square with sides being a little less than nine miles.

All physics formulas describe relations between measurable quantities where "measuring" means comparing with a standardized quantity. It is no wonder that different standards have been defined at different places on earth. While Americans still measure lengths in inches, feet, yards and miles, the Europeans agreed in 1875 on a length standard which they called meter (from the Greek word "metron" for measure). Based on this standard, shorter or longer lengths could then be measured using decimal fractions or multiples of a meter, namely micrometer = meter/1,000,000, millimeter = meter/1,000, centimeter = meter/100, kilometer = 1,000 meters. Before the meter was introduced, a large variety of standards for measuring lengths were used in the different countries of Europe.

When standards for measuring lengths, time durations and masses were originally defined, these standards referred to quantities, related to the human body, which everybody could easily imagine. This reminds us of the old Greek philosopher Protagoras (490-411 BC) who said, "Man is the measure of all things." The unit "meter" corresponds to the length of one human step, a second is close to the time between two heartbeats, and a kilogram is approximately the mass of what a person eats and drinks at a meal. Though these elementary units originally were defined with reference to our daily experiences, their exact definitions deviated more and more from their original references because it became more and more necessary to increase the precision of the measurements. Today, the definitions of the units of time and length, meter and second, seem to be rather strange. The actual definitions are:

> A *second* is 9,192,631,770 times the cycle time of a certain periodic process based on cesium atoms changing their state of energy.

> A *meter* is the 299,792,458[th] fraction of the distance light travels in one second through a vacuum.

Everybody who reads these definitions for the first time wonders where those two high numbers might have come from. But the evolution of these definitions is not as difficult to understand as you might suppose. Let's consider the definition of the second as the unit of time duration. When people began to think about time durations, it was quite natural that they used the time from noon to noon as the standard reference. Noon is defined as the point in time when the sun is at its zenith with respect to an actual location on earth, i.e., the sun has reached its highest position during the day. In the old days, nobody could understand that this time is not an absolute constant, but varies over time. The construction of clocks required that the time from noon to noon be divided into shorter periods. These periods are the hour, the minute and the second. Once a clock was built, it could be tested to see if it really counted 60*60*24=86,400 seconds from noon to noon. Since the scientists wanted to measure time durations with very high precision, they began to search for periodic processes whose cycle times are much shorter than one second and don't vary much over time. Today, they use very complex systems which are called "atomic clocks." The periodic process in these systems is based on changing energy states of cesium atoms (see Chapter 11). Once such a system is available, it is possible to count how many of its cycles are contained in one second as measured by a conventional clock. This conventional clock is not very precise, i.e., its second does not always correspond to the same time duration. Therefore, the number of cycles of the atomic clock which fit into a conventionally measured second will vary over time. But compared with the absolute number of such a count which lies in the range of billions, the variation is very small. Finally, in a purely arbitrary act, the number which lies in the middle of the variation interval was chosen for the definition of the "atomic second." This number is 9,192,631,770.

In the definition of the meter as the standard unit of length, the speed of light is used as a reference. Originally, the meter had been defined as the ten millionth fraction of the distance from the equator to a pole of the earth. But this was only a rough guideline. The original standard meter was realized using a sectional beam which was kept in Paris since 1889. It was made of an alloy of iridium and platinum, and its length at zero degrees centigrade was defined to be one meter. But later, when it became possible to measure durations with very high precision, the unit of length could be defined with reference to the speed of light which is considered to be a natural constant and does not change with time. The meter is now defined as the 299,792,458-th fraction of the distance light travels in a vacuum in one second. This number has been determined in a way similar to the determination of the number in the definition of the second. Using atomic clocks and conventional methods for measuring distances, the speed of light was measured. The results of these measurements were not all the same since the measurements of time and distance could not be absolutely precise. Arbitrarily, the middle of the variation interval of the measured speed of light was chosen. This makes it seem as though this decision has defined the speed of light, but since this speed is a natural constant and cannot be defined, determining a number for this speed merely means defining the unit of length.

My detailed description of the problems of measuring time durations and distances was to emphasize the fact that the foundations of physics consist in definitions of physical quantities and their units of measurement. It was a tremendous achievement of mankind to come up with ideas about how to measure phenomena which are experienced as something that can be more or less. While it is easy to compare a given distance with a standard unit distance, it is not at all obvious how to compare two temperatures or two forces. Everyone can feel that the temperature of the air or water in a container is higher or lower, but nothing indicates how one could map temperature values to a continuous scale of numbers. The Swede Anders Celsius (1701-1744) defined a temperature scale where the freezing-point of water was given the number 0 and the boiling-point was given the number 100. This definition is still used, and the unit centigrade (°C) is associated with this scale. However, with the definition of the two end points of the temperature scale, 0 °C and 100 °C, the problem of measuring temperatures was not yet solved. We want to know which temperature values are assigned to all the numbers between 0 and 100. We may know that a temperature of 25 °C has a distance from the boiling-point of water which is three times the distance from the freezing-point. But how could these distances be measured? In the 18$^{th}$ century, scientists found that gases expand when their temperature rises. Consider a closed balloon: its volume can be changed by putting it into a cooler or warmer environment. It is smaller when it is cooler. Volume also depends on temperature with metals. Unless they are restrained by strong forces, the rails of a railroad track expand in summer and contract in winter. As a consequence, the gap between two adjoining rails is smaller in summer and larger in winter.

Many of the instruments used for measuring temperatures are based on this dependency of volume on temperature. Using this effect even makes it possible to measure temperatures slightly outside of the range defined by the Celsius scale. But measuring temperatures far below the freezing-point of water or far above its boiling-point required completely new methods. At this point, I shall not present the definition of temperature which is actually used; you shall find this in Chapter 10. But at least I can tell you here that this definition allows us to measure extremely low and high temperatures. For example, one now knows that the boiling-point of hydrogen lies at -252.9 °C and that the flame of a welding torch has a temperature of roughly 3000 °C.

## What Sir Isaac Newton Found Out about Forces and Moving Bodies on Earth and in the Sky

When I heard the word 'mechanics' for the first time, I was a child living in the Black Forest which is a mountainous region in south-western Germany. At that time, the region hosted many companies which did precision engineering. The origin of these companies dates back to the time when the Black Forest was famous for its mechanical clocks which were mainly cuckoo clocks. Therefore it is no wonder that I thought that mechanics is a name for this type of industry. For me, the symbol of mechanics was a small gearwheel. Only some years later, I learned from my physics teacher that mechanics is the name of that special area of physics which deals with forces affecting the shapes and the motions of bodies. At least once a year, I still spend some days on vacation in the Black Forest. And on such a day, I was sitting on a bench in front of a farm house, listened to the burbling of a fountain and, lost in thought, looked at its aesthetically shaped jet of water. I imagined being a Greek philosopher who asks himself why the jet has exactly this shape and not a different one. I went into the house, got a measuring stick and measured the characteristic distances of this shape. The result is represented in the left side of Fig. 7.1. Since, at that time, I had learned earlier in high school that the shape of a falling jet of water is a parabola, I defined an adequate xy-coordinate system and developed the formula of the parabola. This formula is included underneath the picture of the fountain.

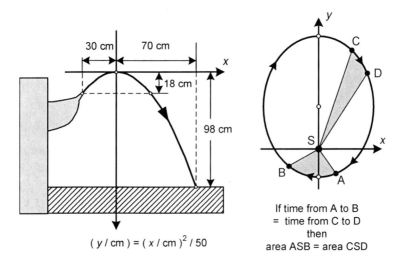

**Fig. 7.1** Parabolic arc of a water jet and elliptic orbit of a planet

Stimulated by the parabolic shape of the jet, it came to my mind that there is another aesthetic shape which also occurs in mechanics and is a relative of the parabola. This other shape is an ellipse, and both the parabola and the ellipse are so-called conic sections. Before I shall explain to you what this name means, I shall tell you where the ellipse can be found. While I could see the jet of water and its parabolic shape, and take a snapshot of it, the ellipse which came to my mind cannot be seen but can only be derived from certain observations. I was thinking of the elliptical shape of the orbit of a planet around the sun, a shape like a slightly deformed circle. Nicolaus Kopernicus believed that the planets orbit the sun in circles, but the astronomer Johannes Kepler (1571-1630) had more precise data and came to the conclusion that the orbits are ellipses and that the position of the sun coincides with one of the two focuses of each of these ellipses. What a focus is shall soon be explained along with the relationship between ellipses and parabolas. Johannes Kepler not only discovered the elliptical shape of the orbits, but also found a law which describes the variation of the speed of the planets on their elliptical courses. Any pair of light beams connecting the sun and the planet determines both a time interval and an area. Kepler's law says that the ratio between the time interval and the area is always the same for any arbitrarily chosen pair of beams. This law is illustrated in the right side of Fig. 7.1. If the two shapes ASB and CSD have the same area, the time the planet needs to go from A to B is the same as the time needed for the way from C to D. From this it follows that the planet must go faster on its way from A to B than on its way from C to D, since the distance AB is longer than the distance CD.

Figure 7.1 gives you an impression of the phenomena that confronted the great British scientist Isaac Newton (1642-1727) when he began his search for the laws of mechanics which explain these phenomena. But before I begin to describe the evolution of these laws, I first must give an explanation which I mentioned earlier. I said that parabolas and ellipses are conic sections. What this means is illustrated in Fig. 7.2. Assume that the cones shown are made of wood and that you cut off a certain piece, keeping the saw-blade in a plane. Depending on the relation between the angle of this plane and the angle of the cone, the border of the cut surface is a particular type of conic section. A parabola is obtained if the two angles are equal. If the two angles differ, the curve is either an ellipse or a hyperbola. While the cutting plane intersects only one cone in the case of a parabola or an ellipse, two cones are intersected in the case of a hyperbola as shown in the figure.

The concept of a focus is illustrated in the bottom section of the Figure 7.2. Assume that the interior of the curves is covered with a reflecting material. Then, each beam which begins at one of the two focuses of the ellipse, will arrive at the other focus. While an ellipse has two focuses, a parabola has only one. Any beam which starts at the focus of a parabola will be reflected in such a way that afterwards it is parallel to the symmetry line of the parabola. And, of course, the direction of such a beam can been reversed, i.e., all beams which arrive as lines which are parallel to the symmetry line will be reflected into the focus. This property is used in a so-called dish antenna such as those used for satellite television reception and whose shape is a rotated parabola. When such an antenna is used as a receiver, all incoming energy is concentrated at the focus. In the case of a transmitting antenna, the energy to be transmitted is inserted at the focus of the antenna.

156                                                                                    7. Mechanics

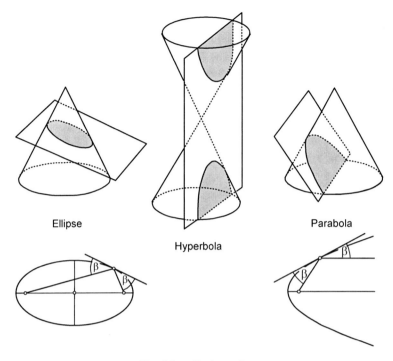

**Fig. 7.2** Conic sections

Now my introductory remarks are over, and we are ready to consider how the phenomena in Fig. 7.1 can be explained, i.e., what laws force the jet of water into the shape of a parabola and force the path of a planet around the sun into the shape of an ellipse. If we were left alone and had to find these laws by ourselves we most probably would be absolutely helpless. It required the mind of a genius to find the right approach, and that genius was the English scientist Isaac Newton (1642-1727). You shall soon see that Newton had to find not only one ingenious approach, but two. His first approach answers a question which had previously been posed by Aristotle, "What keeps a moving stone in motion?" Aristotle believed that a stone cannot stay in motion unless a so-called "mover" is active all the time. In contrast, Newton introduced an abstract property of the stone which later was called "momentum." If the momentum is zero, the stone is stationary, otherwise it is in motion. Surely you will immediately object that there was no need to introduce this new property, since the difference between resting and being in motion had been captured long before Newton by the concept of speed: if the speed is zero, the stone is at rest, otherwise it is in motion. Newton, however, realized that the speed is not sufficient to capture the essence of motion; he was convinced that the mass of the stone also had to be taken into account. Therefore he defined the *momentum* of an object as the product of its speed and mass. With two bodies moving at the same speed, the one with the greater mass has the higher momentum. While the mass which usually is measured in kilograms (kg) is not a directed quantity, the speed not only has a value, but also a direction. Therefore momentum is also a directed quantity which has the same direction as the speed. The diagram in Fig. 7.3 illustrates the steps which lead from the concept of momentum to the other two fundamental concepts of mechanics, *force* and *energy*.

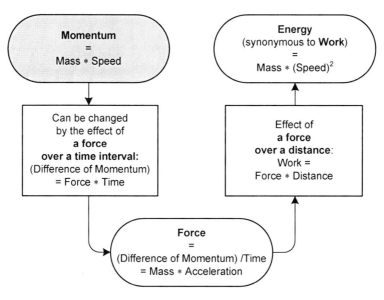

**Fig. 7.3** Concepts of mass motion

The momentum of a body stays unchanged as long as nothing happens which changes its speed or mass. A stone at rest will not begin to move unless it is influenced by some force and correspondingly, a stone which is in motion will not change either the value or the direction of its speed unless a force acts on it. In Fig. 7.1 both the drops of water in the jet and the planet on its orbit around the sun have a continuously changing speed because in both cases there is a gravitational force in effect. In the case of the water jet, the force has a constant value and is directed downwards in the y-direction. In the case of the planet, the value of the force changes continuously, and it is always directed towards the sun. The effect of a force is greater, the longer it is acting.

Besides the effect of a force over time, there is also an effect of a force over a distance. Imagine that someone had to carry a heavy suitcase from the ground floor to the third floor. When the person has reached the final destination, he or she might say, "Believe me, this was hard work." Although during this process the momentum of the suitcase did not remain constant, the momentum at the end of the process is the same as it was at the beginning, namely zero. At the beginning, the suitcase was standing on the ground floor, and at the end it is standing on the third floor. Obviously, the force applied by the carrying person did not cause a difference of momentum but, of course, it still had an effect since now the suitcase is standing high above the ground floor. In order to correctly capture the idea of the effect of a force over a distance, we also must look at a different process. Now consider the case where the suitcase is to be carried horizontally from one house to another house which is one mile away. We will not be surprised if the porter again says that it has been hard work. But in physics, only the first process where the suitcase was carried along a vertical distance is considered as work. On first glance, this might be hard to accept. But there is a reasonable explanation: in the case of the horizontal transportation, the pain could be almost eliminated by using adequate equipment. Think of a cart on wheels which would roll almost by itself to its destination after it has received a small initial momentum by being pushed a little bit. In contrast, there is no adequate equipment which would lift the suitcase to the third floor almost by itself. Using a motor as a substitute for the porter is not permitted, since then the motor would have to provide the same work. Using block and tackle is allowed but would not really reduce the needed work since the product of force and distance would stay the same. If, for example, the number of supporting ropes is four or six, the force to be provided by the pulling person will be one fourth or one sixth of the weight of the suitcase, but the distance the rope has to be pulled will be four or six times the distance between the ground and the third floor, so the product which corresponds to the work would be the same.

The difference between vertical and horizontal transportation is obtained quite naturally by computing the work as the product of the force and the distance. In this case the two factors are both directed quantities since both the force and the distance have their own directions. Therefore it is not sufficient to specify that force and distance must be multiplied, but it must be decided which type of multiplication of directed quantities should be applied. In Chapter 3, two types of products of two vectors were introduced, namely the scalar product and the perpendicular product. When computing the work as a product of a force and a distance, the scalar product must be used, since the work is determined completely by its value and doesn't have a direction. Maybe you still remember that the scalar product depends on the cosine of the angle between the two factor vectors. The function $\cos(\varphi)$ has its maximum value, namely 1, if the angle $\varphi$ is 0, and $\cos(\varphi)$ is 0 if the angle $\varphi$ is 90 degrees. Thus, the work has its maximum value if force and distance have the same direction, and the work is zero if force and distance are perpendicular. When the suitcase is carried from the ground floor to the third floor, force and distance have the same direction and the work has its maximum value. If however the suitcase is carried along the horizontal road, the force points upward while the distance points horizontally, and therefore the work is zero.

The node of the momentum in Fig. 7.3 is shaded grey which indicates that the concept of momentum is the first concept in Newton's mechanics. The standard units used to describe momentum are kilogram (kg) for the mass and meter (m) and second (s) for the speed which is measured in meters per second. If, as shown, the momentum can be changed by a force which acts over time, the force can be defined by the fraction momentum/time. Therefore, there is no need for a specific unit of force, since the force can be expressed using the units kg, m and s. While in the case of momentum, a mass is multiplied by a speed, the unit of the force is the product of a mass times an acceleration, i.e. kg*(m/s$^2$). Going from the force to energy, the force is multiplied by a distance. Thus, the resulting unit of the energy is kg*(m/s)$^2$, i.e., the product of a mass by the square of a speed.

I wouldn't be surprised if all the multiplying and dividing of physical quantities seems strange to you. Maybe you wonder how it could have happened that someone came up with such ideas. Certainly, I myself never would have come up with these ideas – they are the final result of centuries of intensive human thinking, and the genius Newton was the one who came to the final conclusions. Nowadays, we are the lucky ones who harvest these fruits.

Fig. 7.4 shows how the momentum of a drop of water in a jet is changed along its way by the force of gravity which has a constant value and a downward direction. Both the speed and the momentum can be divided into two components, a horizontal one and a vertical one. Since there is only a vertical force, the horizontal component of the momentum or of the speed is not changed, and stays the same all the time. Therefore, the drop moves to the right with a constant speed, and this allows us to interpret the x-axis alternatively as a time-axis or as a distance-axis. I used the symbol $\Delta t$ for the time the drop needs to travel a horizontal distance of 10 cm. The value of $\Delta t$ can be obtained by applying Galilei's law which describes the relation between the distance and the time of a falling body. This law (see page 148) tells us that a drop needs 0.447 seconds to fall a distance of 98 cm. According to Fig. 7.1, the drop needs the same time to move 70 cm to the right. Thus, the value of $\Delta t$ is 0.447/7 seconds= 0.06385 s.

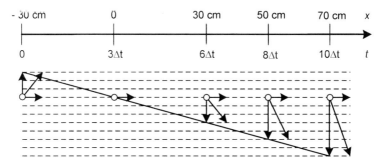

**Fig. 7.4** Momentum of the water jet in Fig. 7.1 over time

Since the force which changes the vertical component of the momentum of a drop is equal to its constant weight, the vertical component constantly increases. This is illustrated in Fig. 7.4 by the straight line which connects the arrowed ends of the vertical components of the momentum. The downward slope of this line is one vertical unit per $\Delta t$, since the horizontal interval of 10 $\Delta t$ corresponds to a vertical interval of 10 units. At the point where the water leaves the pipe and the jet begins, the jet goes diagonally upward, and correspondingly the momentum of a drop has a vertical component which is directed upward. It has a value of three units which are decreased to zero on the way to the maximum height of the jet; this height is reached after 3 $\Delta t$. After that, the vertical component of the momentum points downward.

The arrows in Fig. 7.4 which are interpreted as momentums can also be interpreted as speeds. Then the vertical units, the distances between the dashed horizontal lines, correspond to constant increases of the speed, $\Delta v$. The value of such a unit can be obtained as follows: for the vertical distance of 98 cm the jet needs 0.447 s which corresponds to an average speed of (98 cm/0.447 s)=2.19 m/s. Since the speed at t=3*$\Delta t$ is zero, its value at t=10*$\Delta t$ will be 7*$\Delta v$ which corresponds to an average speed of (7*$\Delta v$)/2. Thus, we know that 2.19 m/s= (7*$\Delta v$)/2 from which we get $\Delta v$=0.626 m/s. An increase of speed per time unit is called an acceleration which in this case is $\Delta v/\Delta t$=(0.626 m/s)/(0.06385 s)=9.8 m/s$^2$. This is the acceleration due to the gravity on the surface of the earth, and it says that the speed of a falling body is growing at a rate of 9.8 m/s per second. In formulas, this acceleration is symbolized by the letter g.

Until now, we took it for granted that there is a force of gravity which pulls bodies downward. But we didn't ask where this force comes from. Finding an answer to this question was the second great achievement of Isaac Newton. He had the idea that the force which pulls bodies down to the floor or to the ground might have the same origin as the forces which keep the moon in its orbit around the earth and the planets in their orbits around the sun. On his search for an adequate law, he had to take into consideration that it should explain Kepler's law which relates areas and times as illustrated in Fig. 7.1. The law he finally found is represented in Fig. 7.5; it is called *Newton's law of gravity*. The letter $\gamma$ stands for the so-called gravitational constant which must have the unit m$^3$/(kg*s$^2$), since its multiplication by the fraction (kg*kg)/m$^2$ must result in the unit of a force which is (kg*m)/s$^2$. The law of gravity says that two bodies attract each other with a force which depends upon the masses of their bodies and upon the distance between them. The force is doubled when the product of the masses is doubled, and the force is halved when the square of the distances is doubled. If, for example, the distance is increased by a factor of four, the force drops to one sixteenth of its former value.

Now I shall show you that this law actually explains the elliptical shape of the orbit of a planet. However, I shall not develop the formula which describes an

$$\text{(Force of attraction)} = \gamma * \frac{\text{(Mass of body 1)} * \text{(Mass of body 2)}}{\text{(Distance between the two bodies)}^2}$$

**Fig. 7.5** Newton's law of gravity

ellipse in a coordinate system, but I shall start with the assumption that the orbit is an ellipse and then show that this assumption satisfies all the conditions described by Newton's laws. For the following explanation, I need the concept of the so-called *field of potential*. When we say that someone or something has a certain potential, we mean that the person or the object we are talking about has the ability to produce a certain effect. Think of a large tile resting on the roof of a house. This tile has the potential to cause severe damage by falling from the roof – it may even kill someone who is walking on the street. Since a tile which is lying on the street does not have the same potential as the tile on the roof, we may say that the potential of a tile increases with its height above the street. The concept of a field of potential is nothing more than the generalization of this idea. We want to assign to each point in the space, a physical quantity which describes the potential a body has when it is located at that point. The term "field" is used in physics when considering the distribution of a physical quantity in a space. The physical quantity might be directed like a speed or undirected like a temperature. In the case of the field of potential, an undirected quantity is assigned to each point in the space. The quantity we are looking for should be related to the energy a body has after it has fallen from the actual point down to the street level, since this energy determines the effect the body might have. But the unit of the potential cannot be that of an energy since the energy of a falling body is not completely determined by the height of the fall, but it also depends on the mass of the body. From this it follows that the unit of potential must be the unit of an energy divided by the unit of a mass. Thus, the unit of the potential is $(kg*(m/s)^2)/kg = (m/s)^2$ which is the square of a speed.

I derived the acceleration of falling bodies, $g = 9.8$ m/s$^2$, in connection with Fig. 7.4. The cause of this acceleration is the weight of the body, i.e., the force which pulls it down. According to Fig. 7.3, this force is equal to the product of the mass of the body and its acceleration. Thus we get *weight=m*g*. Fig. 7.3 also implies that the energy which is needed to move the body from the street level to the height *h* is equal to the product of the weight and the height, *energy = weight*h = m*g*h*. Since we already deduced that the potential is equal to the energy divided by the mass, we can now write *potential = g*h*. This is a product of an acceleration times a distance which results in a square of a speed as is required.

While the potential above the street level is helpful for explaining the parabolic shape of a jet of water, the field of the potential around the sun is helpful for explaining the elliptical shape of the orbit of a planet. While the potential above the street level is proportional to the height $h$, the potential around the sun grows with increasing distance $r$ from the sun. But it is not proportional to r because, according to the law of gravity, the attracting force decreases proportionally with the square of the distance. In the case of the potential above the street level, we could neglect the dependency of the attracting force upon the distance between the earth and the falling body, since in this case the variation of the distance is relatively small. The attracting force between two bodies according to the law in Fig. 7.5 depends not on the distance between the surfaces, but on the distance between the centers of gravity of the two bodies. The radius of the earth is approximately 6,400 km, and this primarily determines the distance which has to be used in the law of gravity. In relation to this big number, the variation by a few meters because of the fall of the body from above the earth's surface is so small that it doesn't have a measurable effect. In the case of a planet in its orbit however, the variation of the distance between the planet and the sun cannot be neglected. In the case of the earth, the distance varies between 147 and 152 million kilometers, and in the case of the planet Jupiter, the variation is even higher, namely between 740 and 816 million kilometers. Although an extreme variation such as shown in Fig. 7.6 does not occur in our solar system, the factor four between the shortest and the longest distance is helpful for my explanation.

Since we are not considering a particular planet but only want to understand why a planetary orbit is an ellipse, I could choose the numbers in Fig. 7.6 to simplify the formula which describes the potential. You may interpret the concentric circles which have the sun as their center as if they were lines of equal height on a map of a hilly region. The numbers assigned to these circles can then be interpreted as heights above the zero level. Whenever a group of circles having the same center occur on a map, this indicates either a conic hill or a conic crater. In the case of Fig. 7.6, it is a crater because the numbers are increasing with increasing radius. The walls of the crater get steeper and steeper as we get closer to the center, and get shallower as we move away from the center. At a great distance from the center, the ground is almost flat.

The arrows on the orbit indicate that the planet is moving clockwise around the sun. Along the left part of its path, the planet must climb up the wall of the crater and will continuously loose speed. Along the right part of its path, the planet rolls down the wall of the crater with increasing speed. Thus, the speed of the planet will have reached its maximum when its distance from the sun is minimal, and the speed will be minimal when the distance has reached its maximum. From Kepler's law, it follows that the maximum speed must be four times the minimum speed since the relation of the corresponding distances from the sun is 1:4. Since the

Newton's Discoveries 163

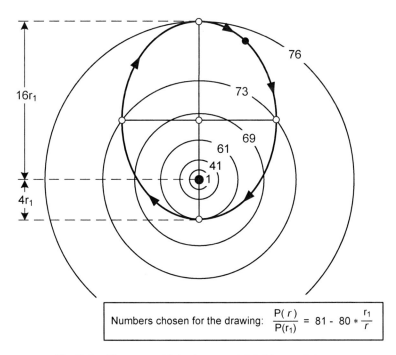

**Fig. 7.6** Planetary orbit in the potential field around the sun

energy is determined by the square of the speed, the maximum energy is $16*\Delta E$ if the minimum energy is $\Delta E$.

The potential of the planet is minimum, namely 61, when its speed is at its maximum, and the potential has reached its maximum, namely 76, when the speed is minimum. On both the left and right halves of the orbit, there is an exchange of energy corresponding to the difference of 15 between the two extreme values of the potential. This corresponds exactly to the difference of $15*\Delta E$ between the minimum and the maximum speed-dependent energy. At the beginning of its way along the left part of the orbit, the planet has an energy of $16*\Delta E$, and when it has reached the upper extreme point, it has lost most of it, namely $15*\Delta E$. However, this difference is not really lost, but has been stored in the potential which the planet gets back on its way along the right part of the orbit. Thus, the total energy of the planet is given by a sum of two components, namely the speed-dependent so-called *kinetic energy* and the position-dependent *potential energy*.

You probably are very familiar with processes where a continuous exchange between kinetic and potential energy occurs, although you might not have been aware of such an exchange. Think of a pendulum clock. When the pendulum moves through its lowest position, its speed and its kinetic energy are at their maximum and the potential energy is at its minimum. When the pendulum has reached the point on the left or the right side where it reverses its direction, its speed and its kinetic energy are zero and its potential energy is maximum.

To me, it seems like a miracle that the rather simple concepts defined in Fig. 7.3 together with the law of gravity enabled us to completely explain the phenomena shown in Fig. 7.1. But there are still some phenomena in mechanics which cannot be understood without the introduction of additional concepts. Until now, we considered only bodies which we could look at as if they were points of mass moving along a path which was either a straight line or a curve depending on whether or not a force was affecting the motion of the body. Now we must expand our view and take into account that the bodies might rotate. In this case, it is no longer adequate to look at these bodies as if they were points of mass. The most general case of the motion of a body is composed of two components, the motion of the center of gravity along a path, combined with a rotation of the body around its center of gravity. Think of a tennis player who, after having won a hard match, is so excited that he throws his racket high in the air. The center of gravity of this racket will move along a path which brings it back to the ground, and while it is moving along this path, the racket will spin around its center of gravity in a rather complex way.

The mechanical system which we shall use as an example to illustrate the problem and the solution concepts for dealing with rotation is shown in Fig. 7.7. A heavy disk is rotating with rather high speed around its horizontal axis. One end of this axis is mounted at the top of a vertical rod using a bearing in such a way that it cannot become unattached from the base, but the bearing does not hinder the axis from changing its angle. Looking at this system, we wonder why the disk, instead of tipping down, keeps moving slowly in the horizontal plane with its center of gravity making a full turn around the z-axis every two seconds. One important application of this strange effect is the gyroscope which for many years has been the principal device for navigation on ships. The same effect makes it possible to ride a bike.

When scientists are confronted with a new problem which they cannot solve using familiar concepts and methods, they first try to slightly modify the tools which have been so effective in the past. If such modifications are sufficient to solve the

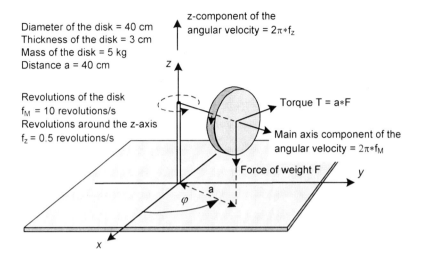

**Fig. 7.7** Motion of a rotating body

new problem, this has the advantage that the scientists need not learn completely new concepts and methods, but still can apply the old formulas with only some changes in interpretations. This has actually been possible in the case of going from moving points of mass to rotating bodies. The starting point is Fig. 7.3 where the physical quantities of distance, duration, mass, speed, acceleration, momentum, force and energy are considered with respect to mass points. A body consists of many mass points which cannot move relative to each other, and if such a body moves without rotating, all its mass points have the same speed. The opposite of this kind of motion is pure rotation which is characterized by the fact that the center of gravity doesn't move at all, while the other mass points of the body move around a straight line, the *axis of rotation*, which contains the center of gravity. As long as there are no forces affecting the rotation, the direction of this axis of rotation and the number of turns per time unit will not change. This corresponds to the fact that neither the direction nor the value of the speed of a mass point will change without having a force applied. Therefore, in the case of rotation, the familiar concept of "speed" has to be modified to become the new concept of *"angular velocity."* Like normal speed, the angular velocity is a directed quantity. Its direction is the same as the direction of the rotation axis. Its value tells how often the body spins around this axis per unit of time. Its speed may be measured in turns per second or by any other fraction with a number as its numerator and a time unit as its denominator. While the direction of the normal speed is defined by the motion of the mass point, the direction of the angular velocity does not follow from the rotation; an additional definition is required. This definition is called the "right hand rule" which says, "if the rotation axis is grasped by the right hand in such a way that four fingers point in the direction of the rotation, the thumb will point in the direction of the angular velocity."

Although both the speed and the angular velocity have a time unit in the denominator, they differ with respect to their numerators. While the numerator in the case of speed is a physical quantity, a distance, the numerator in the case of the angular velocity is a real number. This means that the angular velocity is not a fraction involving two physical quantities. It may sound strange to you that the number of turns is not considered a physical quantity. Doesn't rotation happen in the physical world? Certainly it does, but you shouldn't think that anything which can be expressed by numbers in the physical world is necessarily a physical quantity. A physical quantity cannot be measured without comparing it to an arbitrarily defined standard quantity. Think of a distance which, for example, can be measured in meters or yards. And now think of the number of revolutions a body makes per second. Do you need an arbitrarily defined standard quantity for comparison? Of course you don't, since the concept of a revolution is not a physical, but a mathematical concept. Here you don't ask questions like, "How far is it?" or "How long does it take?," but you ask only, "How often was it?" And that means counting which is a basic concept of mathematics. Instead of saying, "The body made one revolution," we could say, "The body made a turn of 360 degrees," or "The body turned by an angle of $2\pi$." We certainly can express the same fact by using different words, but we cannot refer to different standards as in the case of a distance. While the length of a meter or a yard could have been defined arbitrarily, we cannot arbitrarily define what a revolution should be. We can only define that the number $\pi$ corresponds to half of a revolution and one degree corresponds to the 360-th part of a revolution. When the word "degree" is used as the unit of a temperature, it is a physical unit, but when it is used in connection with an angle, it is not a physical unit but a mathematical term.

The fact that I gave you such a detailed explanation of the difference between the normal speed and the angular velocity should lead you to conclude that the concept of angular velocity is a crucial concept for an adequate handling of rotation. The following is a list of pairs of concepts which shows that partners must be associated with the concepts in Fig. 7.3 in order to obtain the set of concepts for rotational motion:

distance ↔ angle
speed ↔ angular velocity
acceleration ↔ angular acceleration
momentum ↔ angular momentum
force ↔ torque
mass ↔ moment of inertia

When I introduced the concepts in Fig. 7.3, I began with momentum which corresponds to the angular momentum in Fig. 7.8. But my explanation of Fig. 7.8 begins with the concept of torque which is the partner to the concept of force. Imagine that you own a clock or a toy driven by a spring mechanism which must be wound up every now and then in order to provide the energy the mechanism requires. Winding up is done by turning a key having two symmetric flat surfaces. The resistance of the spring can be overcome only by applying a pair of opposite forces to these surfaces. These two forces together with the distance between the two surfaces determine the torque. Its value is obtained by multiplying the value of one of the forces and the distance, and its direction is defined by the right hand rule. In Fig. 7.3, the product of a force and a distance is the undirected energy, while in Fig. 7.8 the product of a force and a distance is the torque which has a direction. How can this difference be explained? In both cases, the physical unit of the product must be the same, namely the product of the unit of force, $(kg*m)/s^2$, and the unit of distance, m, which gives $kg*(m/s)^2$. But it doesn't follow from this that torque and energy are essentially equivalent concepts. The difference comes from the different types of multiplication. Both factors of the product, namely the force and the distance, are directed quantities and, as you know from Chapter 3, there are two types of products of such quantities, the scalar product and the perpendicular product. If an undirected result is desired, the scalar product must be computed; otherwise the product used must be the perpendicular product. While the maximum energy is obtained when force and distance have the same direction, the maximum torque is obtained when force and distance are perpendicular. Therefore, when winding up your clock, you make sure that the forces you apply to the surfaces of the key are perpendicular to the straight line connecting these surfaces.

The introduction of the concept of torque, as given above, corresponds to that path in Fig. 7.8 which leads upward from the force to the torque. But, as you can see, there is a second path leading to the torque, and this path is coming down from the angular momentum. However, we cannot follow this path yet because we do not know what the angular momentum is. Of course, according to the formal correspondence between the structure in Fig. 7.8 and that in Fig. 7.3, the angular momentum must be the product of the partner of the mass and the partner of the speed. I already have introduced the angular velocity as the partner of the speed, but until now, the partner of the mass has been introduced only in the form of its name, *moment of inertia,* which still needs an explanation. The deduction in Fig. 7.9 is based on the fact that in Fig. 7.8 two different paths lead to the concept of torque. The left side of the equation corresponds to the path leading upward from the force to the torque, while the right side of the equation corresponds to the path leading downward from the angular momentum to the torque. Fig. 7.9 tells us that the unit of the moment of inertia is $kg*m^2$ which is the product of a mass and the square of a distance. How can we interpret this result? Assume that you are sitting

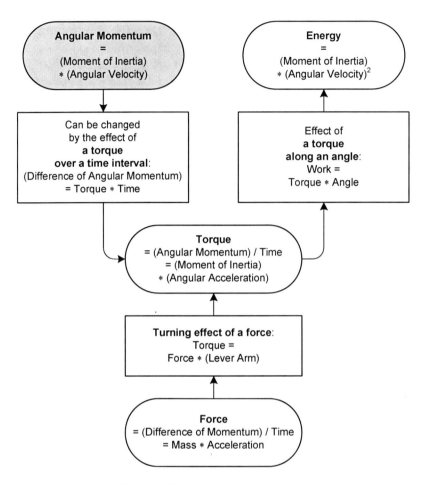

**Fig. 7.8** Concepts of rotational motion

in a circus, e.g., Barnum & Bailey, and watching the impressive performance of a weight lifter. He not only lifts a heavy barbell with two thick disks at the ends, but he makes it rotate above his head. His upright body is the axis of rotation. The effort required depends not only on the total weight of the barbell and the number of turns per second to be achieved, but also on how far the heavy disks at the ends are apart from each other. Actually, it's the square of this distance which determines the energy to be provided. If the weight is left unchanged, but the distance of the disks is doubled, four times the effort will be needed to obtain the same number of revolutions per second. The moment of inertia provides the information about how the parts of the mass are distributed around the axis of rotation.

# Newton's Discoveries

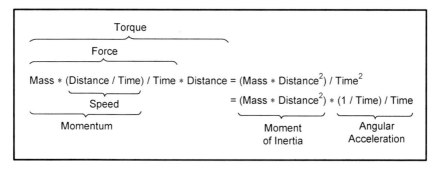

**Fig. 7.9** Two ways to define the concept of torque

If you own a car, you will be familiar with the situation when new tires are needed. With a new tire, it may happen that the distribution of the mass of the tire on the wheel is not symmetric about the axis of rotation which can lead to vibrations at high speeds. This can be rectified by attaching small pieces of metal at the right places on the wheel rim. This procedure is called balancing the wheel.

Now all the partners of the concepts in Fig. 7.3 have been introduced, and we can compare the physical units for each pair. This comparison is shown in Fig. 7.10. The physical units of the partners differ both in the pair (momentum, angular momentum) and in the pair (force, torque). From this it follows that there are essential differences between the partners in these pairs. Such a difference does not exist between the two types of energy which indicates that these two types are equivalent. This becomes obvious by comparing the following two scenarios. In the first scenario, we are expected to accelerate a track vehicle to a certain speed. We do this by pulling a rope which is connected to the front of the vehicle. If the vehicle is going straight in one direction, it may be compared to a

| | |
|---|---|
| Momentum | = Mass∗Speed = Mass∗(Distance/Time) |
| Angular Momentum | = (Moment of Inertia)∗(Angular Velocity) |
| | = (Mass∗Distance$^2$) ∗ (1/Time)     ≠ Momentum |
| Force | = Mass∗Acceleration = Mass∗(Distance/Time$^2$) |
| Torque | = (Moment of Inertia)∗(Angular Acceleration) |
| | = (Mass∗Distance$^2$) ∗ (1/Time)$^2$     ≠ Force |
| Energy of a moving mass point | = Mass∗Speed$^2$ = Mass∗(Distance/Time)$^2$ |
| Energy of a rotating body | = (Moment of Inertia)∗(Angular Velocity)$^2$ |
| | = (Mass∗Distance$^2$) ∗ (1/Time)$^2$   = Energy of a moving mass point |

**Fig. 7.10** Comparing units for moving mass points and rotating bodies

mass point. Acceleration means putting energy into the mass point. In the second scenario, we are expected to make a particular heavy wheel rotate with a certain number of revolutions per second. The wheel's center of gravity cannot move since the axis of rotation is mounted between two fixed bearings. Next to the wheel, on the same axis, there is a drum with a rope wound around it. By pulling this rope, we can make the wheel rotate faster and faster which means that we put energy into the wheel. For the person pulling the rope there is no difference between the two scenarios. It requires a certain force to pull the end of the rope, independent of the question of what happens at the other end of the rope. This indicates that there cannot be any essential difference between the energy of a moving mass point and the energy of a rotating body.

From what I told you about pure rotation, you should realize that the motion of the disk in Fig. 7.7 is not a pure rotation since its center of gravity moves on a circle around the z-axis. Pure rotation requires that the center of gravity of the rotating body doesn't move at all. However, there is still a point in the system in Fig. 7.7 which doesn't move and which is the center of the rotation, although it is not the center of gravity; this point is the top of the vertical rod. Therefore, the concepts of pure rotation can still be applied if the moment of inertia is appropriately modified. When I introduced the moment of inertia, I said that this is the information about how the body's mass is distributed around the axis of rotation. But this is correct only for cases where the direction of the axis of rotation doesn't change over time. In the case of Fig. 7.7, this condition is not satisfied since the axis of rotation changes its direction continuously. In the most general case, the moment of inertia is not a single quantity, but a set of six quantities which are combined in the form of a 3×3 matrix whose elements in the pairs ($J_{jk}$, $J_{kj}$) have equal values if j≠k. There is no need for you to understand this special result. Even most mechanical engineering students have difficulties understanding the corresponding deduction. Fortunately, the symmetries in the system in Fig. 7.7 simplify the problem of determining the moment of inertia. There are not six, but only two different components required for the moment of inertia, one, $J_M$, for the main axis of the disk, and a second one, $J_z$, for the z-axis.

The application of the concepts given in Fig. 7.8 to the system represented in Fig. 7.7 leads to Fig. 7.11. The angular velocity ω has two components, a horizontal one $ω_M$ which is associated with the fast rotation around the main axis of the disk, and a vertical one $ω_z$ which is associated with the slow rotation around the z-axis. Since there are 10 revolutions per second around the main axis and only 0.5 rotations per second around the z-axis, the ratio $ω_M$:$ω_z$ is 20:1. You may wonder why the angular momentum does not have the same direction as the angular velocity. In the case of moving mass points, the momentum always has the same direction as the speed since the momentum is obtained by multiplying the directed speed with the undirected mass. In the case of rotation, however, the angular momentum is obtained by multiplying the directed angular velocity by the moment of inertia which, as I said above, is a matrix. A matrix is a structure of numbers which, in the most general case, is more complex than a vector and cannot be said to have a direction. However, in the case of the simple system in Fig. 7.7, the two components $Q_M$ and $Q_z$ of the angular momentum can be obtained by multiplying the two components $\omega_M$ and $\omega_z$ of the angular velocity by the corresponding components of the moment of inertia: $Q_M = J_M*\omega_M$ and $Q_z=J_z*\omega_z$. The direction of the angular momentum is the same as the direction of the angular velocity only if the factors for the M- and the z-component were the same, i.e., if $J_M=J_z$. This, however, is not the case in the example considered. Here, $J_z$ is actually greater than $J_M$ by a factor of approximately ten. Therefore, the ratio $Q_M:Q_z$ is 2:1 while the ratio $\omega_M:\omega_z$ is 20:1.

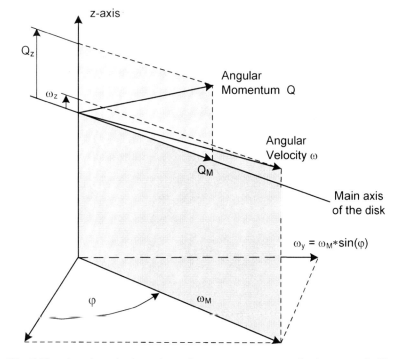

**Fig. 7.11** Angular velocity and angular momentum vectors for the system in Fig. 7.7

According to Fig. 7.3, a momentum can be changed by a force, and the direction of the change is the same as the direction of the force. Correspondingly, Fig. 7.8 says that an angular momentum can be changed by a torque, and that the direction of the change is the same as the direction of the torque. The torque which changes the angular momentum in the system shown in Fig. 7.7 is determined by the weight of the disk and the length of the lever arm. Its direction is given by the right hand rule; it is perpendicular to the plane defined by the two axes of rotation, the main axis and the z-axis. Thus, the torque cannot change the value of the angular momentum, but only its direction, since it is perpendicular to it. This, together with the fact that one end of the main axis is fixed to the top of the vertical rod, makes the disk rotate around the z-axis.

---

Time dependent momentum = $P(t)$

Slope of $P(t)$ = $\dfrac{dP}{dt}$ = Time dependent force = $F(t)$

---

Time dependent angular momentum = $Q(t) = \begin{pmatrix} (J_M * 2\pi f_M) * \cos(2\pi f_z * t) \\ (J_M * 2\pi f_M) * \sin(2\pi f_z * t) \\ (J_z * 2\pi f_z) \end{pmatrix}$

$= 2\pi * \begin{pmatrix} (0.10 \text{ kg}*m^2) * (10/s) * \cos(2\pi * (0.5/s) * t) \\ (0.10 \text{ kg}*m^2) * (10/s) * \sin(2\pi * (0.5/s) * t) \\ (0.86 \text{ kg}*m^2) * (0.5/s) \end{pmatrix}$

Slope of $Q(t)$ = $\dfrac{dQ}{dt}$ = Time dependent torque = $T(t)$

$= (2\pi)^2 * \begin{pmatrix} (J_M * f_M * f_z) * (-\sin(2\pi f_z * t)) \\ (J_M * f_M * f_z) * (\cos(2\pi f_z * t)) \\ 0 \end{pmatrix}$

$= (2\pi)^2 * \begin{pmatrix} (0.5 \text{ kg}*(m/s)^2) * (-\sin(2\pi * (0.5/s) * t)) \\ (0.5 \text{ kg}*(m/s)^2) * (\cos(2\pi * (0.5/s) * t)) \\ 0 \end{pmatrix}$

**Fig. 7.12** Formulas for the rotational system in Fig. 7.7

Fig. 7.12 shows how the behavior of the system in Fig. 7.7 is captured by formulas. This will be much too much detail for most of my readers who may skip Fig. 7.12. My main goal was to show you that in physics – as in other fields – it is sometimes possible to successfully transfer the structure of known concepts and methods to a new but related field. In our case, it was the structure in Fig. 7.3 belonging to the field of moving mass points, which has been transferred to the field of rotating bodies. The corresponding structure is shown in Fig. 7.8.

# Chapter 8
# How Albert Einstein Disregarded Common Sense

Until the turn of the century in the year 1900, so many philosophers, mathematicians and physicists had considered the problems of space, time and motion that almost all experts were convinced that no further interesting concepts or laws could be found in this field. But then, in 1905, Albert Einstein entered the stage and published a paper which wiped out the old ideas about space and time. Even today, most people believe that the ideas which Einstein presented in that paper can be understood only by someone who is a genius similar to Einstein, and that ordinary people shouldn't even try to follow Einstein's path to his results. I know for sure that this belief is a prejudice. It all depends on the quality of the explanation. At the beginning of this chapter, I feel like a mountain guide who takes his group on a rock wall climbing tour which is commonly said to be much too difficult for ordinary tourists. Please trust me. I have climbed this wall very often, and hammered so many hooks into the rocks that you will find it quite natural to follow me.

## How Meters and Clocks Were "Relativized" and the Speed of Light Was Made the Standard Reference

First, you must ask yourself what it means to say, "space and time are absolute" - since this was the belief everybody had until 1905. Someone who says this certainly doesn't believe that the lengths of all meter sticks are exactly the same, and that all clocks are running absolutely synchronously. It has long been known that a meter stick gets longer with rising temperature, and that one clock may run slower or faster than another clock. Absolute space and time means that the simultaneousness of two events and the distance between two simultaneous events are well-defined absolute concepts. It is evident that, if the concept of absolute simultaneousness disappears, the concept of absolute distance will also disappear. And that's exactly what Einstein did – he made these two concepts disappear.

In all the illustrations which I shall present in this discussion, the concept of the so-called "life line" plays a major role. What a life line is shall be explained using Fig. 8.1. Here, we consider three different objects: a black arrow, a white arrow and a rocket flare. The diagram shows these objects in different situations over time. The black arrow is flying with constant speed downward while, at the same time, the white arrow is flying upward at the same constant speed. Time passes from left to right and the objects move vertically along the x-axis. Since all objects are moving on the same straight line, the y- and z-coordinates are not relevant and could have been omitted. Lengths and distances are expressed as multiples of a reference length a, and time durations are expressed as multiples of a reference duration T. Each of the two arrows has a length a and the value of their speed is taken to be a/2T. The rocket flare is considered as a point object flying with the speed a/T, double the speed of the arrows; this will be discussed later. If we choose a = 100 m and T = 1 s, the diagram covers a distance of 400 meters (4 lengths of an arrow) and a time of 4 seconds. In the situation shown at the far left, the two arrows are still apart by exactly one length of an arrow. One second later, the two arrow heads meet, and after one more second, when the time is zero, the two arrows lie in parallel. Then one second later, only the two tails are together. At the far right of the diagram, the distance between the two tails has already reached the length of an arrow.

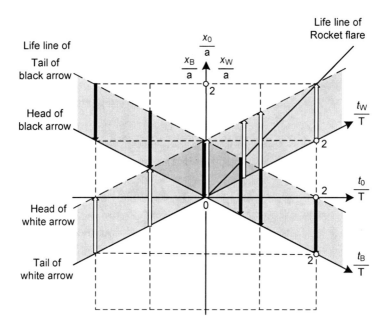

**Fig. 8.1** Life lines for moving objects in absolute time

For any point in time, each point of an object has its own location in the diagram, and the line which connects all these locations for a given point of an object is called the *life line of that point*. The boundaries of the grey shaded bands are the life lines of the heads and the tails of the two arrows, and the fact that these lines are straight indicates that the arrows are moving with constant speed. The third object, the rocket flare which is fired at t = 0, is also moving with constant speed. Since this speed is a/T, the life line leads upward at an angle of 45 degrees. I assumed that the rocket flare is fired vertically from the tail of the white arrow upwards towards its head. Since the rocket flare and the black arrow are flying against each other, the rocket flare needs only two thirds of T to reach the tail of the black arrow. At that point in time, it has travelled only one third of the length of the white arrow, since this arrow is flying in the same direction as the rocket flare. The head of the white arrow is finally reached after a time of 2T.

The arrows at the right side of Fig. 8.1 are for time axes. You may wonder why there are three different axes for the time coordinate. The reason for this is the assumption that there are three different observers having individual views of the system, and that the three axes $t_0$, $t_B$ and $t_W$ are the life lines of these observers. Each of these observers describes the situations he experiences over time in his own coordinate system, i.e., in $(t_0, x_0)$ or $(t_B, x_B)$ or $(t_W, x_W)$ respectively. Since the life line $t_0$ is a straight horizontal line, the associated observer has the speed zero, i.e., this observer doesn't move at all. He sees the black arrow flying downward, the white arrow flying upward and the rocket flare flying upward like the white arrow, but with double speed. The life line $t_B$ belongs to the head of the black arrow, and if this is also the life line of an observer, this observer must be sitting on the head of the black arrow. In the view of this observer, the black arrow doesn't move. This observer sees the white arrow flying upward with the speed a/T and the rocket flare flying upward, too, with the speed 1.5 a/T. For the observer sitting on the black arrow it seems as though the observer having the life line $t_0$ is flying upward with the speed 0.5 a/T. The third observer, i.e., that one with the life line $t_W$, is sitting on the tail of the white arrow, and his experiences are quite different compared to those of the observer on the black arrow.

The concept of life lines is very helpful for explaining processes occurring in space and time, and this is shown quite convincingly in Fig. 8.2. You probably are familiar with the rather amazing sound effect which occurs when you are in a fixed position and a police car, a fire engine or an emergency vehicle passes you at high speed while sounding its siren: the pitch of the siren drops significantly exactly at the moment when the vehicle reaches your position and passes by. Your life line in Fig. 8.2 is the horizontal axis while the straight line leading upward is the life line of the vehicle. The points located at equal distances from each other on the life line of the vehicle indicate the fact that the emitted sound of the siren has a constant pitch. The shorter the distance between the points, the higher is the pitch. Although the distances between the points on the life line of the vehicle are the same on the left and right sides of the vertical axis, the corresponding distances on the horizontal life line are different. They differ by a factor of 3 which is a consequence of my unrealistic assumption that the speed of the vehicle is half the speed of the sound. The first scientist who analyzed and described this phenomenon was Christian Doppler (1803-1853), and in order to honor him this effect is called the "Doppler-effect."

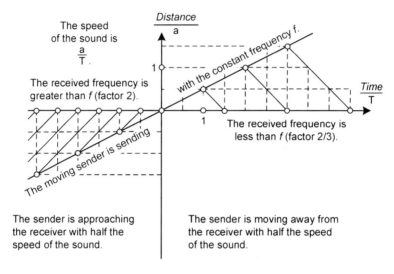

**Fig. 8.2** Doppler effect

In order to follow Einstein's thoughts, we have to go back again to Fig. 8.1. We now assume that instead of the rocket flare, a flash of light is sent towards the head of the white arrow and that the speed a/T is now the speed of the light. Correspondingly, the two arrows now fly with half the speed of the light. If Fig. 8.1 were a correct description of the facts, observers sitting on the arrows would measure different relative speeds of the light. Observers sitting on the black arrow would realize that it takes two thirds of the time T for the light to go the full length a of the arrow, and from this they would conclude that the relative speed of the light is 1.5 a/T. In contrast to this, observers sitting on the white arrow would realize that it takes twice the time T for the light to go the full length a of the arrow, and from this they would get 0.5 a/T for the relative speed of the light. The two relative speeds measured by observers sitting on the two arrows would differ by a factor of 3. This difference cannot be avoided if space and time are really absolute.

However, at the end of the nineteenth century, physicists were confronted with the fact that all recent measurements of the speed of light showed that the measured speed did not depend at all on the question of whether or not the source of light and the observers are moving relative to each other. From this it follows that Fig. 8.1 doesn't describe the facts correctly, and that a new approach was required – at least

# Special Relativity Theory

for the case when the speed a/T is considered to be the speed of light. That was exactly the situation in the year 1905 when Einstein came up with his approach. Of course, he was not the only one searching for a solution to the problem, but he was the first one to come up with the right approach. His idea was that it might well be that space and time are interwoven in such a way that distances and time durations cannot be determined independent of each other, with the interdependence only becoming relevant when the speeds considered get close to the speed of light. As long as the speeds considered are one-millionth or less of the speed of light, Fig. 8.1 describes the facts correctly.

Einstein began his considerations with two assumptions. His first assumption was that there are no experimental phenomena from which it would be possible to decide whether an object is or is not moving. Think of two spaceships flying through space at constant speeds. An astronaut who is sitting in one of these spaceships and looks out of the window, may see the other spaceship passing by, but from his observations he cannot determine whether his spaceship or the other one or both are in motion. The second assumption Einstein made was that two observers moving relative to each other with constant speed will measure the same speed of light. On the basis on these two assumptions, Einstein deduced rather simple mathematical results which I now shall present to you. The top section of Fig. 8.3 shows how physicists handle space and time mathematically in the case of a system where two objects are flying in opposite directions with constant relative speed v, where v is extremely small compared to the speed of light. Each of the two objects is assumed to be connected to its own coordinate system. The origin, i.e., the intersection of the three axis x, y and z of each coordinate system, is located at the tail of the actual arrow. Since the motion occurs only along the x-coordinate, the y- and z-coordinates could have been omitted in the diagram. In order to get a completely symmetric situation, we assume that the two z-axes are pointing upward. With these given directions of the x- and the z-axis, the direction of the y-axis is obtained by applying the right hand rule: the axis $y_1$ points away from the reader into the book, while $y_2$ points into the face of the reader. From these directions we get the relations $y_1=(-y_2)$ and $z_1=z_2$.

The position (x, y, z) and the time t of an event can be described alternatively in either of the two coordinate systems, and once an event has been described in one system, its description as a vector in the other system can be obtained by multiplying the vector for this system by a particular transformation matrix. This simple transformation was known to Galileo Galilei, and he was honored by having it named the "Galilei transformation." The transformation matrix is therefore called the "Galilei matrix."

Now we leave the assumption that space and time are absolute, and therefore I can no longer draw a picture showing two objects flying in opposite directions. Such a picture would suggest that it shows a situation at a certain point in time.

178                                                                            8. Relativity Theory

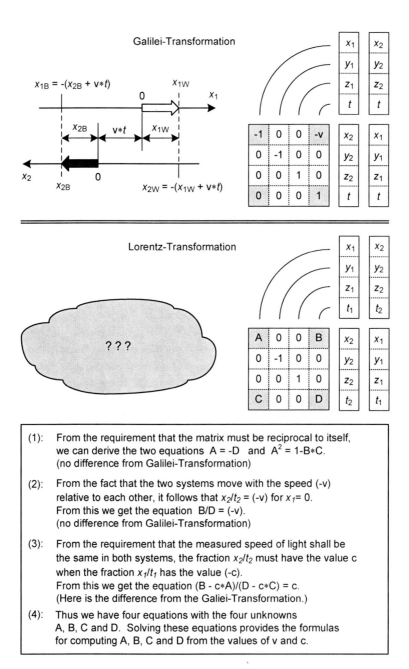

Fig. 8.3  Deducing the Lorentz transformation

Special Relativity Theory 179

This would be misleading to the reader because when space and time are relative, there is no point in time at which two observers, each sitting on his own object, have the same view of the situation. The only thing we know is that each observer has his own devices for measuring time and distance, such as a clock and a meter stick. Unfortunately, some authors try to explain the concept of relativity by using a picture that shows a train with an observer inside and another observer standing next to the track. Such a picture is not at all helpful, since it confuses the essential difference between absolute and relative space and time, and thus makes it difficult if not impossible to reach an adequate understanding. Therefore, in the lower section of Fig. 8.3, you will not find such a picture, but only a cloud with question marks in it.

It would be convenient if the transformation in the case of relativity could also be expressed in form of a matrix, and so we – as Einstein did – will try to find this matrix. As you shall soon see, it really can be found. If this matrix does exist, only the entries in its four corners will differ from the Galilei matrix because these are the positions which describe the relations between distances on the x-axis and the corresponding time durations. Since the motion occurs only along the x-axis, the coordinates y and z cannot be affected by the following considerations. What we are looking for is a set of four formulas which describe how the entries A, B, C and D can be obtained once the relative speed v is given. Certainly, it should be expected that the speed of light will appear in these formulas. The standard symbol for the speed of light used in formulas is the lower-case letter c. As you know from what I told you about equations with unknowns, we need four equations because we have four unknowns. The notes (1) through (3) in Fig. 8.3 explain how these equations are obtained. You should note that three of these equations also must be satisfied in the case of the Galilei transformation.

The solution of the four equations is given by the matrix equation in Fig. 8.4. The coordinate transformation based on this matrix is not called the "Einstein transformation" as one might expect, but the "Lorentz transformation." Hendrik Antoon Lorentz (1853-1928) found this transformation some years before Einstein as a consequence of his analysis of Maxwell's theory of electrodynamics which shall be presented in Chapter 9. Albert Einstein deduced the transformation not by using Lorentz's approach, but merely by thoughts about space and time as presented here.

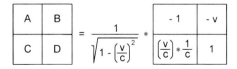

**Fig. 8.4** Lorentz transformation: Solution of the equations in Fig. 8.3

By applying the Lorentz transformation to the scenario in Fig. 8.1, we get Fig. 8.5. While in Fig. 8.1 the observations of the three observers B, 0 and W who are moving relative to each other, could be illustrated by a single diagram, three diagrams are required for the relativistic view, one for each observer. In any (x, t) coordinate system, the lines of simultaneousness are defined as the lines that are parallel to the actual x-axis, and the life lines of objects which don't change their x-position over time are lines that are parallel to the actual t-axis. If, as is the case in Fig. 8.1, all the x-axes are identical, the lines of simultaneousness are the same in all coordinate systems. This means that in Fig. 8.1 each vertical line is a line of simultaneousness, and thus, all observers measure the same time, whether they are in motion or not. This, of course, is not surprising, since Fig. 8.1 has been drawn under the assumption that space and time are absolute. In contrast to Fig. 8.1, the x-axes in Fig. 8.5 have different directions, and consequently, the lines of simultaneousness cannot be the same in the three coordinate systems B, 0 and W. In the 0-system, the lines of simultaneousness are vertical, and in the B- and the W-system, they have the directions of their arrows. Visibility of an object requires that all points of the object have the same time, i.e., that they are located on a line of simultaneousness. Therefore, the arrows as seen by the particular observers, have the direction of the corresponding x-axis.

The diagrams in Fig. 8.5 which are based on the Lorentz transformation could be drawn only under the constraint that the geometric distances in the planes are not consistent with the values of the coordinates in the three different coordinate systems. This inconsistency can easily be seen by computing the length of the white arrow which is sitting on the $x_W$-axis. According to the labeling of this axis, the length of the arrow is 1. In the 0-system, the same distance corresponds to the length of the diagonal of a rectangle, which can be computed by applying the law of Pythagoras with q being the length of one side of a dash-lined square:

$$\text{Length of diagonal} = \sqrt{q^2 + (2q)^2} = q * \sqrt{5} \qquad \text{with } q = \frac{\sqrt{3}}{3}$$

$$\text{Length of diagonal} = \frac{\sqrt{15}}{3} = 1.291$$

The discrepancy between the two different values 1 and 1.291 for the same geometric distance is a consequence of the fact that Fig. 8.5 is a mere mathematical construct obtained by applying the matrix in Fig. 8.4. Later on, I shall show you how this discrepancy can be resolved.

The diagram in the middle of Fig. 8.5 shows the view of an observer having the life line $t_0$ since for him the lines of simultaneousness are vertical. At $t_0=0$ he sees the two arrows side by side in parallel, and for him the two arrows have the same length, namely $a*(\sqrt{3})/2$. He is the only one who sees the two events "white head touches black tail" and "black head touches white tail" at the same point in time. For observers in the systems B and W, these two events are consecutive, and the

# Special Relativity Theory

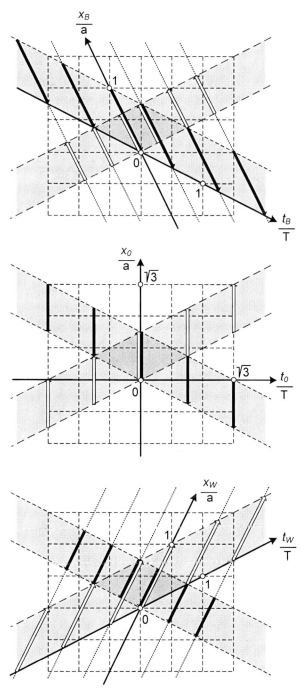

**Fig. 8.5** Relativistic version of Fig. 8.1

observers don't even see them in the same order: An observer in the system B first sees the event "black head touches white tail," and some time later the event "white head touches black tail." From this he concludes that the white arrow is shorter than the black one. For an observer in the system W, it's just the other way around.

Most people, when they heard about Mr. Einstein's ideas, were convinced that his mind had been confused. His results were so obviously opposite to common sense that most people didn't take them seriously. But Albert Einstein had clearly realized that common sense only provides useful results as long as it is applied to the phenomena of every day life, and that it fails completely when scenarios are considered which lie far away from our everyday experiences. You should not forget that Fig. 8.5 is based on the assumption that an observer sees two arrows passing by with half the speed of the light. Nobody ever has experienced objects flying at such a high speed. If we talk about high speeds, we think of rockets or rifle bullets. But their speeds are not even one thousandth of the speed of light, because otherwise they would need only one second to travel the distance from New York City to Boston. Most speeds we are used to from our every day experiences are less than one millionth of the speed of light. In these cases, the differences between the Galilei transformation and the Lorentz transformation are so small that they cannot be detected using standard measuring equipment.

Perhaps you once heard or read somewhere that according to Einstein there is no speed greater than the speed of light. However, this is not an assumption which he made before he began his mathematical deduction, but a result which can be deduced from the Lorentz transformation. This result is the answer to the question of how two speeds which point in the same direction can be added. Think of an observer who is standing near railroad tracks and sees a train passing by in which a passenger is walking towards the front of the train. Let's assume that the train travels at a speed of 50 miles per hour and that the passenger is walking with a speed of 2 miles per hour. If time and space were absolute, the speed of the passenger, as measured by the observer, would be the sum of the two speeds, i.e., 52 miles per hour. But from the Lorentz transformation, it follows that the resulting speed must be computed according to the formula

$$(v_1 +_{relativistic} v_2) = \frac{v_1 + v_2}{1 + \left(\frac{v_1}{c}\right) * \left(\frac{v_2}{c}\right)}$$

where $v_1$ and $v_2$ are the two speeds, and c is the speed of light. If we apply this formula to the speeds of our example, we get the relativistic sum 51.999,999,999,999,988,56 miles per hour which is so close to the arithmetic sum 52 miles per hour that it is impossible to measure the difference between these two values. Now we assume that the two speeds $v_1$ and $v_2$ both have the value of half the speed of the light. The conventional sum of these two speeds is exactly the

Special Relativity Theory                                                                 183

speed of light whereas their relativistic sum is only 80 percent of the speed of light. The formula for computing the relativistic sum of two speeds has an interesting property: the result does not exceed the speed of light as long as neither of the two summands exceeds the speed of light.

Fig. 8.5 illustrates that the relativistic sum of two speeds which both are half of the speed of light is 80 percent of the speed of light. Since the observers in the 0-system sees the two arrows flying with half the speed of light in opposite directions, observers sitting on one of the arrows will see the other arrow passing by with a speed which is the relativistic sum of two summands each having the value 0.5c, with c= a/T. Let's consider observers sitting on the black arrow, one sitting at the head and the other sitting at the tail. The view of these observers is illustrated by the upper diagram in Fig. 8.5. The white head reaches the observer at the black head at the time $t_{Bh}$ = - 0.75T, and the white head arrives at the observer at the black tail at the time $t_{Bt}$ = 0.5T. In the view of the two observers, the white head needed $(t_{Bt}-t_{Bh})$ = 1.25T for the length a of the black arrow, and from this they conclude that the speed of the white arrow is a/(1.25T)=0.8*(a/T)=0.8c.

The left part of Fig. 8.6 illustrates how it is possible for the observers in all three systems B, 0 and W to measure the same speed of light. In all three coordinate systems, distances are expressed as multiples of a standard length a, and time durations are expressed as multiples of a standard duration T with a/T=c. Thus, the life line of a flash of light must be a straight line with a slope of 45 degrees. While the coordinates of the start event are the same in all three systems, namely x=0 and t=0, the coordinates of the arrival event depend on the system. But since the arrival event is a point on the 45 degrees line, the ratio (x/a):(t/T) is 1 in all three coordinate systems, and this means that the speed x/t is a/T=c and does not depend on the coordinate system.

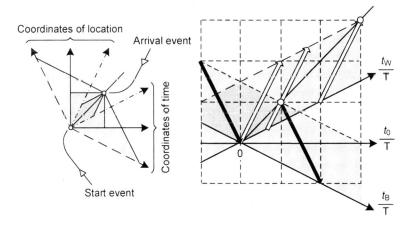

**Fig. 8.6**   Observer-independent speed of light

Fig. 8.1 shows a 45 degrees line which is interpreted as the life line of a rocket flare which is fired at t = 0 from the tail of the white arrow towards its head. The diagram on the right side of Fig. 8.6 also shows the life line of something which is fired at t=0 from the tail of the white arrow towards its head, but now it is no longer a rocket flare, but a flash of light. When this flash of light reaches the end of the black arrow, it has travelled only one third of the length of the white arrow. Thus, the coordinates of this event are

$$(t_B, x_B) = (a, T) \quad \text{and} \quad (t_W, x_W) = (a/3, T/3).$$

When the flash reaches the head of the white arrow, it has already travelled three times the length of the black arrow. Thus, the coordinates of this event are

$$(t_W, x_W) = (a, T) \quad \text{and} \quad (t_B, x_B) = (3a, 3T).$$

In each of the coordinate systems, an event has its own coordinates, and this means that distances and time durations depend on the system in which they are measured. This is illustrated by Figs. 8.7 and 8.8. Fig. 8.7 deals with the question of what is the measured length of the black arrow in each system. This length is the distance between the head and the tail of the arrow and must be measured on a line of simultaneousness. The two points which determine the length must be points on the life lines of the head and the tail of the arrow. Since the direction of the lines of simultaneousness depends on the coordinate system, the measured lengths will differ.

Fig. 8.8 deals with the question of what time duration is measured between two events. The event that the head of the black arrow touches the tail of the white

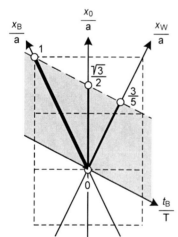

| Speed v of the observers in relation to the arrow | Observed relative length of the arrow |
|---|---|
| 0 | 1 |
| 0,5 c | 0,866 |
| 0,8 c | 0,6 |
| $\left(\frac{v}{c}\right) * c$ | $\sqrt{1 - \left(\frac{v}{c}\right)^2}$ |

**Fig. 8.7** Observer-dependent lengths

arrow is considered as the first event. All three time coordinates of this first event have the value zero. The second event is defined by the coordinates $(t_B, x_B)=(T, 0)$ which means that T units of time have passed since the first event; this corresponds to an "aging" of the head of the black arrow. In Fig. 8.8, three lines of simultaneousness, one for each coordinate system, are drawn through the point which corresponds to this event. The time coordinates which belong to these lines differ, and the corresponding values indicate that the greater the relative speed between an observed object and its observers is, the faster is the process of aging of the observers in relation to the observed object. The factor by which an observer ages faster than the black arrow corresponds to the relative times belonging to the observer's line of simultaneousness. The table in Fig. 8.8 contains the reciprocals of these values since here the observed object, and not the observers, is taken as reference.

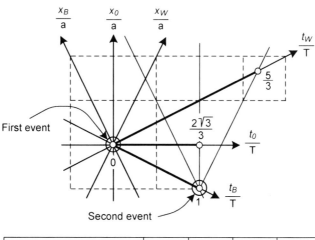

| Speed v of a point object in relation to its observers | 0 | 0,5 c | 0,8 c | $\left(\frac{v}{c}\right)*c$ |
|---|---|---|---|---|
| Aging of the object in relation to its observers | 1 | 0,866 | 0,6 | $\sqrt{1-\left(\frac{v}{c}\right)^2}$ |

**Fig. 8.8** Observer-dependent time durations

Although it is very hard for us to accept these strange results, there are experimental findings through which they have been confirmed. Muons are elementary particles which can be generated using high energy radiation to break up certain atoms. These muons have a very short lifespan, namely only about two millionth of a second, and then they are consumed in the creation of other particles. Muons are generated not only in physics laboratories, but also about 10 km above the earth where high energy radiation coming out of space hits the atoms of the atmosphere and partially breaks them up. Without the effects described by the Lorentz transformation, it would be impossible for these muons to fly from high up in the atmosphere down to the ground. Even if they could fly with the speed of the light, their short lifespan wouldn't allow them to fly a distance longer than 600 m. Travelling the distance of 10 km with a speed close to the speed of light would require 33.5 millionth of a second. But to the great surprise of the physicists, such muons have been detected at ground level. This can be explained only by the fact that for objects moving with a speed which is close to the speed of light, distances and time durations differ very much from those measured by the observers. A distance which we measure as 10 km can well be 600 m for the muon, and a time duration which for us is 33.5 millionth of a second, can be 2 millionth of a second for the muon if it moves fast enough. The relative speed,

as measured by the muon is    $(600 \text{ m})/(2*10^{-6} \text{ s})$,
while the observer measures $(10{,}000 \text{ m})/(33.5*10^{-6} \text{ s})$,
but the value is the same in both cases, namely 300,000 km/s.

Again, I want to emphasize that the results presented in the figures 8.5 through 8.8 could be obtained by formal application of the Lorentz transformation, and that I am in the same situation as you and Albert Einstein: we have to accept the mathematical results, and we cannot reconcile them using our common sense.

In 1905, Einstein published three extraordinary papers, among them the paper containing the ideas presented above. Einstein died in 1955. These were the reasons the German government and some scientific associations declared the year 2005 as "the Einstein year." His famous formula $E=m*c^2$ could be found on many of the posters which invited the public to visit special exhibitions and events. Most people couldn't explain what it means, and from those who could, only a very small percentage could explain how Einstein came up with this formula. Since the basis of nuclear power plants and nuclear weapons follow from this formula, it should be expected that any person who considers himself well-educated knows this formula and can tell which thoughts led to it. Again I repeat what I have said before more than once: it required a genius to find the chain of thoughts which led to the results presented here, but any somewhat intelligent person can understand these thoughts if they are explained well enough. If you could follow me to this point, you certainly will have no problems following me on the rest of the way.

In Fig. 7.3, the concept of energy was introduced as work which can be delivered, stored or consumed. The unit of energy was derived as the product of a mass and the square of a speed. The only new aspect which is provided by the formula $E=m*c^2$ is that it contains the speed of light c. As long as you don't know the details behind this formula, you might think that this formula determines the energy which has to be provided for accelerating an object until it is flying at the speed of light. But from what I told you about the relativistic sum of two speeds, you know that it is impossible to accelerate an object to the speed of light. Then what happens if we keep applying a force to a body which is already flying with very high speed? Even if we cannot accelerate it any more, we keep putting more energy into it. From this we may conclude that, if the energy cannot be increased anymore by increasing the speed, we actually increase the energy by increasing the mass. From our every day experiences, we don't know any processes where we increase a mass without adding matter. Until now, mass has always been for us a property of matter, a property which doesn't change unless matter is added or taken away. But now we have to face the possibility that mass cannot only be changed by adding or removing matter, but also by adding or removing energy. Again, this can be deduced from the Lorentz transformation.

Assume that we have two heavy cubes of exactly the same size and material. They differ only with respect to their color, where one cube is white and the other one is black. In between these cubes, we place a strong spring which we initially compress and lock. Thus, we have a symmetrical system. As long as we do not unlock the spring, all three objects, i.e., the two cubes and the spring, move together and have the same life line. At time t=0 we unlock the spring which now will push the two cubes in opposite directions. Since the system is completely symmetric, the value of the acceleration of the two cubes will always be the same in the view of an observer sitting at the spring. After a very short time, the spring will no longer touch the cubes which from then on will fly with constant speed. We assume that this final speed is half the speed of light. From this point on, there are three life lines, namely the two life lines of the two cubes and the life line of the spring which also is the lifeline of the observer. These three life lines correspond to the axis $t_B$, $t_0$ and $t_W$ in the two diagrams in Fig. 8.9. The left diagram shows the view of the observer who is sitting at the spring and who sees the two

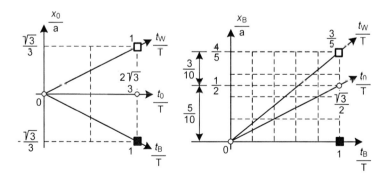

**Fig. 8.9** Symmetric versus asymmetric view of the system used for deducing relativistic mass

cubes flying away in opposite directions. The life line of this observer is horizontal which indicates that his relative speed is zero. In the right diagram, the horizontal life line belongs to the black cube which indicates that this diagram shows the scenario in the view of an observer sitting on the black cube.

While the left diagram shows that the observer with life line $t_0$ sees a symmetrical scenario, the right diagram shows that same scenario is unsymmetrical in the view of an observer with life line $t_B$. He sees observer 0 flying upward together with the spring with half the speed of light, and the white cube flying upward with 80 percent of the speed of light. In the view of the observer B, observer 0 is not in the middle of the distance between the two cubes. But he knows that the spring had been in between the two cubes and that the forces of the spring affecting the two cubes had been the same, only in opposite directions. Therefore the values of the *momentums* of the two cubes should be equal. From this, the observer B concludes that the mass of the white cube must be greater than the mass of the black cube, since only then is it possible for the two products (white mass)*(white speed) and (black mass)*(black speed) to be equal. The ratio (white speed):(black speed) relative to the spring in the view of the observer B is (0.8c – 0.5c):(0.5c) = 3:5, and therefore, observer B concludes that the ratio (white mass):(black mass) must be 5:3.

While we previously had to accept that distances and time durations depend on the relative speed between the observers and the system whose properties are measured, we now are confronted with the fact that this speed dependency also exists with respect to masses. An observer who measures the mass of an object which, in his view, doesn't move, will get the mass $m_0$, while an observer who sees the object moving with the relative speed v will measure a greater mass m(v), which is the mass of the object as a function of its relative speed. By generalizing the scenario in Fig. 8.9, the formula can be deduced which describes the relationship between $m_0$, v and m(v). This formula is shown in the upper left corner of Fig. 8.10. If we had the special disposition of Albert Einstein, we now would begin playing around with this formula, and one result of our manipulations would be the series expression which says that m(v) is equal to the sum of an infinite number of weighted powers of (v/c). A relativistic formula can be correct only if, for cases where v is extremely small compared to c, it provides the results according to the Galilei transformation. Therefore, we check what the series expression in Fig. 8.10 provides for the case v << c, and we get m(v) = $m_0$ + $m_0$*(v/c)$^2$/2. Someone who is familiar with formulas in the field of mechanics would see immediately that the second summand in the reduced series strongly resembles the formula for the kinetic energy of a body moving with speed v. According to Newton's laws of mechanics, this energy is m*v$^2$/2. Therefore, it is not a far-fetched idea to multiply the series expression by $c^2$; the result is represented in the lower section of Fig. 8.10.

# General Relativity Theory

$$m(v) = \frac{m_0}{\sqrt{1-\left(\frac{v}{c}\right)^2}}$$

$m_0$ is the mass at relative speed zero,
$m(v)$ is the relative mass, depending on the relative speed $v$.

$$m(v) = m_0 * \left(1 + \frac{1}{2}*\left(\frac{v}{c}\right)^2 + \frac{1*3}{2*4}*\left(\frac{v}{c}\right)^4 + \frac{1*3*5}{2*4*6}*\left(\frac{v}{c}\right)^6 + \ldots\right)$$

$$m(v)*c^2 =$$
$$= m_0*c^2 + \frac{1}{2}m_0*v^2 * \left(1 + \frac{3}{4}*\left(\frac{v}{c}\right)^2 + \frac{3*5}{4*6}*\left(\frac{v}{c}\right)^4 + \frac{3*5*7}{4*6*8}*\left(\frac{v}{c}\right)^6 + \ldots\right)$$

Motion independent | Motion dependent fraction of the energy

**Fig. 8.10** Deducing the equivalence of energy and mass

While this equation could be deduced formally from the Lorentz transformation, its interpretation was mere speculation, at least until it could be confirmed by experiments. Einstein was convinced that the formula not only provides a more accurate expression for the energy of a body in motion, but that there is already energy contained in a body which doesn't move, and that this energy is $m_0*c^2$. It took over 30 years until Einstein's theory, which he published in 1905, got confirmed experimentally. In 1938, the three scientists Otto Hahn (1879-1968), Lise Meitner (1878-1968) and Fritz Strassmann (1902-1980) detected and explained the existence of nuclear fission. They showed that the sum of the masses of the fragments after the fission was less than the mass of the uranium nucleus before the fission. The difference of the mass corresponds to the energy which was dissipated as heat.

According to the formula $E = m_0*c^2$, one kilogram of matter corresponds to the amount of electrical energy which a big power plant produces in one year. Unfortunately, no methods have yet been found to transform a common material like sand in such a way that its entire mass is turned into energy. If this were possible, half a ton of sand per year would be sufficient to satisfy the demand for energy for all nations on earth.

## How the Beautiful World of Mr. Newton Got Bended

After he had finished the theory which is presented above and which later became called "Special Relativity Theory," Einstein then found it quite natural to ask what two observers would observe if their relative speed doesn't stay constant, but changes over time. He finally found the answer to this question, and this answer is called "General Relativity Theory." At the time when Einstein was developing his Special Relativity Theory, he was an employee of the Swiss patent office in Bern, and it was only in the evenings and over weekends that he could find time to think about finding a consistent theory. Nevertheless, it took him less than a year to come up with his solution which was published in 1905. At that time, he was only 26 years old and had not yet received his Ph.D. degree which he received a few months later. The papers he had published in 1905 were so extraordinary that he immediately became well known in the academic world. In 1909 he obtained a position as associate professor at the Institute of Technology in Zurich, and from then on he could spend most of his time searching for the solution to the problem of general relativity. But it took him six more years to finalize this theory. In comparison to the development of the General Relativity Theory, the development of the Special Relativity Theory can be considered child's play. Otherwise it would not have been possible to present the basic mathematical deduction of this theory within the scope of this book. In contrast to this, it is absolutely impossible for me to go into similar detail with respect to General Relativity Theory. From the moment of its first publication, this theory got the reputation that only an extreme minority of scientists could understand it. This has been expressed by the following anecdote. A journalist wanted to write an article about Einstein's new theory and went to a physicist of high reputation to ask him for help. As an introduction to their conversation, the journalist told a joke he had heard that there were only three people on earth who really had a profound understanding of this theory. Hereupon the scientist said, "I wonder who the third one might be."

But I shall not restrict myself to telling you that this theory is very difficult. At the very least, I want to tell you about the problem this theory deals with, and what the characteristics of the solution are, since both the problem and the characteristics of the solution are not too difficult to be understood by most people. The extreme difficulty lies only in the mathematical deduction which is so difficult that even Einstein had to ask a friendly mathematician (Marcel Grossmann, 1878-1936) for help. That the related mathematical problems are extremely challenging can be seen, too, by the fact that the great mathematician David Hilbert (1862-1943) also spent some time searching for adequate solutions.

Most people, including Albert Einstein, begin to look for new explanations as soon as they encounter facts which are inconsistent with their present view of the world. The fact that motivated Einstein to search for the Special Relativity Theory was the independence of the speed of light from the relative speed between the source of the light and the measuring observers. This was inconsistent with the Galilei transformation which, until then, had been believed to be absolutely correct. Einstein's motivation to search for the General Relativity Theory originated from his wondering about the fact that the concept of mass is used in the explanation of two extremely different phenomena. On one hand, the mass of a body is used to explain the effect of inertia which determines the relation between a force and the resulting acceleration. On the other hand, mass is used to explain gravitation,

General Relativity Theory 191

i.e., the attracting force between two bodies. The sameness of the so-called *inertial mass* and the *heavy mass* had been introduced by Isaac Newton in his laws of mechanics and gravitation. And since no experimental results had been found which contradicted the assumption of this sameness, nobody saw any reason to ask further questions concerning this subject. Everybody was satisfied with the situation as it was, i.e., that the inertia of a body is doubled or tripled when its weight is doubled or tripled. Einstein, however, found it very strange that nobody had wondered about this sameness and nobody had begun to search for an explanation.

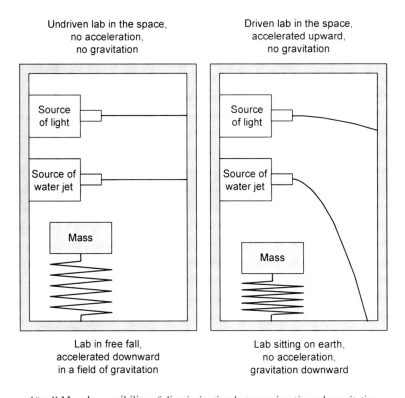

Fig. 8.11   Impossibility of discriminating between inertia and gravitation

I hope to make you wonder in the same way Einstein wondered using Fig. 8.11 and 8.12. Fig. 8.11 shows two scenarios which could occur in a physics laboratory. In the scenario on the left side, the beam of light and the jet of water are exactly horizontal and the spring is not compressed. In the contrasting scenario on the right side, the beam and the jet are bent downward and the spring is compressed. What you should wonder about is the fact that it is impossible to conclude from these observations whether or not the laboratory is actually accelerating. You might object that the impossibility of a conclusion is only a consequence of the fact that the laboratory doesn't have any windows. But even if it had windows and we could see objects passing by with constant or time variant speed, this wouldn't enable us to determine whether the laboratory is accelerating. In the case of an observed relative acceleration, there is no way to decide which object is accelerating, the observer or the observed object or both. In most situations we are convinced we know exactly whether or not we are accelerating, but this is merely a subjective opinion and cannot be proven by objective observations. A roofer who unfortunately is falling from the roof will find it absurd to assume that it is not he who is accelerating towards the earth, but that it is the earth which is approaching him with increasing speed. Nevertheless, the two views are absolutely equivalent.

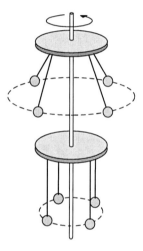

**Fig. 8.12**  The problem of relative rotational motion

While Fig. 8.11 concerns the question of whether an object is accelerated in a certain direction, Fig. 8.12 concerns the question of rotation. When I was drawing Fig. 8.12, I was thinking of a ride on a chairoplane in an amusement park. I put two such systems one above the other on the same axis. From what you see, you certainly will say spontaneously that the upper system is rotating while the lower one is standing still. You will argue that the centrifugal forces which pull the balls outward are a clear indication of rotation. Until now, you probably have not yet realized what it is you should wonder about. And of course, I also hadn't been aware that there is something to wonder about here before I became familiar with Einstein's considerations. His arguments were as follows: "There are no observable phenomena from which one could conclude whether an object is moving or not. Motion is always relative and is a relation between two objects." With respect to the two systems sitting on one common axis in Fig. 8.12, it is not justified for us to say that one is rotating and the other one is standing still; the only thing we can say is that they are

General Relativity Theory 193

rotating relatively to each other. Therefore, an explanation must be found why the balls of the upper system are pulled outward while the balls of the lower system are hanging straightly down. If the system shown in Fig. 8.12 were the only object in the universe, the situation in the drawing could not possibly occur. But since the universe contains not only the system in Fig. 8.12, but unimaginably many galaxies with unimaginably great masses, the situation shown somehow must be the consequence of the distribution of masses in the universe.

Fig. 8.13 also deals with rotation, but in contrast to Fig. 8.12, the dimensions of the system shown are so big that it cannot be thought as possible to build. We now assume that the circumferential points move with 80 percent of the speed of light. If the diameter of the wheel is big enough, even such a high speed does not require an extremely high number of revolutions per second. Think of the rotation of the wheel of a water mill; such a wheel rotates rather slowly and it may well need up to ten seconds for one revolution. In the case of Fig. 8.13, I assumed that the wheel needs five seconds for one revolution, and from the requirement that its circumferential speed is 80 percent of the speed of light it follows that the radius must have a length of 190,000 km, which is half the distance between the earth and the moon.

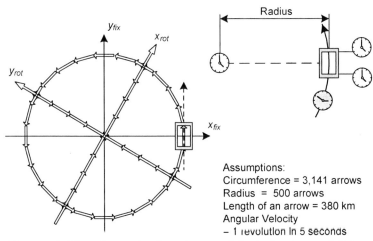

Fig. 8.13  Relativistic view of a rotating wheel

Although the two coordinate systems in Fig. 8.13 rotate relatively to each other – the fix-system clockwise and the rot-system counterclockwise – the subscript *fix* indicates that this system is assumed to be connected to the page of the book while the rot-system is connected to the wheel. The wheel is thought of as being built of arrows which are welded together. Thus, the circumference is a regular polygon with 3,141 corners. In the view of an observer in the fix-system, all arrows have approximately equal length, namely 380 km, with the circumferential arrows being longer by a few meters compared to the arrows of the radius. The window shown at the right side of the circle is assumed to be fixed in the fix-system, and its height is such that a circumferential arrow fits into it exactly. When an observer in the fix-system measures the diameter and the circumference of the wheel and then computes their ratio, he will get the number $\pi$ as expected. And by measuring the height of the window and the time an arrow head needs from the moment it enters the window until the moment it leaves it, he can compute the circumferential speed and get a value which is 80 percent of the speed of light.

Now we assume that the observer in the fix-system is familiar with Special Relativity Theory. Then he must conclude that an observer in the rot-system who is sitting on the wheel and rotating with it will measure a smaller height of the window. For the rot-observer, the window will not be high enough for an arrow to fit into it entirely. For him, the height of the window will only be 60 percent of the length of a circumferential arrow. That a fix-observer and a rot-observer measure different lengths of an arrow is only true with respect to the circumferential arrows, but not to the radius arrows. Their length does not have the same direction as the relative speed between the two observers, but is perpendicular to it, and therefore the Lorentz transformation does not apply to the radius arrows. Thus, the ratio of the arrow lengths (length of a radius arrow):(length of a circumferential arrow) is 1:1 for the fix-observer and 1:(1/0.6) = 0.6 for the rot-observer. From this it follows that the ratio between the circumference and the diameter, which is $\pi$ for the fix-observer, is $\pi$:0.6 for the rot-observer. This strange result says that the laws of plane geometry do not apply in the world of the rot-observer.

A detailed consideration of the time relations will provide another surprise. We assume that there are the four clocks shown in the upper right part of Fig. 8.13. The shaded clock is fastened to the wheel and is rotating with it while the other three clocks belong to the fix-system. For the fix-observer, these three clocks always run synchronously and he can read the actual time from any of them. For measuring the circumferential speed, the fix-observer uses the two clocks which are located at the entrance and the exit of the window. For him, an arrow head needs the time $\Delta t_{fix}$ to travel through the window. The rot-observer reads the shaded clock when the arrow head enters the window and again when it leaves it. Then, he computes the time difference and gets $\Delta t_{rot}$. Because of the Lorentz transformation, the ratio between the two times is $\Delta t_{rot}:\Delta t_{fix}=0.6$. The fact that $\Delta t_{rot}$ is less than $\Delta t_{fix}$ corresponds to the fact that the height of the window $h_{rot}$ as measured by the rot-observer is by a factor of 0.6 less than the height $h_{fix}$ measured by the fix-observer. Thus, both observers measure the same relative speed, since $h_{fix}/\Delta t_{fix}$ is equal to $h_{rot}/\Delta t_{rot} = (0.6*h_{fix})/(0.6*\Delta t_{fix})$.

# General Relativity Theory

Until now, we have not considered the clock which is sitting at the center of the wheel. We assumed that it isn't rotating, i.e., it belongs to the fix-system. But even if it were fastened to the wheel, it would still run almost synchronously with the other two clocks in the fix-system since it would stay in the center of the wheel and rotate with one revolution in five seconds. Everybody knows that the time he reads from a watch does not depend on whether he reads it while he is riding a merry-go-round or while he is standing outside and watching his child going around. Thus, we are confronted by the surprising fact that two clocks which don't move relatively to the rot-observer cannot run synchronously although they are built exactly the same. The clock at the circumference runs slower by a factor of 0.6 compared to the clock in the center.

Summing up our considerations we may say that only the fix-observer experiences a "normal world" where Euclid's laws of geometry do apply, and where all clocks which do not move relatively to each other can be synchronized. In contrast to this, the rot-observer experiences a world where the laws of conventional geometry do not apply and where the running speed of a clock depends upon its location. The cause for these different worlds cannot be the mere rotation, since each observer rotates relatively to the other one with the same angular velocity. The cause must be the same as in the case of Fig. 8.12: only one of the observers experiences centrifugal forces. Einstein himself wrote about this [EIN]:

*The observer who is sitting on the wheel may consider the wheel as the system in relation to which all phenomena are described. He may do so because of the relativity principle. In his view, the forces which affect him and other objects which are fastened to the wheel can be considered as caused by a field of gravitation though the values and directions of this field cannot be explained by Newton's law of gravitation. But this doesn't bother him since he believes in the general relativity principle. He is hoping, and rightly so, that a general law of gravitation will be found which does not only explain correctly the motion of the objects in the sky but also the forces he is experiencing on his wheel.*

These considerations motivated Einstein to search for a generalization of Newton's theory of gravitation. In Newton's law which describes the attracting force between two bodies, Newton assumed that a force can be caused just by the fact that another body is located at a certain distance. In contrast to this, Einstein was convinced that such instantaneous effects over a distance do not exist. According to his view, the cause of a force of gravity which affects a given body can be found only in the direct local neighborhood of the body. Of course, he didn't deny the fact that the force somehow depends on other bodies being somewhere; but he stated that the influence of the distant masses must be explained by certain changes in the space between the bodies.

From the finding that the standard laws of geometry could not be applied with respect to all observations in the universe, Einstein concluded that the solution of this problem could be found only in the form of an adequate mathematical description of the geometry of the space. He was convinced that the "bent space" and the problem of non-synchronizable clocks must have the same cause. As a result of his Special Relativity Theory, he had already found that space and time were not quantities which were independent from each other, but had to be considered as a composite structure. Therefore, the new laws of geometry he was searching for had to be laws for the four-dimensional space with the coordinates $x$, $y$, $z$ and $t$. Unlike the time in 1905 when he developed his Special Relativity Theory, he now could apply the results which the mathematician Hermann Minkowski (1864-1909) had published in 1908. Maybe you still remember the problem of inconsistent distances which I described in my comments about Fig. 8.5. There I showed that the law of Pythagoras provides different results depending upon the choice of the coordinate system. Mr. Minkowski was quite unhappy with this inconsistency, and he was convinced that it could be eliminated. As Albert Einstein had done before, Mr. Minkowski disregarded common sense and therefore finally found a surprising solution: he introduced imaginary values for the time coordinate.

When I presented the history of the creation of numbers and introduced the imaginary numbers as mirrored real numbers, I tried to convince you that there was no reason to be scared by these numbers. Here again, I emphasize the fact that everybody, even the most genius mathematician, has to accept imaginary numbers as pure formal constructs which have no further meaning beyond the fact that their squares are negative numbers. Nothing more is required since it is enough to know how to use them in arithmetic computations. In contrast to me, the British physicist Stephen Hawking (born in 1942) has an absolutely contrary opinion. He wrote [HA 2]:

> *I would like to make it clear that the imaginary time is a concept which we have to accept. It is a mental leap of the same type as the discovery that the earth has the shape of a sphere. The day will come when we shall consider the imaginary time as evident as the fact that the earth is a sphere.*

I object to this opinion by pointing out that disks and spheres are shapes which we observe every day and which we can easily imagine scaled to extremely small or big sizes. The problem of imagining the earth being a sphere did not originate from an inability of imagining a very big sphere, but from the fact that gravity had not yet been observed as a force which can have different directions. In contrast to this, the concept of an imaginary time is a formal mathematical concept from the outset and will stay so forever. Of course we can write $t = 5i$ seconds and conclude from this that then $t^2$ must be minus 25 seconds$^2$, but this doesn't give us any reason to expect that one day we will be able to associate an actual picture with this imaginary time.

General Relativity Theory

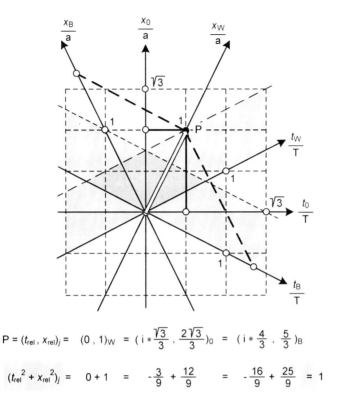

$P = (t_{rel}, x_{rel})_j = (0, 1)_W = (i * \frac{\sqrt{3}}{3}, \frac{2\sqrt{3}}{3})_0 = (i * \frac{4}{3}, \frac{5}{3})_B$

$(t_{rel}^2 + x_{rel}^2)_j = 0 + 1 = -\frac{3}{9} + \frac{12}{9} = -\frac{16}{9} + \frac{25}{9} = 1$

**Fig. 8.14** Invariance of distance in the space-time continuum using imaginary time

In the case of the ball-shaped earth, however, the real picture exists, not only in our imagination, but even by looking out of the window of a spaceship.

The example in Fig. 8.14, which refers to Fig. 8.5, illustrates the fact that the dependency of distances from the choice of the coordinate system really disappears by using imaginary values for the time coordinate. The distance between the intersection of the t- and x-axes and the point P is found by applying the law of Pythagoras; the square of the length is $t_j^2 + x_j^2$. If we would not take imaginary values for the time coordinates $t_j$, the result of this computation would not be the same in all coordinate systems. Only by using imaginary values for $t_j$, can we make the result independent from our choice of the coordinate system; in this case, the resulting length is always one.

In the example in Fig. 8.14, the resulting length is a real number since its square is a positive number. This is always the case when the angle between the horizontal axis and the line considered exceeds 45 degrees. If, however, this angle is less than 45 degrees, the square of the length is negative and the resulting length is imaginary. If the angle is exactly 45 degrees, the square of the length is zero. In this case, the line is the life line of a flash of light. Life lines of real objects – think of the head of the white arrow – always have an angle which is less than 45 degrees. Distances on such life lines are always imaginary which indicates that the distance is to be interpreted as a time duration. A positive distance can never be part of a life line, but always lies on a line of simultaneousness and describes the distance between two points in space.

The Lorentz transformation and Minkowski's imaginary time are the concepts which define the structure of the four-dimensional space-time continuum of Special Relativity Theory. However, you already know that this structure cannot be applied by an observer sitting on the wheel in Fig. 8.13. Einstein suspected that the strange phenomena which occur in connection with rotation could be explained by the assumption that the four-dimensional space-time continuum were somehow bent, and that this bending is determined by the distribution of the masses. This certainly sounds rather strange, but instead of being scared away you should remember what I told you when I introduced the concept of higher dimensional spaces: we always look first at structures in the real three-dimensional space, describe these by formulas, and then extend these formulas by formally adding further dimensions. That's exactly how we shall now proceed. We look at a bent two-dimensional space, i.e., a bent surface which is embedded in the three-dimensional space. Fig. 8.15 shows a bent surface which is sitting on a plane. It reminds me of a sun-helmet sitting on a table, but it also might be a hill out in the country. Now we consider the task of drawing a circle having its center at the top of the surface. This circle which is shown in the left part of the figure runs at a certain height along the hillside, and its radiuses are the lines to the top of the hill, along which drops of water would run down from the top. The ratio between the circumference and the diameter of this circle is less than $\pi$ since the diameter is longer than it would be if the circle had been drawn in a plane. This is a consequence of the condition that all lines must be drawn on the bent surface. We are not allowed to drill a tunnel through the hill in order to obtain shorter diameters.

The right part of the figure illustrates the fact that the shortest connection between two points can be a bent line. Think of a rubber string whose two ends are fixed by hammering nails into the table at the rim of the sun-helmet. Since all points of the string have to touch the bent surface the string will necessarily run along a bent line, and among of all possible such bent lines, there is exactly one which is the shortest.

The reason we have no problems seeing the bent surface in Fig. 8.15 is the fact that it is embedded in the three-dimensional space. If we could see a four-dimensional space, we presumably would have no problems seeing a bent three-dimensional space embedded in it. But since four-dimensional spaces are mere formal mathematical constructs which nobody can see, the question of whether the

General Relativity Theory 199

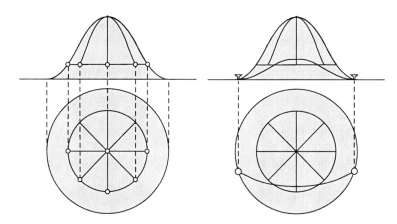

Fig. 8.15 A circle and a shortest distance on a bent surface

three-dimensional space we are living in is bent cannot be determined by looking at it. The problem of answering the question about bent spaces actually had been considered long before Albert Einstein came up with it. It was the German mathematician Carl Friedrich Gauss who, while he was involved in a project surveying Northern Germany starting in 1818, had the idea that it might be possible using certain measurements to find out whether or not our three-dimensional space is bent. The idea is quite simple, since it only requires a formal transfer of certain geometric phenomena from the case of bent surfaces to the case of our three-dimensional space. In the case of a bent surface, the ratio between the circumference and the diameter of a circle can be less than $\pi$, and the sum of the angles of a triangle can be more than 180 degrees. While these phenomena can be detected by measuring lengths and angles on very small bent surfaces, they cannot be detected by measurements in small sections of our three-dimensional space. From this, Gauss concluded that the bending of the three-dimensional space, if such a bending exists at all, could be detected only by measuring edges or angles of triangles which are much bigger than those we have access to on earth. It took about 100 years until the ideas of Gauss could be applied and the bending of space was proven on the basis of observations of astronomers.

Fig. 8.16 illustrates that a bending cannot be detected if the section within which the measurements are performed is too small. Here, a bent one-dimensional space, i.e., a curve, is partitioned into a sequence of intervals, and the lengths of these intervals is chosen such that within an interval no bending can be detected.

We now take this bent one-dimensional space as the starting point for our way which finally shall lead us to the mathematical description of bent four-dimensional spaces. This is the reason the formula for computing the distance $\Delta s$ between the entrance point of an interval, i.e., its origin, and a point within seems

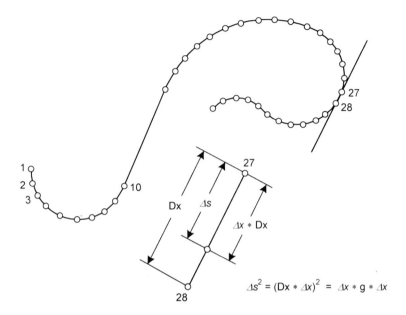

**Fig. 8.16**   Piecewise linear view of a bent one-dimensional space

to be unnecessarily complicated. Each interval has its individual total length Dx, and each point within the interval is determined by its relative coordinate $\Delta x$ which has the value $\Delta s/Dx$. You will not understand why the formula has been structured in this way until we proceed to higher dimensions.

The step which leads us from one dimension to two dimensions will enable us see how the formula must be structured in order to apply to four-dimensions. Since the time when mankind realized that the earth has the shape of a globe, mathematical methods for dealing with bent two-dimensional spaces have been developed. By drawing a net of lines onto a surface, this surface can be partitioned into a set of meshes. If the distances between the lines are small enough, each mesh can be considered approximately either as a quadrangle or as a triangle in a plane for which Euclid's laws of geometry apply. In the case of the globe of the earth, the partitioning lines are the meridians containing the poles, with the circles of latitude being concentric with the equator.

Fig. 8.17 shows a mesh of an assumed grid on a bent surface. The lines of the grid are assumed to have natural numbers assigned to them, with the x-lines and the y-lines being enumerated separately. The mesh in Fig. 8.17 lies in between the x-lines 34 and 35 and the y-lines 13 and 14. In Fig. 8.16, each interval had a specific factor $g = (Dx)^2$ assigned to it for being used in the formula $\Delta s^2 = \Delta x*g*\Delta x$. Now we now try to find a corresponding factor G assigned to the mesh

# General Relativity Theory

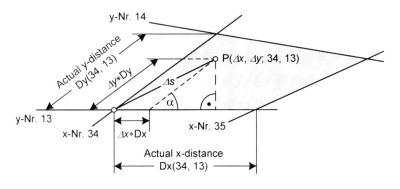

**Fig. 8.17** Determination of a distance in a mesh of coordinates

in Fig. 8.17 to produce the formula $\Delta s^2 = (\Delta x, \Delta y) * G * (\Delta x, \Delta y)$. Such a factor G cannot be a simple number but must be a "two-dimensional factor." Fig. 8.18 shows that the factor G is a matrix having the elements $g_{jk}$ according to Fig. 8.17. The formulas in Fig. 8.17 are obtained by applying the law of Pythagoras together with the definitions of the two functions $\cos(\alpha)$ and $\sin(\alpha)$. As can be expected from Fig. 8.16, in Fig. 8.17 only the variables $\Delta s$, Dx and Dy represent real physical distances while the variables $\Delta x$ and $\Delta y$ represent the ratios $\Delta s/Dx$ and $\Delta s/Dy$, respectively, and have values which are pure numbers. The computation in the bottom part of Fig. 8.18 is based on the assumption that the two distances Dx and Dy have the values 10 km and 9 km and that the angle $\alpha$ is 36.87 degrees; then $\cos(\alpha)=0.8$. With these values the point P in Fig. 8.17 has the coordinates $(\Delta x; \Delta y)=(0.25; 0.65)$.

The structure shown in Fig. 8.18 is exactly what enables us to perform the last step which brings us to our goal. It doesn't require a genius to conclude that, while the bent two-dimensional space could be described by two-dimensional matrices, the description of a bent four-dimensional space will require four-dimensional matrices where each section of the space has its own individual matrix assigned to it (see Fig. 8.19). Since each entry $g_{jk}$ of the matrix is equal to the entry $g_{kj}$, we need ten entries to fill the 4×4-matrix. In the simplest case, the matrix can be the unit matrix, and in this case the distance $\Delta s$ is obtained by simply applying the law of Pythagoras $\Delta s^2 = \Delta x^2 + \Delta y^2 + \Delta z^2 + \Delta t^2$, as shown in Fig. 8.19. It is always possible to partition a bent surface in such a way that one or more sections get the unit

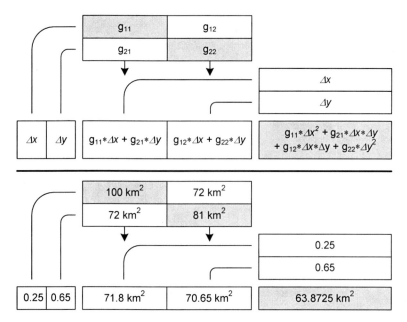

**Fig. 8.18** Computing a distance using a mesh-specific matrix

matrix assigned them, but it is impossible to find a partition where the unit matrix can be assigned to all sections. Think of the surface of the globe being partitioned by the meridians and the circles of latitude: the meshes near the equator can be treated approximately as squares, but the nearer we get to a pole, the more the borders Dx and Dy of the meshes are no longer perpendicular; i.e., the angle $\alpha$ then is no longer 90 degrees and the value of $\cos(\alpha)$ is no longer zero.

I believe that the mathematical structures which I have introduced to this point have not yet been too difficult to be presented to non-expert readers. But this was only the simplest part of the tour to the top of the mountain which is called General Relativity Theory. Actually, we have only walked from the valley over a hiking trail to a mountain shelter at the bottom of the steep rock wall. Climbing this wall really is much too difficult for us. This is the rock wall which I had in mind when I said at the beginning of this section that the mathematical deduction of the General Relativity Theory could not be presented in this book. Soon after Einstein had published this theory, the German physicist Max Born (1882-1970) wrote a book in attempt to explain the concepts of this theory to an interested non-expert audience. In this book, I found the following statement [BOR]:

*It has been Einstein's idea that the field which is generated by a body produces a repercussion on the body itself and thus affects its life line. This is an extremely difficult mathematical problem the structure of which we cannot even adumbrate.*

# General Relativity Theory

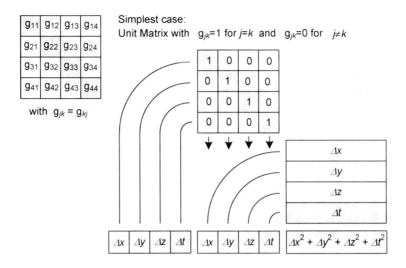

**Fig. 8.19** Extending the concept from Fig. 8.18 to the four-dimensional space-time continuum

Another indication of the difficulty of this problem is the fact that, in most university physics curricula, General Relativity Theory is not a required subject, but only an elective.

In his book, Max Born characterizes the results of the theory as follows:

*The metric field (which determines distances) and the field of gravity (which determines forces) are two different aspects of the same thing and both are represented by the ten quantities of the 4×4-matrix (see Fig. 8.19). Thus, Einstein's theory is a miraculous merger of geometry and physics, a synthesis of the laws of Pythagoras and Newton.*

In the two-dimensional space-time-continuum in Fig. 8.5, the lifelines of the objects considered are straight lines which indicate that the motion of these bodies stays unchanged as a consequence of their inertia. The lines of simultaneousness are also straight lines. In a plane, the shortest distance between two points is a straight line. Bending a surface has the effect that the shortest distances are no longer straight lines (see Fig. 8.15). Nevertheless, these shortest distances still have the same interpretation as in a plane: either they represent the life lines of bodies which are not influenced by any forces besides those caused by their inertia, or they are lines of simultaneousness. The square $\Delta s^2$ of the distance between two points on such a shortest connection still can be positive or negative.

With respect to the basic concepts of General Relativity Theory, we may say that most of them have already been introduced by Special Relativity Theory. The only new concept is the bending of the space-time-continuum which is due to the distribution of the masses. A four-dimensional space-time-continuum was already a basic concept of Special Relativity Theory, but there it is not bent and does not depend on the distribution of the masses.

Figures 8.15 and 8.16 show examples of bent spaces, the first being two-dimensional and the second being one-dimensional. These illustrations of bent spaces were possible only because, in both cases, the bent space could be represented as a continuous subspace of a space which has one additional dimension: the line is a subspace of a (two-dimensional) plane, and the surface is a subspace of the three-dimensional space. From this, we can conclude that an illustration of a bent four-dimensional space would be possible only if we could represent it as a subspace of a five-dimensional space which, however, is impossible. But in order to give you an idea of a space whose bending depends on a distribution of masses, I suggest that you think of a trampoline. As long as no bodies are placed on the jumping surface, it can be considered approximately as a plane. Now we assume that balls of different sizes and weights are placed randomly on the sheet which will cause the sheet to be bent. As a consequence of this bending, some of the balls may begin to move which will change the bending, and this again will have an effect on the motion of the balls. Obviously, the distribution of the balls influences the bending and the bending influences the distribution of the balls. If I had to find a mathematical description of a four-dimensional bent space whose bending depends on the distribution of masses and where the bending influences the distribution, I would begin with the problem of describing the case of the trampoline, and if I found a solution I would try to expand it formally into higher dimensions.

Since General Relativity Theory describes the interdependence of space, time and the distribution of masses, it is no wonder that this theory became of great importance for astronomy and cosmology. Astronomy is the science of the positions of bodies in the universe as they actually are and how they change over time, and cosmology is the science of the emergence and evolution of the universe. General Relativity Theory became relevant only recently for the development of systems which we use in our every day life. These systems contain clocks which are moving with high speed relative to each other, with some of them being on earth while the others are in satellites out in space. The high relative speeds between these clocks requires the application of Special Relativity Theory while the fact that the influence of gravity is not the same for all of the clocks, requires the application of General Relativity Theory. According to this theory, the time measured by a clock sitting on the ground differs slightly from the time measured by a clock in an airplane flying at an altitude of 30,000 feet. But this difference is so small that it could not be measured during Einstein's lifetime. But in the meantime, it became possible to compare the times measured with very high precision, and actually the difference has the value predicted by Einstein's theory. Since according to Newton's law of gravity (see Fig. 7.5) the force of gravity on ground level is stronger than in the height where the airplane is flying, the clock in the plane runs faster than that on the ground. Modern navigation systems based on ground positioning systems (GPS) which are installed in cars and tell the driver when to turn right or left are based on the comparison of the time measured by a clock in the car to the times measured by clocks in different satellites. Based on this information, these systems can deduce the actual position within an interval of a couple of yards. If the engineers who developed the systems would not have applied Einstein's theories, the uncertainty interval for the position would be much greater.

# Chapter 9
# How a Few Frog Legs Triggered the Origin of Electrical Engineering

When we were discussing historical events from the beginning of modern technology, a friend of mine used to say, "That was in the days when electricity was still made from frog legs!" With this statement, he referred to the accidental observation by the Italian Luigi Galvani (1737-1798) which actually triggered the tremendous development of electrical engineering. But before I discuss this observation in more detail, let's have a closer look at the time when all this happened. In those days, all of the scientists who were trying to find new laws for explaining physical phenomena were naturally well familiar with Newton's findings about mechanics. Newton's book had been published about one hundred years before Galvani made his observation. In the decades just before this observation, chemistry had made great progress. Until about 1750, most chemists still believed that the air is a simple element which does not play any role in chemical reactions. But within the next twenty years, it was found that air is a composite of different gases. Carbon dioxide was detected as its first component, and soon afterwards the other two essential components, nitrogen and oxygen, were found. At the same time, hydrogen was found. It is reasonable to assume that all scientists who made major contributions to the progress of knowledge about electricity and magnetism were familiar with actual knowledge in the fields of physics and chemistry. But in Galvani's days, there was not yet much knowledge about electricity and magnetism, and it was still completely unknown that these two are closely interrelated. As in the time of the old Greeks, certain phenomena were well known from every day experience. Certain blocks of iron attract needles and other small iron objects. And by rubbing bodies of certain materials together, attractive or repulsive forces between the bodies can be produced, or sparks can jump from one body to the other. Perhaps you had the experience that you winced because a spark jumped from your hand to a door knob after you walked over a certain flooring material, or your hair stood on end after you took off your woolen hat or used a certain comb.

You probably know that Benjamin Franklin (1706-1790) invented the lightning arrester. He had the bright idea that lightning is nothing but an extreme form of the sparks which jumped from his hand to the door knob. Nevertheless, electrical and magnetic phenomena were still mysterious and unexplained when Galvani made his observation. From that event, it took only eighty years until all electrical and magnetic phenomena and their interrelations were completely understood and the corresponding laws had been expressed by mathematical formulas of surprising clarity. Even the work of today's electrical engineers is based on a theoretical foundation which was completed in 1860. The rock wall which I shall now tackle with you begins with Galvani making his observation in the year 1780, and ends with Maxwell writing his equations in 1860. And at the bottom of this rock wall, we pass a plate with the inscription: "Nine years from now, the French revolution will begin."

## The Tremendous Consequences of Accidental and Simple Observations

Luigi Galvani was a professor of anatomy and gynaecology at the University of Bologna which is one of the two oldest universities in Europe. Galvani performed experiments with frog legs because he wanted to find out how muscles are controlled by nerves. One day, he accidentally observed that the frog legs twitched whenever they touched the iron grid of the window where they were hanging from copper hooks. Galvani published his observation in a journal, and thus Alessandro Volta (1745-1827) got to know about it. Volta was a professor of physics and had already done many experiments with electricity produced by rubbing suitable materials together. He had the idea – which later proved to be correct – that the cause of the twitching of the frog legs was purely physical. In order to check whether his assumption was correct, Volta made experiments with a set-up which contained only those elements from Galvani's structure, which Volta believed were the relevant ones. Instead of a frog leg, he used a piece of cardboard which he dipped into salt water, and instead of the iron window grid and the copper hook he used disks of these metals. The sandwich composed of the two metal disks outside and the cardboard inside showed the same effects as a body which had been charged by rubbing. Volta could intensify the effects by piling up more of such sandwiches and by replacing the original metals with others. He had invented an electrical battery. It worked perfectly, but nobody could explain why. It took more than one hundred years until the explanation was found: it is based on quantum theory.

Of course, Volta published his results, and from then on many scientists performed experiments using such batteries. One of them was the Danish scientist Hans Christian Oersted (1777-1851). One day in the year 1820, there was, accidentally, a magnetic compass lying next to his experimental set-up – perhaps Mr. Oersted owned a sail boat. He observed that whenever he closed the circuit and allowed the current to flow, the magnetic needle of the compass jumped to a new direction and did not return to its original direction until the current was turned off. You certainly can imagine that Mr. Oersted was quite surprised when he realized that his set-up for experiments with current from a battery could have an effect on the compass, although there was no connection; there was only air in between. Oersted, of course, realized that his observation was of great importance, since it indicated that there was a strange interrelation between electricity and magnetism. Therefore he published his findings immediately. Among those who read this publication with great interest were the Frenchman André Marie Ampère (1775-1836) and the Englishman Michael Faraday (1791-1867). Mr. Ampère had done many experiments before using Volta type batteries, and he had observed that there is a force between two parallel wires when currents are flowing through both of them. Whether the force was attractive or repulsive depended on whether the currents in the two wires went in the same direction or in opposite directions. This effect is the basis of the actual definition of the unit Ampere for the electric current – I shall come back to this later.

I think it is helpful for you to realize how little was known at that time about the essence of electrical current. Nobody could see the current flowing, and nobody could know what it might be that was flowing. Somehow, the scientists had the vision that a battery works like a pump which presses something through a closed circuit. In this analogy, the wires correspond to pipes through which a gas or a liquid can flow. Although nobody knew how a battery could work as a pump, the analogy between an electric circuit and a closed system with flowing water was quite useful. However, in contrast to flowing water where we can see in which direction it is flowing, no information was available for determining the direction of flow of the electric current. But the analogy to flowing water required that the electric current had a direction, and therefore the direction was defined arbitrarily. Such a definition had to refer to something unsymmetrical in the electrical circuits. The different metals at the two ends of the battery were easy to discern and formed the basis for specifying the current direction. Referring to the iron and the copper which previously had caused the twitching of the frog legs at Mr. Galvani's window, the arbitrary definition says that the current leaves the battery at its copper end and enters it at its iron end. Today, however, we know that the electric current is a stream of flowing electrons, and they don't flow according to the definition, but opposite to it. When this was detected – which happened eighty years after the original definition was chosen - it had to be decided whether the original definition should be reversed or retained. Since a reversion would have caused much confusion, the original definition of the direction of the electric current was retained. This was done by defining the electrical charge of the electrons as being negative. It is interesting to know that the term *electricity* was used before the existence of electrons was discovered – electron is the Greek word for amber which is one of the materials that can be electrically charged by rubbing, and this was known even at the time of Socrates.

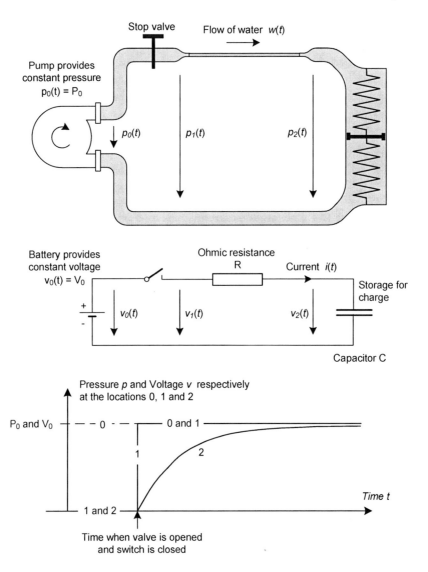

**Fig. 9.1** Analogy between the flow of water and the flow of electrical current

Fig. 9.1 illustrates the analogy between flowing water and electric current. This analogy is not only helpful in teaching, but determines the view of professionals in their daily work. In the upper part of the figure, you see a system filled with water. In the middle section of the figure is represented an electric circuit whose structure is in all details analogous to the water system. At first, we consider the flow of the water. You should assume that the system is horizontal in order to eliminate the influence of gravity. If the system stood vertically, gravity would affect the pressure

Accidental Observations                                                                                      211

in the system. The pump tries to make the water flow clockwise through the closed system which is possible only if the stop valve is open. There are two components in the system which prevent the water from flowing with a speed which is determined only by the pump, and these components are the thin resistor pipe which sits next to the valve and the cylinder with a piston which sits between two springs. These two springs are assumed to be exactly of the same type such that the piston will sit in the middle of the cylinder if the difference of pressure between the two entrance points is zero. This situation will be in effect as long as the stop valve is closed. When the valve is opened, the water will begin to flow. The water leaving the thin pipe will enter into the cylinder, the piston will leave its middle position and water will leave the cylinder through the cylinder's opposite opening. One spring will be stretched, while the other will be compressed. This will cause an increase in the pressure $p_2$ compared to $p_1$ until the value of $p_2$ has reached the pressure $p_0$ of the pump. Then, the water will stop flowing since the pressure difference between the two ends of the thin pipe, $p_1$-$p_2$, will be zero. In this situation, the values of $p_0$, $p_1$ and $p_2$ will be equal.

In the electrical circuit, the battery corresponds to the pump, the ohmic resistor corresponds to the thin pipe and the two metallic plates which are standing very close to each other correspond to the cylinder with the piston and the springs. Although the symbol for the ohmic resistor looks more like a thick short wire, you should think of a long and thin wire through which the electric current has to be forced. The term ohmic resistor refers to the German physicist Georg Simon Ohm (1789-1854) whose name is known because of the so-called *Ohm's law*. Amazingly, many people, even those who have degrees in electrical engineering, confuse Ohm's law with the definition of an ohmic resistor. You will not make this mistake if you never forget that a law of physics is something completely different than a definition of a physical concept. Laws are found as results of experiments, whereas definitions are the result of arbitrary human decisions. An ohmic resistor is defined as an electrical circuit element having two connectors where the current which enters the resistor through one connector and exits through the other connector is proportional to the voltage between the two connectors. I will discuss the concept of voltage later. That means that a doubling of the voltage leads to a doubling of the current. The symbol R, taken from the word resistor, is used as the factor which determines the relation between voltage and current. Thus, the equation R=$v/i$ is not Ohm's law, but the definition of an ohmic resistor. Mr. Ohm found "his" law by experimenting with batteries and wires. The law says that long thin wires behave approximately like ohmic resistors whose value R can be doubled by doubling their length or by halving their cross-sectional area. That's the reason for taking a long thin pipe as the analog of an ohmic resistor.

In professional language, the structure consisting of two plates facing each other with a very small separation is called a *capacitor*. Like the piston in the cylinder which doesn't allow a continuous flow of water in one direction, the gap in between the two plates doesn't allow a continuous flow of electrons in one direction. In the case of the cylinder, a certain amount of water can enter on one side of the piston only if at the same time the same amount of water leaves the cylinder on the other side of the piston. This applies in an analogous way to the capacitor: a certain number of electrons, i.e., a certain amount of electrical charge, can be brought onto one of the plates only if, at the same time, the same amount of electrical charge is flowing away from the other plate. The voltage between the two plates can differ from zero only if the two plates contain different amounts of charge. Now we assume that the voltage $u_2$ is zero when the switch is closed, i.e., that there is no charge on the capacitor. Closing the switch will cause a current to flow which brings charge to one plate and takes charge away from the other plate. This will result in an increase of the voltage $u_2$ which will continue until the capacitive voltage has reached the voltage of the battery. Then, all three voltages $u_0$, $u_1$ and $u_2$ will be equal, and no current can flow any more because the voltage difference $(u_2 - u_1)$ which forces the current through the ohmic resistor is zero.

It should have been clear to you that my descriptions of the processes in the water system and the electric circuit were almost the same. Only the flowing medium was different, with water in the one case and electrons in the other case. Therefore, the driving forces had to be different, namely the pressures $p_0$, $p_1$ and $p_2$ for pushing the water, and the voltages $u_0$, $u_1$ and $u_2$ for pushing the electrons. Thus, you will not be surprised that the curve at the bottom of Fig. 9.1 applies to both the water system and the electric circuit. The horizontal line labeled "0" represents the constant pressure $P_0$ of the pump or the constant voltage $U_0$ of the battery. As long as the valve stays closed or the switch stays open, the pressures $p_1$ and $p_2$ behind the valve, as well as the voltages $u_1$ and $u_2$ following the switch, are still zero. This is a consequence of our assumption that, at the beginning, the two compartments of the cylinder, as well as the two plates of the capacitor, are symmetrically loaded. At the moment when the valve is opened or the switch is closed, the pressure $p_1$ as well as the voltage $u_1$ jumps to the value provided by the pump and the battery, respectively. But the pressure $p_2$ as well as the voltage $u_2$ cannot jump to a higher value since this would require that, in zero time, the cylinder and the capacitor, respectively, get unsymmetrical fillings. But this is inhibited by the thin pipe and the ohmic resistor, respectively. The initial slope of the curve in Fig. 9.1 is given by the initial flow of water and the initial current, respectively, but these initial values will immediately begin to decrease because of the rising values of $p_2$ and $u_2$. This will cause a decrease in the differences $(p_2 - p_1)$ and $(u_2 - u_1)$, respectively, which are the driving forces of the flow through the thin pipe and the ohmic resistor, respectively. Therefore, the pressure $p_2$ and the voltage $u_2$, respectively, will increase with decreasing slope according to the line labeled "2" in Fig. 9.1.

Although the explanations given above make it plausible that the curve representing the rise of the values $p_2$ and $u_2$, respectively, looks as shown, we have to go into some mathematics if we want to know the exact formula for this curve. Fig. 9.2 shows how this formula can be deduced. Since the water system is closer to our daily experience, I did the calculations with the letter $p$ for the pressure, but alternatively, I could have chosen the letter v for the voltage and still come to the same structure for the formula. At the left of the diagram, you find an equation whose two sides were obtained by simply translating our knowledge about the behavior of the thin pipe and the cylinder with the piston and the springs from natural language into mathematical expressions. The left side expresses the fact that the higher the pressure $(P_0 - p_2)$ is which presses the water through the thin pipe, the more water will flow. The higher the value of R, i.e., the longer or thinner the thin pipe is, the less water will flow through per time unit. The right side of the equation expresses the fact that the greater the amount of water entering the cylinder per time unit, the greater the increase of the pressure $p_2$ per time unit. The symbol C is determined by technical characteristics of the cylinder, e.g., the surface of the piston and the stiffness of the springs. The letter C refers to the word capacitor. This word might mislead you to assume that such a capacitor has a fixed capacity, but what is really meant is that the more you force water to flow, the greater the volume of water gets into it. I recommend that you associate C with the idea of elasticity: the higher the value of C, i.e., the more elastic the springs are, the more water is needed to get a certain increase of the pressure $p_2$.

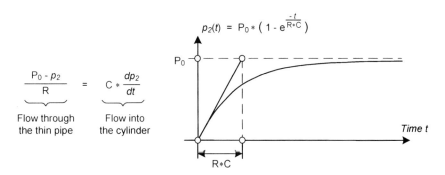

**Fig. 9.2** Mathematical properties of the curve in Fig. 9.1

The equation is a differential equation which does not explicitly represent the desired formula for the curve, but is only a statement of certain facts from which the desired formula follows. I assume that you haven't learned how to solve such differential equations; at least you cannot have learned it from this book. But I have learned it, and therefore I wrote the result at the top of the diagram in Fig. 9.2. From this formula, an interesting feature of the curve can be derived: the initial slope of the curve is given by the fraction $P_0/R*C$. Once the curve has been found by adequate experimentation, the product R*C can be obtained by simply measuring the corresponding distance, and from this, the value of C follows if the value of R is already known. This procedure was performed almost two hundred years ago by the physicists who performed experiments with Volta type batteries. They discovered that two plates facing each other with a very small separation distance have the behavior described above. And they found out that the value of C can be doubled by doubling the surface area of the plates or by halving the distance between them.

In the paragraphs above, I used the term *voltage* without giving an explanation about what that is. The only hint I gave was the analogy between pressure and voltage. When the physicists at the time of Volta thought about voltage, which at that time was not yet called voltage, but electrical force, they referred to the force between the two plates of a capacitor. While the force between two parallel wires depends on the distance between the wires and on the current which flows through them, the force between the two plates of a capacitor depends on the distance between the plates and the voltage across them. Only five years after the twitching of the frog legs had been observed by Mr. Galvani, the French nobleman Charles Auguste de Coulomb discovered that the force between two bodies which carry electrical charges can be expressed by a formula which has the same structure as Newton's law of gravity (Fig. 7.5). The so-called *Coulomb's law* is

$$\text{Force of repulsion} = \chi * \frac{\text{(Charge on body 1)} * \text{(Charge on body 2)}}{\text{(Distance of the two bodies)}^2}$$

In the case of the law of gravity, the fraction never can have a negative value, since masses are always positive. But electrical charges can be positive or negative, and therefore, the result of Coulomb's law can be positive or negative which means that the force can be that of attraction or repulsion. While at Volta's time, the definition of the unit of voltage referred to the force between two bodies carrying electrical charges, today the unit of voltage is coupled elegantly to the unit of current. But before I can explain this to you, we still have to climb to a higher point.

The analogy between flowing water and flowing electrical charge is helpful only as long as magnetic phenomena are not be taken into account. Now, we shall consider the relationship between electrical and magnetic phenomena. I mentioned earlier that it was a Mr. Oersted from Denmark who accidentally detected that there is such a relationship. When Michael Faraday became aware of this, he began a long series of experiments which finally lead to our present understanding of electromagnetic phenomena. Fig. 9.3 illustrates the essence of Faraday's experiments.

You see a wire through which a current is flowing upward. This wire is cutting through the grey shaded horizontal plane on which have been drawn concentric

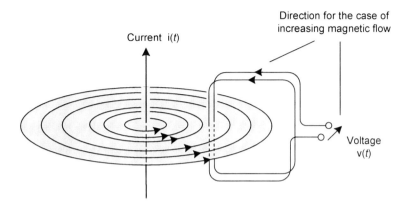

**Fig. 9.3** Experiment showing the connection between electrical and magnetic phenomena

circles. The center of the circles is at the point where the wire cuts through the plane. In Mr. Faraday's experiment, the plane was a piece of cardboard onto which he had scattered iron filings before he turned on the current. At first, these filings were lying without forming any regular pattern. At the moment when the current was turned on, the filings immediately adjusted themselves and formed the circles shown. From this observation, Faraday concluded that the current generates a so-called *magnetic field* which lies concentrically around the wire. He used the term *field* for the filling of the space with invisible properties, and by this ingenious idea he laid the groundwork for a new way of thinking and talking for physicists, a way which since then hasn't lost its importance.

It is quite plausible that the magnetic force which affects the direction of the filings depends on their distance from the wire through which the current is flowing, and that this force is lower at greater distances. The idea to perform the experiment with the filings didn't require a genius mind since it suggests itself as a consequence of Oersted's observation. But Faraday also performed another experiment which doesn't suggest itself at all. The right part of Fig. 9.3 shows a coil which is in such a position that some of the circles of the magnetic field run through its center. Faraday was very much surprised when he found that a short pulse of voltage appeared between the two ends of the coil whenever the current was turned on or turned off. The direction of the voltage depended on whether the current was turned on or off, and the value of the voltage was higher when the coil had more turns. Of course, these observations motivated him to try variations of the experiment, and thus he found that he could leave the current turned on and still obtain voltage pulses: he only had to move the coil, either by pushing it closer to the wire or pulling it away from it, or by changing the angle between the plane of the coil and the plane of the magnetic field (the horizontal plane). The more turns the coil has, the higher was the value of the voltage. Today, the law which describes all these effects and which already was expressed correctly by Faraday himself is called "the law of induction." It says that the voltage "induced" in a coil is proportional to the speed at which the magnetic flow through the coil is changed. However, at this point in my description, I cannot yet tell you the definition of a magnetic flow; I shall present this later.

From the fact that currents, voltages and the related forces are directed quantities, it follows that the magnetic field must have a direction, too. In Fig. 9.3, I indicated the direction by placing arrows on the concentric circles. But how could I decide what the direction was to be? There is no "magnetic something" which we could see flowing. Therefore, as was the case with the direction of the electrical current, the direction of the magnetic field can be defined arbitrarily. The definition has been based on the right-hand-rule which is known from mechanics where it is used to describe the relationship between the directions of the torque, the angular speed and the angular momentum. Applied to a current which is surrounded by a magnet field as shown in Fig. 9.3, the thumb of the right hand must point in the direction of the current. Then the remaining four fingers point in the direction of the magnetic field.

The direction of the voltage which is induced in a coil by the change of the magnetic flow is also obtained by the right-hand-rule. If the thumb of the right hand points in the direction of the increasing magnetic flow, the four fingers point in the opposite direction of the current which would flow if the coil were closed. The induced voltage which is measured between the two ends of the unclosed coil has the same direction the current would have if the two ends were connected together to allow a current to flow.

Between 1831 and 1838, Michael Faraday published many reports describing his experiments and the corresponding conclusions. Later, he summarized the individual papers in two thick volumes with a total of more than 2,000 pages [FA] which contain many precise drawings of the experimental set-ups and results, but not a single formula. Faraday had no college education and was not familiar with higher mathematics. A later author once even made the derogatory remark that Faraday hadn't got beyond the rule of three. But this view is opposite to the high esteem the ingenious James Maxwell had for Faraday when he wrote that Faraday's ideas went far beyond the visions of narrow-minded mathematicians.

## How Mr. Maxwell Transferred His Ideas from the Bath Tub to Free Space

It was quite clear that it wouldn't take long until someone would successfully express in mathematical language the laws which Faraday had described in natural language. Actually, it was James Clerk Maxwell (1831-1879) who expressed these laws in a form which, even now, most physicists and electrical engineers think is absolutely perfect [MA]. The Austrian physicist Ludwig Boltzmann (1844-1906) was so excited about the elegance of Maxwell's equations that he began a lecture on this subject with a citation from "Faust," the best known piece of poetry from the German poet Goethe: "Was it a god who wrote these symbols?"

In the preface to his book which was published in 1873, Maxwell praised the great importance of Faraday's preparatory work. In contrast to some malicious authors of later times, Maxwell didn't mock Faraday's lack of education in higher mathematics. In fact, he wrote, "When I made progress at studying Faraday's papers, I became aware that his way of looking at the phenomena actually was mathematical, although the ideas were not written in the usual form of formulas. I saw that his methods were well suited for being expressed in standard form with mathematical symbols. With his mind's eye, Faraday saw lines of force running throughout all of space where, up until then, mathematicians had seen only the effect of forces over distances. Faraday saw a medium where the others saw only a distance." With these statements, Maxwell emphasized the importance of the change represented in Faraday's viewpoint. Since Newton had published his law of gravity, it was quite common to accept a remote action, i.e., to believe that forces can have effects over great distances. Now this concept was replaced by the so-called *close-range effect*.

After these preliminary remarks, we have reached the point where we can start to climb Maxwell's rock wall. When the gentlemen Volta, Ampère and Ohm were searching for laws describing the relationship between current and voltage, they could base their considerations completely on the analogy between flowing electricity and flowing water. They considered only currents flowing through wires, and according to the analogy, the elements corresponding to the wires were the pipes. But Maxwell had to consider phenomena which occur in the entire three-dimensional space in which wires do not exist except at some particular locations. Look again at Fig. 9.3; there are only two wires, the vertical one with the current and the other one which has the form of a coil. Motivated by Faraday, Maxwell saw something streaming along all the lines to which I added an arrow, although in some cases, it couldn't be a current. Looking at Fig. 9.3, Maxwell would have seen three different kinds of flow, namely the current flowing upward through the vertical wire, the magnetic flow along the concentric circles, and a flow along the arrows belonging to the coil. For this third kind of flow, which is called *dielectrical flow*, it doesn't matter that the coil is not closed; it can flow as if there were a bridge between the two ends of the coil.

Even before Maxwell began his study of Faraday's papers, he was familiar with the mathematical methods for describing streaming media, although at that time the media he considered were fluids. Following Maxwell's way upward towards the top of the mountain called *electromagnetism*, we also have to acquire some knowledge about the methods for describing streaming media in three-dimensional space. Therefore, for a while, we now leave the world of electromagnetic phenomena and restrict our consideration to a space filled with streaming water. If the water doesn't stream at all, each particle has a fixed location which it never leaves. If there is some kind of streaming, some particles may still stay at their initial location, but at least there will be some which move at certain speeds in certain directions. Of course, there is the possibility that all particles move at the same speed in the same direction, but this special case is not what we are interested in. If some of the particles carried small lights, we could see the paths along which they move. As long as no particles enter or leave the space considered, the paths cannot have beginnings or ends. In this case, the streaming field is said to be without sources or sinks. This, however, is not the case either with a swimming pool or with the ocean. In both cases, there is rain and evaporation. In addition, there are pipes connected to the pool for adding or removing water, and there are rivers which flow into the ocean. Therefore, the question of where the sources and sinks are located in the space is considered of great importance. The question of whether there are turbulences, where they are and how fast they are spinning is of similar importance.

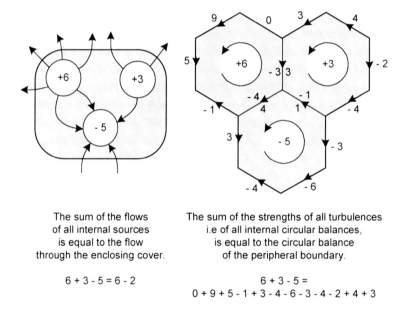

The sum of the flows
of all internal sources
is equal to the flow
through the enclosing cover.

6 + 3 - 5 = 6 - 2

The sum of the strengths of all turbulences
i.e of all internal circular balances,
is equal to the circular balance
of the peripheral boundary.

6 + 3 - 5 =
0 + 9 + 5 - 1 + 3 - 4 - 6 - 3 - 4 - 2 + 4 + 3

**Fig. 9.4** Fundamental properties of sources, sinks and turbulences

Figure 9.4 illustrates the mathematical concepts for describing sources and turbulences. Assume that you are looking from above down to the surface of the water in a pool. The two shaded areas shown in Fig. 9.4 do not represent the entire pool, but are only sections which have been selected arbitrarily. Thus, the borders are not real walls; they just define the part of the pool which actually interests us. In the left half of the figure, sources and sinks are considered. Here three pipes ending at the bottom of the pool have the relative yields (flow rates) of 6, 3 and -5. A source with a negative flow is called a sink. The sum of these yields is 4 which means that more water is entering through the two pipes than is leaving through the sink. Now the border of the whole shaded area is considered the opening of a bigger pipe, and the total yield or flow of this pipe will be the sum of the flows of the inner pipes.

Turbulences are circular motions of water and are considered in the right half of the figure. In order to focus your view on the essentials, I made some simplifying assumptions. I assumed that the paths of the particles could have sharp bends, and that their speed could change abruptly. Although this doesn't correspond to reality, the concepts presented are still correct. Whether or not there is turbulence within a closed border can be decided by computing the so-called *circular balance*, i.e., the sum of weighted speeds along the closed border which makes up the outside border or periphery. If the speed were constant along the entire periphery, the circular balance would be the product of this speed and the length of the periphery. The areas considered in Fig. 9.4 are hexagons where all six edges having the same length. Thus, I could assume that the speed stays constant along an edge, and the balance corresponds to the sum of the numbers written next to the edges. Since the speeds are directed, it was necessary to decide which direction should be used as the positive one. I decided that turbulences spinning counter-clockwise would have a positive balance. Then the vector of the angular velocity, which is determined by the right hand rule, will point into our face, since this is where the thumb is pointing when the remaining fingers are wrapped counter-clockwise. A speed belonging to a border between two hexagons cannot have the same sign in the balances of both hexagons, since for one hexagon it points clockwise and for the other one it points counter-clockwise. By combining the three hexagons into a larger area, we get a new enclosure for which we can also compute the circular balance. This is equal to the sum of the balances of the hexagons.

Until now, we assumed that the locations and directions of all sources, sinks and turbulences were such that everything could be seen by looking onto a plane from above. But now, we have to overcome this simplification and be more realistic since sources and sinks can be distributed all over the three-dimensional space, and turbulences can have any directions. The most general concept for describing a distribution of sources and sinks within a three-dimensional space is the so-called *divergence* which is an abstract property of the points in the space. Divergence is the limit of the ratio between the flow through an enclosing surface and the volume of the space enclosed within the surface for the case when the volume becomes infinitely small. Since the flow has the unit "unit of the flowing media per time", e.g., gallons per second, the divergence which represents a flow per volume will have the unit "unit of the flowing media per time and volume." If the divergence has a negative value, it describes a sink. A detailed analysis, which I shall omit, leads to the result that the divergence is equal to the sum of the derivatives of the three components of the streaming vector ($S_x$, $S_y$, $S_z$) with respect to their own directions, i.e., div(S) = $\partial S_x/\partial x + \partial S_y/\partial y + \partial S_z/\partial z$. You may wonder why the derivatives in this formula are not expressed with the normal letter d, but with the strange round $\partial$. This indicates that the function which is derived is not only a function of the variable for which the derivative is taken, but of at least one additional variable which is considered to be a constant in the derivative. In streaming fields, all three components of the stream are functions of four variables, i.e., the values of $S_x$, $S_y$ and $S_z$ depend on the coordinates $(x, y, z)$ of the point in the space to which the actual stream components belong, and of the actual point in time $t$.

While the distribution of sources and sinks can be captured by an undirected property of the points of the space, the distribution of turbulences requires that a directed quantity be assigned to the points. A turbulence has an angular velocity, and this has a direction which is defined by the right hand rule: when the fingers go around the axis of rotation in the direction of the turbulent flow, the thumb points in the direction of the angular velocity. Look again at Fig. 9.3 and assume that the concentric circles show the direction of a turbulence in a pool; then the wire would be the axis of rotation, and the direction of the angular velocity would correspond to the direction of the current. The directed quantity which is assigned to the points in order to capture the distribution of turbulences is called *rotation*. It has the direction of the angular velocity, and its value is the limit of the ratio between the circular balance of the flow along the circumference of a circle around the axis of rotation and the area of the circle whose radius becomes infinitely small. The direction for computing the circular balance can be chosen arbitrarily since the actual direction of the angular velocity will be indicated by the sign of the result: if the result is positive, the right hand rule applies; otherwise, the direction is opposite.

Since a turbulence can have any direction in the space, the rotation vector must be described as a set of three components, one for each of the three directions of the axes x, y and z of a coordinate system. Fig. 9.5 illustrates the process of computing these three components, each of which is perpendicular to a square face of the cube. The circular balance is the sum of the values belonging to the four edges of this square. The right half of the figure is of interest only for those readers who really want to know in detail how the components of the rotation vector are computed. This computation uses differential equations, and from this it follows that we have to assume that the edges of the cube are infinitely short. The stream cannot change abruptly from one point to another nearby point, and therefore, the difference between the streams along two parallel edges of the cube will decrease when the distance between these edges is decreased. If the stream differences of both pairs of parallel edges of a square were zero, the circular balance would be zero, too, since then the positive contribution of one edge would always be cancelled by the negative contribution of its parallel partner. The stream difference between a pair of parallel edges, related to the area of the square, is equal to the derivative of the stream component along the edge with respect to the distance between the two edges. As an example, we look at the vertical edge in the rear of the grey shaded square. It has the distance dy from the z-axis. Its contribution to the circular balance of the grey shaded square is determined by the derivative of the z-component of the stream in the direction of the distance dy. At the bottom of Fig. 9.5 you find the sum which we get if we compute the circular balance along the edges of the grey shaded square. The result is the value of the y-component of the rotation vector.

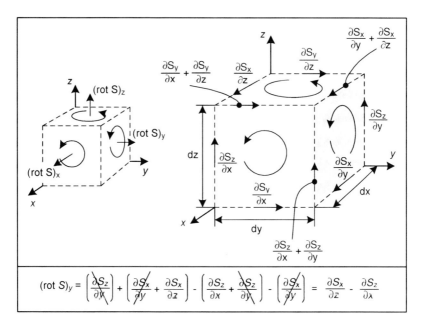

**Fig. 9.5** Computing the three components of the rotation vector

There is a third concept besides divergence and rotation related to streaming fields. Although it is not used in Maxwell's equations, I shall present it here in order to complete the subject. I presented the concept of potential (see Fig. 7.6) in connection with the law of gravity. There I said that a potential field can be considered like a mountainous region which is characterized by two types of lines, i.e., the lines which connect points of equal height, and the lines which connect the points of maximal slope. The lines of the second type are those along which water would flow downhill. The potential field is not directed, i.e., the potential p(x, y, z) is an undirected quantity assigned to the points of the space. But the maximal slope, which is well defined for each point of the space, has both a value and a direction and therefore can be considered a streaming field. The slope has three components $(S_x, S_y, S_z)$ and is called the *gradient* of the potential p. If the potential is given, the gradient can be obtained by computing the derivative of the potential in the three directions of the coordinates, i.e., grad(p) = $(S_x, S_y, S_z)$ = $(\partial p/\partial x, \partial p/\partial y, \partial p/\partial z)$.

At this point, I must remind you that we have not yet returned to the subject of electromagnetism, but are still strolling through the area of flowing media of any kind. The diagram in Fig. 9.6 shows the relationships between the fields which have been discussed in the paragraphs above. There are two types of fields of undirected quantities and two types of fields of directed quantities. The grey shaded node represents the streaming field. The other three fields have been introduced in reference to the streaming field. If the potential field is given, the streaming field is obtained by computing the gradient. And from a given streaming field, both the field which describes the distribution of the yields of the sources and the field which describes the velocities of the turbulences can be derived.

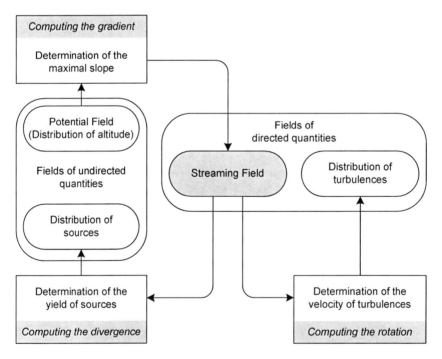

**Fig. 9.6** Relations between a streaming field and related fields

# Maxwell's Laws

As has been shown, the formulas for computing rotation and divergence contain the derivatives of the three components of the streaming field ($S_x$, $S_y$, $S_z$). Since there are three different derivatives for each component, corresponding to the directions of the coordinates $x$, $y$ and $z$, there are all together nine different derivatives to be considered. Three of them occur in the formula for the divergence, the other six in the formula for the rotation according to Fig. 9.5. Like everyone else, mathematicians try to avoid unnecessary work, and therefore they looked for a formal scheme which would allow them to obtain the formulas for computing gradient, divergence and rotation without much thought. They finally found a rather ingenious combination of the symbolism for derivatives and the structure of matrix multiplication. This is presented in Fig. 9.7. If the cells of the vectors and the matrix were filled with numbers, the figure would represent at the left the multiplication of a vector with a single number, in the middle the scalar multiplication of two vectors, and at the right the computation of the perpendicular product of two vectors (see Fig. 3.9). However, the structures in Fig. 9.7 cannot really represent multiplications since the rectangle with the thick border, which has the position of the first factor in the multiplication structures, is not a vector. The entries in the three cells of this rectangle are not numbers but requests to compute certain derivatives. The functions from which the derivatives must be computed can be found by following the 90 degrees curves to the cells in the structures which have the position of the second factor. The derivatives then have

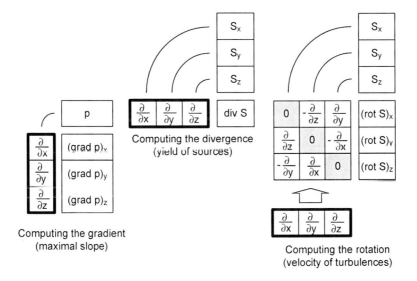

**Fig. 9.7** Computing the gradient, divergence and rotation using the matrix multiplication pattern

to be added as if they were the component products in a real matrix multiplication. It is amazing how this modification of the interpretation of the matrix multiplication formalism simplifies and unifies the formulas for computing the gradient, the divergence and the rotation.

I can imagine that the wall we have just climbed was more difficult for you than the ones you had to cope with earlier. There is still a short distance ahead of us before we reach the plateau of Maxwell's laws of electromagnetism, but we have already reached a rather high rest area where we can linger for a moment. A mountaineer who reaches such a rest place will certainly look back to the lower regions from which he came. Our position is already so high that we have to use binoculars in order to see the frog legs of Mr. Galvani or the magnetic compass of Mr. Oersted which are far down in the valley. From our rest area, we now take the same route to the top which James Maxwell took. All mountaineers who came later saw no reason to search for a different route.

After having discussed Fig. 9.3 which illustrates the essential experiments by which Michael Faraday found the most important relationships between electrical and magnetic quantities, we left the area of electromagnetism and considered concepts for capturing three-dimensional streaming fields of fluids or gases. Now we return to the area of electromagnetism. In my comment about Fig. 9.3, I said that if Maxwell would have looked at this figure with his mind's eye, he would have seen three different kinds of flow, namely the flowing current through the vertical wire, the magnetic flow around this wire and the dielectric flow along the coil. Maxwell had the idea that any streaming field requires a field of "pressure" in analogy to water which will flow through a pipe only if it is pressed through the pipe. When the pressing field is given, the strength of the streaming field will depend on the *permeability* of the space with respect to the flowing media. Permeability is just a physical constant whose unit compensates for the difference between the units of pressure and flow.

In the simple example of streaming water, the unit of flow is either "volume per time and area" or "mass per time and area", and the unit of pressure is "force per area." Then the unit of the permeability which is (unit of flow)/(unit of pressure) will be either "volume per time and force" or "mass per time and force." When the standard units of meter, kilogram and second are used, the unit of permeability for flowing water will be either $(m^2*s)/kg$ or $s/m$, respectively. You should not try to find plausible interpretations of these units; even experts just use them in their computations and rely on the fact that they have been deduced correctly.

Now we are ready to interpret Maxwell's equations which are written in the shaded fields of Fig. 9.8. The equation $S = \delta*P$ in the top field does not belong to the set of Maxwell's equations; I introduced this equation as a pattern for the interpretation of the first three of Maxwell's equations which are represented in the two frames underneath. The letter $J$ stands for the density of the electrical current,

# Maxwell's Laws

and its unit is "current per area." The symbol for the unit of the current is A which is the first letter of the name Ampère, and with this, the unit of $J$ is $A/m^2$. The pressing field which forces $J$ through the space is called the electric field and its symbol is $E$. However, this field $E$ not only provides the force for moving electrical charges through wires or space, but also forces the dielectrical stream $D$. The permeabilities which express the relation between the pressing field $E$ and the resulting streams $J$ and $D$ are symbolized by the Greek letters $\gamma$ and $\varepsilon$, respectively. You should not brood over the question of what kind of stream $D$ might be. It is an abstract concept called *dielectrical flow* which Maxwell introduced, since without it, he could not completely capture all the phenomena which occurred in Faraday's experiments.

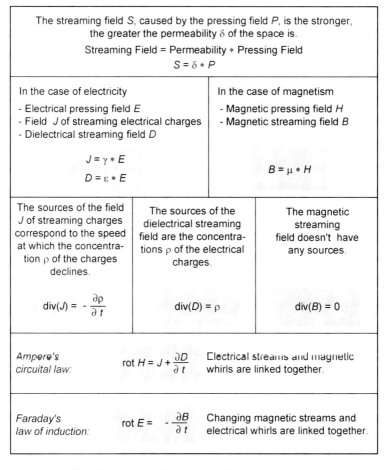

**Fig. 9.8** Maxwell's equations for electromagnetism

In the case of magnetism, there are no different kinds of streams, but only one magnetic flow which is symbolized by the letter $B$ and which is forced through the space by the field $H$. The ratio $B/H$ is given by the magnetic permeability µ.

The three equations just considered do not represent physical laws which could be checked by experiments, but are definitions which Maxwell introduced as a basis for the equations which finally will describe the real laws. Once the streaming fields $J$, $D$ and $B$ have been introduced, the question comes up about what the sources of these streams might be, i.e., what div($J$), div($D$) and div($B$) might be. The answers are given by the three equations in the middle of Fig. 9.8. The equation on the left side states a quite logical fact: a small enclosed part of the space which contains electrical charges in an actual concentration ρ can be the source of a flow of charges through its enclosure only if the concentration ρ decreases over time. Such a logical explanation does not exist with respect to the equation describing the divergence of $D$. This equation is just an ingenious approach by which Maxwell achieved the completeness of the theory. With respect to the magnetic flow $B$, no experimental results had been observed which would have required the introduction of magnetic sources. Therefore, the divergence of $B$ is zero everywhere and at all times.

The laws which describe the relationship between electricity and magnetism as they follow from Faraday's experiments are expressed by the two equations in the frames at the bottom of Fig. 9.8. These two laws have already been described in natural language in my comment concerning Fig. 9.3. In an interpretation which refers to Figure 9.3, the first equation, which is called Ampere's circuital law, captures the relationship between the current $i$ and the magnetic field $H$. Additionally, this law says that it is not absolutely necessary to have a real flow of electrical charges, but that the same effect is obtained by an increase of the dielectrical flow $D$ over time. With respect to Fig. 9.3, the second equation, which is called the law of induction, captures the relationship between the voltage induced in the coil and the speed at which the magnetic flow through the area of the coil decreases over time.

Some years ago, when I was walking through Manhattan, I came by a construction site where a whole block of houses had been torn down in order to put up a new building. At the time, no part of the new building was yet to be seen except for a big and deep hole. At the bottom of the hole, construction workers were busy making the foundation, using tons of concrete and iron. The reason I am telling you this is the fact that Fig. 9.8 shows a foundation which has a much greater significance than all the foundations underneath buildings. Fig. 9.8 actually shows the foundation for the entirety of modern electrical technology. While the foundations of buildings deteriorate over time – the concrete crumbles and the iron rusts – the foundation shown in Fig. 9.8 doesn't show any signs of wear and tear although it already has lasted one hundred and fifty years.

As in the case of foundations of real buildings, the foundation in Fig. 9.8 doesn't tell us anything about the building which finally has been erected above. This building consists of two very high towers, one which is the tower of electrical power technology, and the other one which is the tower of information technology. But before I provide you with a closer look on these towers, I have to get back for a moment to the Lorentz transformation which I discussed in detail in Chapter 8 on relativity theory. There I told you that Antoon Hendrik Lorentz found this transformation some years before Einstein. By a quite formal analysis of Maxwell's equations, Lorentz discovered that the form of these equations stays the same when the variables $x$, $y$, $z$, $t$ and $\rho$ are substituted by $x'$, $y'$, $z'$, $t'$ and $\rho'$ which he obtained by a certain transformation from the original variables. Einstein later discovered that this transformation has a much more general interpretation and tells us something about the essence of space and time.

## How the Feasibility of High Voltage and Radio Waves Became Evident without Experimenting

Whenever electrical current is used to generate heat, light or motion, the systems are said to belong to the area of electrical power engineering. Since sparks caused by static electricity have been observed even since the time of Socrates, the physicists who experimented with Volta type batteries were not really surprised when they found that a thin wire gets hot when a current is flowing through it. And when the current and the heat with it are increased, the wire begins to glow and emit light. The problem of inventing the light bulb consisted mainly of finding a metal which doesn't melt too quickly. In addition, it was necessary to prevent the hot wire from reacting with the oxygen of the air and burning up. Therefore, the wire had to be placed in an evacuated bulb, in a vacuum where the oxygen had been removed.

Evidently, Mr. Maxwell's theory is not needed for generating heat and light with electricity. The findings of the gentlemen Volta and Ohm are quite sufficient. The only connection between their experiments and Maxwell's equations consists in the permeability $\gamma$. After Ampère and Oersted found out how electrical current could be used to generate magnetic forces, enough knowledge was available for the invention of electrical motors. Thus, the knowledge of how to use electrical current to generate heat, light and motion already existed before Maxwell developed his equations. But generating the tremendous amounts of heat, light and motion which we use today requires much more electrical current than was available in those days. Then, the only way to get electrical current was to use batteries, and their energy density, i.e., the ratio between the deliverable electrical power and their volume, is rather low. If the electrical power which is consumed today in a big city had to be provided exclusively by batteries, these batteries would take up about ten percent of the volume of all the buildings in the city. Furthermore, the batteries would have to be replaced by new ones every few weeks. A big step forward was made when someone saw the possibility to generate almost unlimited amounts of electrical power with rather little effort. In 1866, the Englishman Charles Wheatstone (1802-1875) and the German Werner von Siemens (1816-1892) independently came up with the same great idea. At that time they could not yet have read Maxwell's book since it wasn't published until 1873.

You probably are familiar with the so-called chicken or egg problem – which came first? In order to get chickens, there must be eggs, and in order to get eggs, there must be chickens. It's a dynamic cycle, and we ask how it got started. By looking at the two laws at the bottom of Fig. 9.8, we may be reminded of the chicken or egg problem: if we have a current we get a magnetic field, and if we have a magnetic field, we can get an electric current. If we could succeed in getting this dynamic cycle started, we would have a method to generate electrical power, as much as we want. Of course, mechanical energy would be required for turning a coil through the magnetic field (see Fig. 9.3). The chicken or egg problem is not a difficult problem in this case because we already have currents from batteries and magnetic fields coming from magnetic blocks of iron before the cycle is started. To get the dynamic cycle started, the weak magnetic field which is provided by the steel components of the generators is sufficient. This cycle is the basis of all generators which run in our power plants; their rotation is produced by steam turbines. Once the principle of rotating electrical generators had been invented, it took only a few years until the first cities installed electric street lights and the first electric street cars were running.

But even before the generator was invented, electromagnetism was utilized. Because of the restricted availability of electrical power from batteries, the purpose of the electrical systems was not the production of heat or light, but providing communication over long distances. Since it had been discovered that currents produce magnetic forces, the forces could be turned on and off by turning a current on and off. A simple system could be built with a pen which moved up and down and made dots or dashes on a continuously running paper tape. Thus it was possible to transmit messages in Morse code (see Fig. 6.4) between two locations if they were connected via a cable. In the following paragraphs, I shall show you how the basis of wireless long distance communication follows from Maxwell's equations.

We now assume a space which doesn't contain any electrical charges and where, consequently, no electrical currents can flow. In addition, we assume that the permeabilities $\varepsilon$ and $\mu$ for the dielectric flow and the magnetic flow, respectively, do not depend on location but have the same values everywhere in the space. Think of a space which is empty except that it is filled with air. The simplified version of Maxwell's equations which is shown in Fig. 9.9 follows from these assumptions.

# Generators and Radio Waves

> We assume that there are no stored or flowing electrical charges i.e. $\rho = 0$ and $J = 0$, at all locations and at any point in time. We also assume that the permeabilities $\varepsilon$ and $\mu$ do not depend on the location or on the time.
>
> Then, Maxwell's equations are simplified as follows:
>
> $$\operatorname{div} D = \operatorname{div}(\varepsilon * E) = \varepsilon * \operatorname{div} E = 0$$
>
> $$\operatorname{div} B = \operatorname{div}(\mu * H) = \mu * \operatorname{div} H = 0$$
>
> $$\operatorname{rot} H = \frac{\partial D}{\partial t} = \varepsilon * \frac{\partial E}{\partial t}$$
>
> $$\operatorname{rot} E = -\frac{\partial B}{\partial t} = -\mu * \frac{\partial H}{\partial t}$$

**Fig. 9.9** Simplified Maxwell's equations

Although this system of equations looks rather simple, you should still be aware that the field quantities depend on the four variables $x$, $y$, $z$ and $t$, although only the latter appears explicitly in Fig. 9.9. The coordinates $x$, $y$ and $z$ will become relevant in the computation of the rotation as in Fig. 9.5. If we assume that the directions of the fields are restricted as shown in Fig. 9.10, where $E$ has only a y-component and $H$ has only a z-component and both depend only on $x$ and $t$, we get the simplest possible form of the rotations. This assumption takes into account that Faraday's experiments had shown that the fields $E$ and $H$ are perpendicular and cannot have the same direction.

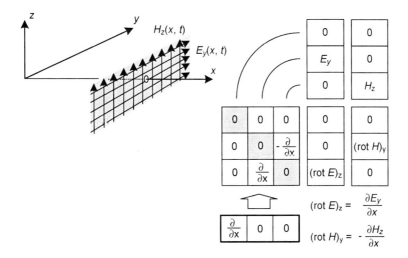

**Fig. 9.10** Simplest case of the rotation equations

By applying the assumptions from Fig. 9.10 to the equations in Fig. 9.9, we get the equations in Fig. 9.11. A professional mathematician would not have any difficulty solving the system of two differential equations at the left side of the arrow and find the solution given in the shaded rectangle at the right. There is no reason to feel bad because you don't know the methods which lead to this result. You may wonder why the two constants $\varepsilon$ and $\mu$ which occur in the differential equations are not contained in the solution, but that doesn't mean that they are irrelevant. The information about these values is now contained in the new constants $E_0$, $H_0$ and v, i.e., the values of $\varepsilon$ and $\mu$ determine the values of $E_0/H_0$ and v. The solution leaves open the value of $E_0$ and the function f; they may be chosen arbitrarily.

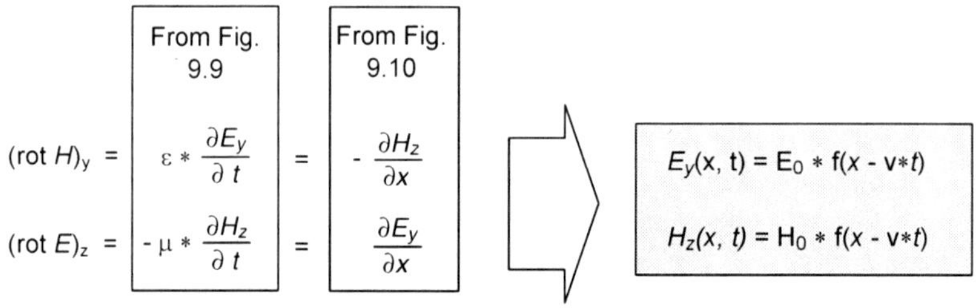

$$(\text{rot } H)_y = \varepsilon * \frac{\partial E_y}{\partial t} = -\frac{\partial H_z}{\partial x} \quad\Rightarrow\quad E_y(x, t) = E_0 * f(x - v*t)$$

$$(\text{rot } E)_z = -\mu * \frac{\partial H_z}{\partial t} = \frac{\partial E_y}{\partial x} \quad\Rightarrow\quad H_z(x, t) = H_0 * f(x - v*t)$$

**Fig. 9.11** Consequences of the assumptions in figures 9.9 and 9.10

The function f stands for a distribution of values along the x-axis which moves in the direction x at the speed v. Fig. 9.12 illustrates this. Imagine that the thick line represents a rope having a length of about ten meters whose right end is fastened to a wall and whose left end is held in the hand of a strong man. Since this man has given a short vertical pulse to his end, the rope is no longer a horizontal line, but contains a specifically-shaped curve which is now moving to the right. The thick line shows the situation at the time $t_a$, and after the time duration $\Delta t$, the shape has moved to the right by the distance $v*\Delta t$ as indicated by the dashed line. The time doesn't change the shape of the rope, only its horizontal position. Therefore, such wave propagation can be described by a function $f(s)$ which originally has only one argument $s$ and describes the static shape. Using the substitution $s=x-v*t$ captures the moving shape, because now many different pairs of values of $x$ and $t$ provide the same argument for the function f. In Fig. 9.12, the two pairs (a, $t_a$) and (a+v*$\Delta t$, $t_a+\Delta t$) are considered. They both provide the same argument $s=a-v*t_a$ and give the same value to the function: $f = b$.

Because the function f appears in the solutions for both $E_y$ and $H_z$ in Fig. 9.11, it follows that the electric field and the magnetic field have the same shape, although physically they are quite different. But the difference between these two fields is expressed only by the two constants $E_0$ and $H_0$. Now I shall outline the way how Fig. 9.11 leads to the formulas which describe the way $E_0/H_0$ and v can

# Generators and Radio Waves

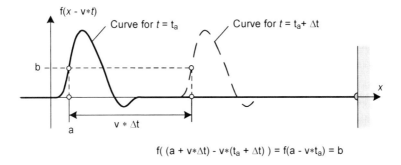

**Fig. 9.12** Example of a wave running along a rope

be computed from the values of ε and μ. However, those readers who are not interested in the details of this mathematical excursion may skip the next paragraphs and join me again two pages from here – like readers of a detective story who skip the description of the detailed investigations and jump immediately to the last pages of a book in order to find out who the murderer was.

Since the function f actually has only one argument $s$ which is substituted for by an expression containing $x$ and $t$, there is a simple relationship between the two derivatives of f in the $x$ and $t$ directions. The derivative in the x direction when $t$ is assumed to be constant gives the slope of the shape f at time $t$. In the case of Fig. 9.12, this means that the shape of the rope and its slope are determined for all points along the x-axis. Correspondingly, the position $x$ is assumed to be constant when the derivative in the t direction is determined. In this case, you have to assume that you were sitting at a certain position on the x-axis, e.g., the position a, and had to measure the speed of change of the actual vertical deviation of the rope at your position. Since the shape on the rope is running from left to right, the deviations which in Fig. 9.12 are farther right, i.e., at higher values of x, will come to you earlier at your position than those which are farther left. Thus, you would observe the small negative deviation before the large positive deviation. The relation between the two kinds of derivatives is represented in the big arrow pointing downward in Fig. 9.13. Above this arrow, you see again the differential equations from Fig. 9.11, but now the variables $E_y$ and $H_z$ have been substituted using the expressions from the shaded rectangle. By applying the relationships contained in the big arrow pointing downward to the differential equations which are standing above this arrow, we get the two equations below the arrow. Here, the derivative $\partial f/\partial x$ occurs as a factor on both sides of both equations, and therefore it can be deleted. This is a consequence of the fact that the function f can be chosen arbitrarily and is irrelevant to the structure of the solution of the differential equations in Fig. 9.11. After the derivatives of f have been eliminated from the equations, the remaining equations describe arithmetic relationships between the five values $E_0$, $H_0$, v, $\varepsilon$ and $\mu$. An equation which does not contain the constant v is obtained by dividing the upper equation by the lower one, i.e., by dividing both the left sides and the right sides of these equations. The resulting equation finally provides the ratio $E_0/H_0$ as it is represented in the shaded rectangle in the lower left corner of the figure. The other equation in this rectangle tells us how to get the value of v from $\varepsilon$ and $\mu$. This formula, too, is obtained by dividing the two equations at the lower right, but this time the two sides of one of these equations must be interchanged before the division. Only then do we obtain an equation which no longer contains the two constants $E_0$ and $H_0$.

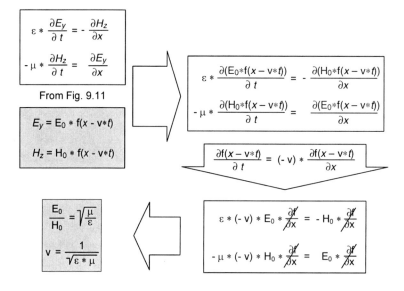

**Fig. 9.13** Relations between the constants in Fig. 9.11

Here ends the section which you could have skipped without missing any important information.

The formula in the lower left of Fig. 9.13 tells us how to get the value of v from $\varepsilon$ and $\mu$, but we cannot apply it yet since I did not yet provide enough information about the constants $\varepsilon$ and $\mu$. Although it is quite clear that the letter v stands for the propagation speed of an electromagnetic wave and therefore has the unit "distance per time," e.g., m/s, we did not yet consider the units of $\varepsilon$ and $\mu$. The formula in Fig. 9.13 requires that the product $\varepsilon*\mu$ has the unit of the reciprocal of the square of a speed, e.g., $(s/m)^2$, but from this we cannot conclude what the units of the factors are. Therefore, we must deal with the questions about what the units are of all the quantities and constants which have been introduced on the way to Maxwell's equations.

## What We Get by Multiplying or Dividing Volts, Amperes and Similar Things

Each letter which appears in our formulas represents a variable physical quantity or a constant. Distances and times are quantities we are used to in our daily life, and it is well understood that they can be measured in meters and seconds, respectively. But what could the units be for measuring the strength of the magnetic field $H$, the magnetic flow $B$ or the dielectric flow $D$? Probably you will be surprised to hear that besides the three units which are needed in the laws of mechanics, namely meter (m), kilogram (kg) and second(s), only one additional unit is required to cover the entire area of electromagnetism. From Fig. 7.3 which represents the basic concepts of mechanics and their interrelations, you know that neither momentum nor force nor energy have their own elementary units. Instead, the units of these quantities are arithmetic expressions composed of meter, kilogram and second – kg*(m/s) for the momentum, (kg*m)/s$^2$ for the force and kg*(m/s)$^2$ for the energy. Similarly, most of the quantities in the area of electromagnetism don't have their own elementary units.

After Mr. Volta invented the battery and many scientists used batteries of this type, they wanted to measure the new quantities which they associated with their observations. In analogy to the pressure and the stream flow in systems where water is forced through pipes, they introduced the concepts of voltage and current although they didn't know what was flowing and what caused this flow. You shouldn't think that we know much more about this today. Of course, we now know that there are extremely small components of atoms called electrons and protons which have a certain property called "electrical charge," where this charge is negative in the case of the electrons and positive in the case of the protons. We know that the current through a wire corresponds to moving electrons - but we still don't know what electrical charge really is. The concept of an electrical charge was invented as an explanation for the attractive or repulsive forces which occurred between bodies of certain materials after they had been rubbed together. As far as our not knowing what a certain quantity really is, there is no difference between the electrical charge of a body and the mass of a body. The concept of mass was introduced by Newton as an explanation of his observations of moving bodies and the related forces – but even today, we don't know what mass really is. The forces between bodies which carry electrical charges are captured in Coulomb's law. These forces could have been taken as a basis for a definition for the unit of charge. Alternatively, the forces between two parallel wires through which currents are flowing could have been taken as a basis for a definition of the unit of current. It was an absolutely arbitrary decision to choose the current to have an elementary unit based on forces, and not on charge. If the charge had been chosen, the unit of the current would have been a fraction "unit of charge / unit of time."

But with the current having the elementary unit, the unit of the charge now is the product "unit of current * unit of time," since the longer a current is bringing charge to a body, the more charge the body will have.

The existence of electrons was discovered eighty years after the need to define a unit for current came up (see Chapter 10). If however the existence of electrons had already been discovered before a unit for current was defined, the definition could have referred to the charge of a single electron. Then the unit of the current would be "a certain number of electrons per second." The definition which is used today was agreed upon in 1948, and although this definition refers to different experimental phenomena than the older ones, the defined unit *Ampere* (A) is approximately the same as before. The actual definition says:

> One Ampere is defined to be the constant current which per meter of length will produce an attractive force of $2*10^{-7}$ (kg*m)/s$^2$ between two straight parallel conductors of infinite length and negligible circular cross section placed one meter apart in free space.

This corresponds to

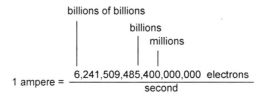

That means that if the current through a wire is said to be one Ampere, approximately 6.24 billions of billions of electrons per second are flowing through the cross section of the wire.

In the table in Fig. 9.14 you will find all the units for electrical and magnetic quantities which follow directly from the unit of current. However, the units in the last two rows of the table cannot be understood without reference to Maxwell's equations in Fig. 9.8 and to the definitions of div and rot as shown in Fig. 9.7. From Fig. 9.7, it follows that the unit of div($D$) is equal to "the unit of $D$ per distance", and from this we may conclude that the unit of $D$ is equal to the product "unit of div($D$) * meter." The unit of div($D$) can be determined from Maxwell's equation div($D$)=$\rho$, which requires that the unit of div($D$) be equal to the unit of the density of charge $\rho$ which is given in Fig. 9.14 as A*s/m$^3$. By multiplying this with the unit meter, we get the unit of $D$.

Ampere's law in Fig. 9.8 says that the unit of rot($H$) is the same as the unit of $J$ which is A/m$^2$. From Fig. 9.7, it follows that the unit of rot($H$) equals "the unit of $H$ per distance" from which we may conclude that the unit of $H$ is equal to the product "unit of rot($H$) * meter", i.e., A/m.

Units of Physical Quantities

| Electrical current | | A | $i$ | Ampere | |
|---|---|---|---|---|---|
| Density of current | = Current per area | $\frac{A}{m^2}$ | $J$ | | |
| Electrical charge | = Current times time | $A*s$ | $Q$ | Coulomb | C |
| Density of charge | = Charge per volume | $\frac{A*s}{m^3}$ | $\rho$ | | |
| Density of dielectrical flow | = Charge per area | $\frac{A*s}{m^2}$ | $D$ | | |
| Strength of magnetic pressing field | = Current per distance | $\frac{A}{m}$ | $H$ | | |

**Fig. 9.14**  Physical units derived from the unit of the electrical current

Although the table in Fig.9.14 contains many of the quantities occurring in Maxell's equations, there are still some more to be considered, namely the electrical field $E$, the magnetic flow $B$ and the permeabilities $\gamma$, $\varepsilon$ and $\mu$. These quantities and constants have been introduced in the upper part of Fig. 9.8, and they are related to the quantities $J$, $D$ and $H$ whose units are given in Fig. 9.14. Thus, we already know the units of the fractions $J/\gamma$, $D/\varepsilon$ and $B/\mu$, but we cannot conclude what the units of the numerators and denominators are from these. Therefore, we need some inspiration which helps us to find an adequate approach. Fortunately, someone else already had the ingenious idea to transfer the concept of potential from the area of mechanics to the area of electricity. When I introduced Coulomb's law, I emphasized the fact that this law has exactly the same structure as Newton's law of gravity. And since this law of gravity led us to the concept of mechanical potential which relates mechanical work to masses, it is no wonder that this concept can be transferred to the area of electricity where it relates mechanical work to charges. Fig. 9.15 illustrates this transfer. In both cases, there is a product which has a difference of potentials as its second factor, and in both cases, the result of the product is an energy. Therefore, we obtain the unit of the corresponding potential by dividing the unit of energy by the unit of the first factor of the corresponding product. From this we obtain the unit of electrical potential as $(kg*m^2)/(A*s^3)$. This looks rather strange and certainly has no evident interpretation. But this doesn't matter as long as we know that the result has been deduced correctly. And there is no need to keep this expression in mind; it is sufficient to know the name which has been given to it, namely *Volt*, and that this is the unit of a difference of electrical potentials. When you are told that the voltage at the power sockets in your home is 110 Volts, this means that the two contacts in a socket have different electrical potentials and that the difference between these is 110 (kg*m²)/(A*s³).

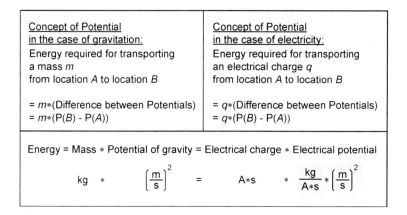

**Fig. 9.15** Concepts of potential in gravitation and electricity

The fact that names are assigned to units which are arithmetic expressions of elementary units could have been seen in Fig. 9.14 where the product A*s, which is the unit of charge, got the name Coulomb (C). It is important for you to realize that elementary units must have names – like meter or second – but that in contrast it is only a possibility and not a necessity that a composed unit have a name. Let's look at the composed units in mechanics (see Fig. 7.3): the unit kg*(m/s) of the momentum has no name, but the unit (kg*m)/s² of the force was given the name Newton (N), and the unit kg*(m/s)² of the energy has been called Joule (J). The unit of power, i.e., energy per time is (kg*m²)/s³, and it was given the name Watt (W).

The names for units are always chosen in reference to scientists who made major contributions to the field. The name Ampere for the unit of current refers to André Marie Ampère (1775-1836), who detected the force between parallel wires through which a current was flowing. The name Volt for the unit of voltage was chosen in honor of Alessandro Volta (1745-1827), the inventor of the battery. Charles Augustin de Coulomb (1736-1806) had already been honored by assigning his name to the law he found; thus, naming the unit of charge after him is a second honor. Sir Isaac Newton (1642-1727) is the father of modern mechanics and certainly is worthy of being honored. James Prescott Joule (1818-1889) was the first who studied the relationship between electric current and heat development. Many people believe that James Watt (1736-1819) invented the steam engine, but this is not correct; he only made major improvements which considerably increased the output of the engine.

The world of electricity was connected to the world of mechanics by defining the unit of voltage to be a difference of potentials. Fig. 9.16 illustrates this connection.

Units of Physical Quantities                                                               237

The shaded area contains the quantities from mechanics which were introduced in Fig. 7.3 where the units of momentum, force and energy were defined as arithmetic expressions on the basis of the elementary units of distance, mass and time. The white area in Fig. 9.16 contains the electrical and magnetic quantities with their corresponding units which were not introduced in Fig. 9.14. By comparing the unit of mechanical power to the unit of voltage, we see that the two units differ only by the unit Ampere (A) which appears in the denominator of the unit of voltage. From this we can conclude that the product of a voltage and a current is a power, i.e., Watt=Volt*Ampere. This is one of the two most important equations of electrical engineering; the other is Volt=Ampere*Ohm. When you read the information "110 V, 60 W" on a light bulb, you can conclude that, in operation, the current which flows through the bulb is 0.5455 A because (60 V*A)/110 V=0.5455 A.

| Momentum | = Mass times speed | $p$ | $\dfrac{kg*m}{s}$ | | | |
|---|---|---|---|---|---|---|
| Force | = Momentum per time | $F$ | $\dfrac{kg*m}{s^2}$ | Newton | N | |
| Energy (Work) | = Force times distance | $E$ or $W$ | $\dfrac{kg*m^2}{s^2}$ | Joule | J | |
| Power | = Energy per time | $P$ | $\dfrac{kg*m^2}{s^3}$ | Watt | W | |
| Voltage | = Difference of electrical potentials = Energy per charge | $v$ | $\dfrac{kg*m^2}{A*s^3}$ | Volt | V | $\dfrac{W}{A}$ |
| Strength of electrical pressing field | = Voltage per distance = Force per charge | $E$ | $\dfrac{kg*m}{A*s^3}$ | | | $\dfrac{V}{m}$ |
| Density of magnetic streaming field | = Force per speed of charge | $B$ | $\dfrac{kg}{A*s^2}$ | Tesla | T | $\dfrac{V*s}{m^2}$ |

**Fig. 9.16** Connections between physical units in mechanics, electricity and magnetism

The units of the field quantities $E$ and $B$ are given in the last two rows of the table in Fig. 9.16. In order to find the unit of $E$, we consider the equation $J=\gamma*E$ which is one of Maxwell's equations in Fig. 9.8. It says that the density $J$ of the current, measured in A/m$^2$, is determined by the strength of the electric field $E$ at an actual point in space. Since we have shown that current must be driven by voltage, $E$ must contain voltage. But it cannot be a pure voltage because a voltage is the difference of the potentials between two different points, and $E$ is defined at a single point. This problem can be resolved by defining the unit of $E$ to be Volt per meter.

Once we have found the unit of $E$, we can obtain the unit of $B$ from the law of induction in Fig. 9.8. This law requires that the unit of rot($E$) is the same as the unit of $\partial B/\partial t$ which is the unit of $B$ per second. From Fig. 9.7 we know that the unit of rot($E$) is equal to the unit of $E$ per meter. Thus we have the equation "unit of $B$ per second = unit of $E$ per meter." From this it follows that the unit of $B$ is equal to the unit of $E$ multiplied by the fraction s/m.

From the experiments of the gentlemen Ampere and Faraday, it is well known that the force which affects an electrical charge in an electric field is determined by the forcing field $E$ and not by the streaming field $D$. Those experiments also showed that the force affecting an electrical charge in a magnetic field is determined by the streaming field $B$ and affects only moving charges. The higher the speed of the charge, the stronger this force. This shows up in the unit of $B$ which corresponds to the ratio force/(charge*speed).

In Fig. 9.17, I attempted to make you see the twin relationships between electricity and magnetism which no one had the faintest idea about one hundred years before Maxwell developed his equations. You see that any unit on one side can be obtained from the corresponding unit on the other side just by interchanging Volt and Ampere. Fig. 9.17 also shows the units of the permeabilities $\varepsilon$ and $\mu$ which correspond to the fractions $D/E$ and $B/H$, respectively. This is a consequence of the corresponding equations in Fig. 9.8.

| Electrical quantities | | | Magnetic quantities | |
|---|---|---|---|---|
| $E$ | $\frac{V}{m}$ | Strength of the pressing field | $\frac{A}{m}$ | $H$ |
| $D$ | $\frac{A*s}{m^2}$ | Density of the streaming field | $\frac{V*s}{m^2}$ | $B$ |
| $\varepsilon$ | $\frac{A*s}{V*m}$ | Permeability for flow | $\frac{V*s}{A*m}$ | $\mu$ |

**Fig. 9.17** Symmetrical relations between electrical and magnetic quantities

Mountaineers sometimes shout "Hooray!" when they finally reach the top of the mountain, and that's what you may shout now, since we really have reached the top of our mountain which makes us understand the theory of the pair of twins called electricity and magnetism. One of the results which we can see by looking downhill was represented in Fig. 9.13. When you look at this figure again, you may conclude that it is still incomplete because, although it represents the formulas for computing

Units of Physical Quantities

the ratio $E_0/H_0$ and the speed v of an electromagnetic wave from the permeabilities $\varepsilon$ and $\mu$, it does not give us any real values. I didn't want to introduce the values of the constants $\varepsilon$ and $\mu$ before their units had been deduced. Now these units are given in Fig. 9.17, and thus it's time to apply the formulas from Fig. 9.13.

from Fig. 9.13

$$\varepsilon = 10^{-11} \frac{A*s}{V*m}$$

$$\mu = 10^{-6} \frac{V*s}{A*m}$$

$$\frac{E_0}{H_0} = \sqrt{\frac{\mu}{\varepsilon}} = 316 \frac{V}{A} = 316 \text{ Ohm} = 316 \, \Omega$$

$$v = \frac{1}{\sqrt{\varepsilon * \mu}} = 316.000.000 \frac{m}{s} = 316.000 \frac{km}{s}$$

**Fig. 9.18** Resistance and speed of electromagnetic waves

Even before Maxwell developed his equations, approximate values of $\varepsilon$ and $\mu$ in free space had been found experimentally. Although much more accurate values have been measured in the meantime, we use the approximations in Fig. 9.18 to get values for the ratio $E_0/H_0$ and the speed v. You should note the fact that the unit of the ratio $E_0/H_0$ is the same as the unit of the ratio voltage/current, which is the unit of ohmic resistors and was given the name Ohm. Therefore, the ratio $E_0/H_0$ is called the wave resistance in free space. This ratio is not restricted by the assumptions shown in Fig. 9.9 which we made in order to simplify the deduction of the formulas in Fig. 9.13. In any electromagnetic wave, the ratio between $E$ and $H$ at any point in space and at any point in time is the same and has the value 316 $\Omega$. Maybe you noticed the fact that all other units which have been introduced in this chapter, e.g., A (Ampere), V (Volt), W (Watt) etc., are abbreviated by the capital first letter of the name of the person to be honored, and that the unit Ohm is the only exception to this rule. The reason for this is that the first letter of the name Ohm can be easily mistaken for a zero; therefore the Greek letter $\Omega$ has been taken which corresponds to the Latin letter capital O.

When Mr. Maxwell applied the formula for the computation of the value of the propagation speed of electromagnetic waves (see Fig. 9.18), he got a result which was a great surprise to him. This result is approximately three hundred thousand kilometers per second and is equal to the speed of light which the Danish astronomer Olaf Roemer (1644-1710) had measured about two hundred years before Maxwell performed his computation. From this equality, Maxwell concluded that light is an electro-magnetic wave. Isn't this amazing? Just by performing some computations on the basis of his equations, Maxwell not only discovered the existence of electromagnetic waves, but he also came to the conclusion that light is such a wave. Maxwell's book, which contains his equations and conclusions, was published in 1773. Thirteen years later in 1886, the German physicist Heinrich Hertz (1857-1894) confirmed Maxwell's theory by experiments. He was the first who succeeded in generating electromagnetic waves in a transmitter and detecting them in a receiver. Only fifteen years later in 1901, the Italian physicist and engineer Guglielmo Marconi (1874-1937) established the first system for wireless transatlantic telegraph communication, which is based upon Maxwell's theory.

# Chapter 10
# Small, Smaller, Smallest – How the Components of Matter Were Found

Alhough we are not always aware of it, our experience makes us classify all phenomena into two categories, discrete and continuous. Those which are not discrete must be continuous. Being discrete means being countable, and undoubtedly we see around us a world of objects which we could begin to count. Besides this discrete world, we also experience a continuous world, for the flow of time, the motion through space and the changes of temperature or daylight are perceived as continuous processes without any granularity. In mathematics, the discrete world is captured by the natural numbers which are used for counting, while the continuous world is captured by the real numbers which are used for measuring. While philosophers never questioned the existence of the discrete world of countable objects, there were philosophers, even at the time of Socrates, who speculated that the continuous world does not really exist, but is only a mental construction as a consequence of the restricted resolution of human perception. In Chapter 7 on mechanics, I mentioned the Greek philosopher Protagoras (490-411 BC) who said, "Man is the measure of all things." In today's language of science, this means that the restricted resolution of our sense organs determines our world view. But in spite of that, we are able to talk and write about magnitudes which are far beyond our capability of perception. We just express extremely great or small numbers by using powers of ten as shown in Fig. 10.1. Nobody can really imagine how short a picosecond is, but of course we can define it as the millionth part of the millionth part of a second. Using these powers of ten, we can enter into the world of extremely long distances and time durations which become relevant when considering the universe. But we can also enter into the world of extremely short distances and time durations which are typical in the areas of electronics and nuclear processes. This is the world which we shall now consider.

| $10^{+3}$ | $10^{+6}$ | $10^{+9}$ | $10^{+12}$ | $10^{+15}$ | $10^{+18}$ |
|---|---|---|---|---|---|
| Kilo | Mega | Giga | Tera | Peta | Exa |
| Thsd | Mio | Bio | Trio | Quadrio | ??? |

| $10^{-3}$ | $10^{-6}$ | $10^{-9}$ | $10^{-12}$ | $10^{-15}$ | $10^{-18}$ |
|---|---|---|---|---|---|
| Milli | Micro | Nano | Pico | Femto | Atto |
| 1/ Thsd | 1/ Mio | 1/ Bio | 1/ Trio | 1/ Quad | 1/ ??? |

**Fig 10.1** Powers of ten with their names

## How the Age-Old Assumption That Matter Is Composed of Atoms became Experimentally Relevant

The question of whether some or all processes which we perceive as being continuous actually have a granularity can only be answered separately for each type of process. Until now, no experimental results have been found which would force us to accept that the flow of time is granular. But granularity was discovered for matter and energy. I shall now describe the ways which led to these findings.

Even small children have the idea that material objects consist of something. They watch their mother baking a cake or a pie and see how she prepares the dough with flour, eggs, milk, butter, salt and other ingredients. And they have no problem accepting the fact that the ingredients cannot be separated once the dough has been prepared. You probably know that there was a time when people believed that gold could be made in a way similar to dough, except that the recipe had not yet been found.

The question "What is this made of?" is not only asked by children, but is a concern of philosophy from its very beginning. Someone who asks himself about the existence of a small set of elements, from which all kinds of matter could be composed, puts himself in the role of a creator who, by an ingenious idea, tries to simplify his job of creation. For if he succeeds and finds the elementary components, his job is almost done, since then the creation of a certain kind of matter can be done just by defining a specific combination of elementary components. About 500 years before Christ, some Greek philosophers, among them Heraclitus and Empedocles., came up with the idea that all phenomena which humans could observe are produced by the interaction of fire, water, earth and wind. About one hundred years later, another Greek philosopher, Democritus, was convinced that all matter is composed of extremely small particles which cannot be divided into smaller parts. Thus, Democritus is the father of the hypothesis of the existence of atoms as indivisible components of matter; the Greek word for "indivisible" is atomos. It took more than 2000 years until it could be proven experimentally that Democritus was right. But before that happened, many philosophers believed he was wrong. The Dutch philosopher Baruch de Spinoza (1632-1677) tried to logically deduce that atoms cannot exist [SPI]. And the German philosopher and mathematician Gottfried Wilhelm Leibniz (1646-1716) wrote [LEI], "Matter can be divided infinitely."

The era when the hypothesis of the existence of atoms became experimentally relevant began around the year 1775, which is about the time when the American Declaration of Independence was written. We may say that the transition from alchemy to chemistry occurred in those years. The main progress consisted of the use of very precise scales for weighing the input and output substances of chemical reactions. The French nobleman Antoine Laurent de Lavoisier (1743-1794), especially, must be mentioned here because he contributed most to the new methods [LA]. But this was not really appreciated by his fellow Frenchmen who resented that he wasn't just a scientist, but also a member of the Financial Council of the French government when the French revolution began. Therefore, he was sentenced to death and executed on the guillotine in 1794. But this couldn't stop the process of establishing chemistry as a serious science. Soon, everyone who performed chemical experiments knew that it is absolutely necessary to measure the exact weights of the substances involved.

The British scientist John Dalton (1766-1844) very carefully analyzed the proportions of these weights, and this led him to the conclusion that a chemical reaction is nothing but a transition from the original grouping of the atoms involved to a different one. Let's assume that the three different kinds of atoms A, B and C are involved in a chemical reaction, and that the proportion of the numbers of atoms involved is a:b:c=3:3:4 where a, b and c are the numbers of the atoms of the kinds A, B and C, respectively. A chemical reaction which corresponds to these assumptions then could be

$$AC_4 + A_2B_3 \rightarrow A_3BC + B_2C_3$$

The formula $AC_4$ which describes one of the original substances says that this substance contains four times as many atoms of the kind C than of the kind A, and that is doesn't contain any atoms of the kind B. By letting the two original substances ($AC_4$ and $A_2B_3$) react, we get two different resulting substances ($A_3BC$ + $B_2C_3$). The total numbers of the atoms involved are not changed by the reaction, but the grouping of the atoms is changed.

The proportions measured by Mr. Dalton allowed him to conclude that the weight of a hydrogen atom is lower than the weights of any other atoms. Therefore he defined the relative weight of the hydrogen atom to be 1. He could do this without knowing what the real weight of a hydrogen atom is; in those days, nobody knew how to determine the real weight of an atom. The relative weights of atoms of other kinds correspond to the factor by which such atoms weigh more than hydrogen atoms. Dalton's idea of using relative weights of atoms has kept its importance until the present time, although in the meantime the actual values of these relative weights had to be partially corrected because of more accurate measurements. During Mr. Dalton's lifetime, the concept of atoms was not yet commonly accepted. Even in 1887, when the concept had been accepted by most scientists, the Englishman Henry Roscoe (1833-1915) made the following mocking remark, "Atoms are small wooden balls invented by Mr. Dalton."

However, even during the lifetime of John Dalton, additional support for his hypothesis came based on a new type of experiments. Alessandro Volta invented the battery, and Michael Faraday used it in his experiments. He wanted to know if there were liquids through which an electric current could flow. He dipped two metal plates apart from each other into a container which contained a liquid he had chosen, and then he connected the metal plates to the terminals of a battery. He made many such experiments by varying the particular liquid and the metals of the plates. In 1832, Faraday coined the term "electrolysis" in order to characterize this type of experiments. His experiments showed that the liquid itself, or substances being dissolved in the liquid, were electrically decomposed into two kinds of components, where one went to one plate and the other went to the other plate. Even pure water was decomposed into its oxygen and hydrogen components, with the oxygen going to the metal plate which had been connected to the positive terminal of the battery and the hydrogen going to the other plate. The ratio of the weights was always hydrogen:oxygen=1:8, and the ratio of the volumes of the gases was always hydrogen:oxygen=2:1. At the time when Faraday did his experiments and measured these ratios, neither he himself nor anybody else could give a reasonable explanation for these results.

The assumption of the existence of different atoms which can be grouped to form molecules was a reasonable explanation for the tremendous number of different kinds of matter, but it was not yet sufficient to explain all the weight ratios which had been measured in chemical experiments. An additional idea was required, and this was the idea that each kind of atom has a specific valency. The simplest view of a valency is that it is the number of bonds which an atom can use to connect to other atoms. Each hydrogen atom has one bond, and each oxygen atom has two. Since water consists of only hydrogen and oxygen atoms, the water molecule can be built from one oxygen atom and two hydrogen atoms where there are two bonds from the oxygen atom, with each connecting to one hydrogen atom. A large part of the concept of valencies was developed by the German chemist Friedrich August Kekulé von Stradonitz (1829-1896). The idea of atoms having valencies made it possible to represent molecules in the form of structure diagrams, examples of which are shown in Fig. 10.2. Once the concept of valencies

Chemistry 245

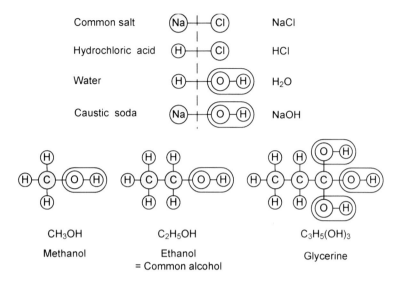

**Fig. 10.2** Chemical structures of common substances

was established, Mr. Kekulé saw that atoms could be connected not only in the form of chains, but also as rings, and that the bonds between two atoms must not necessarily be restricted to one pair of atoms forming a bond, but could consist of two bonds (see Fig. 10.3). The chemical structure diagrams represent the molecules as lying flat in a plane, but of course, this does not correspond to reality where a molecule is a three-dimensional structure. At the time of Kekulé, there was no way of getting any information about the three-dimensional structures of molecules. It took about half a century before such information became available.

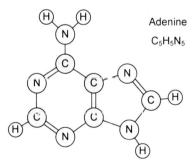

**Fig. 10.3** Chemical structure of the organic molecule adenine

Around the year 1865, which was the time when Maxwell produced his equations describing the theory of electromagnetism, chemistry had become an established science. In those days, some chemists began to wonder whether it might be possible to structure the "zoo" of the known kinds of atoms not only as a linear ordering according to their relative atom weights, but according to additional criteria. Maybe they thought of the Swede Carl von Linné (1707-1778) who had developed a systematic ordering scheme wherein all plants and animals had their proper places. Comparing the tremendous number of different plants and animals with the rather small number of different atoms, it might seem to be a rather simple problem to find an adequate structuring scheme for the less than one hundred chemical elements which were known at that time. But the problem was not simple at all since the desired scheme had to contain empty positions for those elements which had not yet been discovered. The Russian Dimitri Iwanowitsch Mendelejew (1834-1907) and the German Lothar Meyer (1830-1895) independently developed the same scheme which is called the "periodic table of chemical elements," but the results of the Russian were published in 1869, two years before those of the German.

I shall now use Fig. 10.4 to explain the concept of the periodic table of chemical elements and how it was developed. None of the so-called noble or rare gases, which are represented in the right-most column of the scheme, had been discovered before 1894, and therefore these elements could not be taken into account when the periodic table was developed. Since these gases have zero valency, i.e., their atoms have no ability to bond with other atoms, they are almost never involved in any chemical reaction. Besides that, they are, as their name "rare gases" says, extremely rare. Although they are contained in air, their percentage is extremely small. Therefore it is no wonder that it took so long before they were discovered. In 1869, all chemists were convinced that the shaded cells contained the complete set of all 15 elements with the lowest relative atomic weights, i.e., that no element exists which is not present in Fig. 10.4 and which has a relative atomic weight lower than 35.5, that of chlorine. By ordering these 15 elements according to their relative atomic weights and by grouping them according to the number of their valencies and taking into account certain chemical similarities, seven groups were found which correspond to the seven shaded columns. Because of the fact that the pattern 1-2-3-4-3-2-1 of the number of valencies is repeated periodically in the sequence of the elements, the scheme has been called the *periodic table*.

Once the seven groups of elements corresponding to the shaded columns had been found, an interesting question arose concerning whether all other elements which were known at the time would also fit into these groups. Some of them did, but there were others which didn't. The first elements which didn't fit were in positions 21 through 30 in the sequence of all elements ordered according to their relative atomic weights. This required an expansion of the periodic table by inserting additional columns between the elements calcium at position 20 and gallium at position 31. The need for inserting additional columns occurred again when

Chemistry                                                                 247

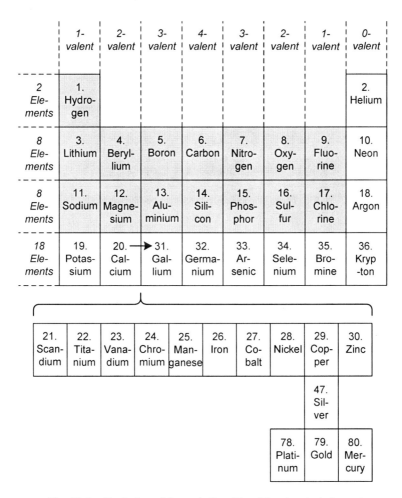

Fig. 10.4  Evolution of the periodic table of the chemical elements

elements with a relative atomic weight in the interval $58 \leq w \leq 71$ were discovered; again these elements didn't fit into the existing columns. The final structure of the periodic table is shown in Fig. 10.5. The neighbors to the right of uranium are called transuraniums, most of which do not occur outside of laboratories where they can be produced by certain experiments. Their lifetime is extremely short, i.e., they decompose soon after they have been produced. At the time of Mendelejew, the transuraniums and many other elements which are known today had not yet been discovered. Beginning in about 1940, one transuranium element after another was discovered, and names were assigned to them. The freedom of choosing names was a good opportunity to honor scientists who had made major contributions to the field. Thus, elements were called Curium, Einsteinium, Fermium,

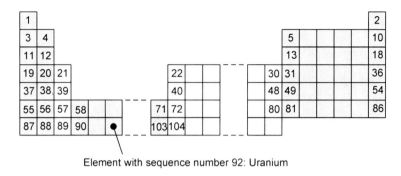

**Fig. 10.5**   Completely expanded periodic table

Mendelevium, Nobelium, Rutherfordium, Bohrium and Meitnerium. However, in technology, these elements are completely irrelevant.

When the periodic table was found, nobody could explain the origin of its structure. It took about half a century before an explanation was found. This was when the internal structure of atoms was discovered and the energy states of electrons in the atomic shell could be described by means of quantum theory. This made it possible to understand the nature of different types of valencies. The essence of quantum theory will be presented in Chapter 11.

In 1870, many indications had been found which made it very likely that the hypothesis of the existence of atoms was correct, although no information whatsoever was known about the size and mass of these atoms. Besides this, nobody could say whether atoms really are as indivisible as their Greek name indicates, or whether they consist of specific components.

## What Can Be Deduced from the Assumption That Gases Are Small Balls Flying Around

Although quantum theory later showed that the assumption of atoms being small balls is far from reality, this assumption nevertheless leads to some rather realistic conclusions. Despite the fact that nothing was known about the size and mass of an atom, interesting physical laws could be deduced theoretically and later confirmed experimentally. However, these laws did not say anything about solid matter or liquids, but only about gases. A gas was thought to be made up of a tremendous number of small elastic balls flying around in complete disorder and bumping into each other every now and then. What happens with such collisions is described by Newton's laws of mechanics. Obviously, knowledge about the behavior of billiard balls was just transferred to the world of atoms.

# Theory of Gases

In addition to knowledge about mechanics, another ingenious idea could be applied concerning the concept of heat and the definition of the unit of temperature. In 1738, the Swiss scientist Daniel Bernoulli (1700-1782) wrote, "The springiness of the air can be increased not only by compression, but also by increasing its temperature, since it is a fact that increasing heat corresponds to an increasing motion of the particles." When the Englishman Benjamin Thompson (1753-1814) observed that both the drill and the barrel got hot when drilling gun-barrels, he said, "I find it difficult if not impossible to imagine that, in this process, anything different from motion could be produced and dissipated." These ideas, which were so evident to the gentlemen Bernoulli and Thompson, were not at all evident to their contemporaries. Even Mr. Lavoisier, who was the one who began measuring the weights of the substances involved in chemical reactions, believed at the beginning of his career that there is something which might be called *fire substance*. He once wrote, "The same substance can consecutively pass through all three states (solid, fluid, gaseous), and in order to make this happen, a certain quantity of fire substance must be added or withdrawn." Finally, about the year 1800, it was well established that heat is nothing but the disordered motion of atoms or molecules; kinetic energy of this motion corresponds to the energy of heat, and temperature is a linear measure of the mean value of this energy.

This definition of temperature laid the groundwork for a new area in physics which is called statistical thermodynamics. It is amazing that something useful could be said about the energy of the small balls, even though their mass was absolutely unknown. By applying Newton's laws of mechanics, one can compute how a collision of two elastic balls changes the ratio of their speeds, depending on the ratio of their masses. These computations show that after many collisions have occurred, all balls will have approximately the same kinetic energy regardless of their masses. The gas then has a well-defined temperature only when this equilibrium state has been reached.

The law which I shall deduce in the following paragraphs is – at least in my opinion – the most important law of gas theory. The physical quantities which are related in this law are measured in the system shown in the upper part of Fig. 10.6. The volume of the gas in the container can be varied by shifting the closing piston in or out. The gas produces a force which tries to push the piston out, and therefore a counterforce is needed to keep the piston in its actual position. The following considerations will help us to find the pressure of the gas as a function of the volume and the temperature:

(1) The speed v of a particle has the three components $v_x$, $v_y$ and $v_z$, from which the value of the speed is obtained by the application of the law of Pythagoras: $v^2 = v_x^2 + v_y^2 + v_z^2$. Since none of the three directions is preferred, we may write $v_x^2 = v^2/3$.

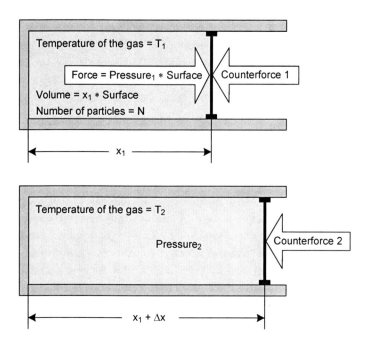

**Fig. 10.6** Experiments with an enclosed gas

(2) We now consider all of the particles which hit the piston within the time interval $t_1 \leq t \leq t_1+\Delta t$. At the beginning of this time interval, i.e., at $t_1$, their distance from the piston could not have been greater than $v_x*\Delta t$ since otherwise they would not have been able to reach the piston within the given time interval. Therefore, they were all contained in a slice sitting left of the piston and having the volume $v_x*\Delta t*$surface. The factor "surface" means the surface area of the piston. The number of particles in this slice is $(v_x*\Delta t*$surface$)*(N/V)$ with V being the total volume and N being the total number of particles in the volume.

(3) Only half of the number of particles in the slice of (2) will hit the piston within the given time interval because the other half has a $v_x$ leading away from the piston.

(4) Since the piston is held in its position by the counterforce, the particles will hit a fixed wall and be reflected, i.e., the sign of the speed $v_x$ will be reversed while the number of particles will stay the same. This corresponds to a change of the momentum from $(m*v_x)$ to $(-m*v_x)$. Thus, the difference of momentum caused by this reflection is $2m*v_x$.

Theory of Gases 251

(5) The total difference of momentum, which results from all particles being reflected within the given time interval, is the product of the momentum difference for one particle with half the number of particles in the slice of (2). This product is $N*m*v_x^2*\Delta t*\text{surface}/V$. According to Fig. 7.3, each difference of momentum corresponds to the product of a force and a time duration. Thus we get a force by dividing our difference of momentum by the length $\Delta t$ of the corresponding time interval. This force is $N*m*v_x^2*\text{surface}/V$.

(6) The pressure is the ratio between force and surface, and this is $N*m*v_x^2/V = N*m*(v^2/3)/V$. Since the temperature T has been defined to be a measure for the mean value of the kinetic energy of a particle, $m*v^2/2$, we can write $T \propto m*v^2$ where the symbol $\propto$ is read as "is proportional to." With this substitution, the pressure formula can be written as $p \propto N*T/V$ which can be transformed to the more common form $p*V \propto N*T$.

Although the number N of particles in the container is unknown, the relation deduced can nevertheless be tested. For a first experiment, an arbitrary quantity of gas is enclosed in a container as shown in Fig. 10.6 and the functional relation between the product p*V and the temperature T is determined by varying the volume and the temperature of the system. The volume can be varied easily by shifting the piston in or out, but before we can vary the temperature we have to find a method for measuring it. The definition which says that the temperature is a linear measure of the mean value of the kinetic energy of the molecules doesn't give us any hint about how we could measure it. It is quite obvious that we cannot measure the speed and the mass of the molecules. At least we can begin with two well-defined temperatures which were used by Mr. Celsius when he thought about measuring temperatures. He assigned the value 0 to the freezing point of water and the value 100 to its boiling point. Fig. 10.7 illustrates a method for producing precise temperatures of water in the interval between 0 and 100 degrees Centigrade. We just mix freezing and boiling water in an adequate ratio. If, for example, we want the gas in Fig. 10.6 to have a temperature of 60 °C, we can start with a bathtub full of water at 60 °C and place the container with the gas into the bathtub. Then we can add more water of the same temperature until the whole system has reached its equilibrium state at 60 °C.

Once we have completed the experiment with the original quantity of gas, we can repeat this experiment with twice the original quantity. Although we don't know what the original number N of molecules was, at least we know that in the second experiment the number is 2N. Fig. 10.8 shows the results of our two experiments. The fact that the product p*V varies linearly with the temperature T as long as the quantity of gas is kept constant confirms the formula $p*V \propto N*T$

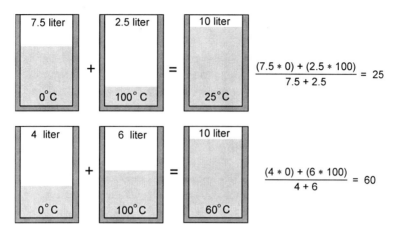

**Fig. 10.7** Producing temperatures between the freezing and boiling points of water

concerning the dependency on T. The fact that the ratio $(p*V)_1/(p*V)_2$ is equal to the ratio $N_1/N_2$ is a confirmation concerning the dependency on N.

The two straight lines in Fig. 10.8 meet at a point which corresponds to the temperature − 273 °C. It is no wonder that lower temperatures do not exist, since the mean value of the kinetic energy of the molecules cannot be less than zero. If this had been known earlier, nobody would have defined a temperature scale with the lowest possible temperature not having the value zero. Unfortunately, different temperature scales were defined before the findings described in Fig. 10.8 became available. Therefore, it became necessary to agree upon a standard way of representing temperatures. In honor of William Thomson (1824-1907), later known as Lord Kelvin, who made major contributions to thermodynamics, this standard has been called the Kelvin temperature. With this standard, the interval between the freezing point and the boiling point of water is still subdivided into one hundred units, as Mr. Celsius had defined, but now the freezing point no longer corresponds to the value 0, but to the value 273 Kelvin. The lowest possible temperature is now 0 degrees Kelvin which is the same as − 273 °C.

In the upper left corner of Fig. 10.8, you again find the formula $p*V \propto N*T$ which has been confirmed by the experimental results represented in the diagram. This formula contains the number of molecules N which are enclosed in the volume V. In 1811, the Italian physicist Amadeo Avogadro (1776-1856) concluded that the number of gas atoms or molecules within a given volume does not depend on the actual type of gas, but is the same for all gases. This could be concluded from the fact that the ratio of the weights of different gases in a given volume corresponds exactly to the ratio of the relative atomic or molecule weights of these gases. This also explained the different ratios of the volumes and the weights of

Theory of Gases

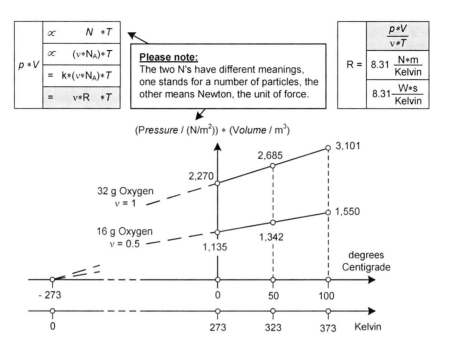

**Fig. 10.8** Experimental confirmation of the relationship between the volume, pressure and temperature of gases

hydrogen and oxygen which result from electrolysis of water. With water, the volume ratio is 2:1 while the weight ratio is 1:8. This corresponds to the fact that the ratio of the relative atomic weights of hydrogen and oxygen is 1:16. I should mention here that the particles in these gases are not atoms but molecules, since an atom uses all possible bonding positions when bonding with other atoms. A hydrogen atom has one bonding position which forms a bond with a second hydrogen atom. And an oxygen atom has two bonding positions, which forms bonds with a second oxygen atom. Since both the hydrogen and the oxygen particles are pairs of atoms, the ratios of the volumes and the weights for a given number of particles is the same as if these particles were atoms.

Although Avogadro's idea was correct, it didn't provide any information about the actual number of particles in a given volume. Only ten years after Mr. Avogadro's death, the first approximate value of the number of particles per unit volume was determined by the Austrian chemist Josef Loschmidt (1821-1895). He found that one cubic meter of gas at a temperature of 0 °C and a pressure of 10 N/cm$^2$ contained approximately $27*10^{24}$ particles. In his honor, this number was later called the "Loschmidt number." From this number, the so-called "Avogadro number" which corresponds to the number of particles of such a quantity of a substance whose weight measured in grams has the value of the relative atom or molecule weight, was derived. For example, let's consider the substances gold and water. The relative atomic weight of gold is 197, and for hydrogen and oxygen, the components of water, it is 1 and 16, respectively. In the case of gold, the particles are atoms, in the case of water they are the molecules $H_2O$. Thus, according to Avogadro, 197 grams of gold and 18 grams of water consist of the same number of particles, namely $6*10^{23}$ particles. Once the Avogadro number $N_A$ had been determined, the number N of gas particles in the volume V could be expressed as a multiple of $N_A$. The factor ν in the relation $N = ν*N_A$ corresponds to the ratio between the weight of the gas in the volume V and the relative atomic or molecule weight, both measured in grams. The two straight lines in the diagram in Fig. 10.8 show the results of experiments with 16 and 32 grams of oxygen. The relative atomic weight of oxygen is 16, and the particles in an oxygen gas are molecules $O_2$. Thus, 16 grams of oxygen correspond to ν = 0.5.

A formula containing the symbol ∝ cannot be used for computations; it has to be transformed into an equation. This can be done by introducing a constant factor which is then called a *proportionality constant*. In the case of our formula $p*V ∝ ν*N_A*T$ we have the choice of either using the constant k or the constant R, since we can either include the constant $N_A$ in the proportionality constant or leave it out. The determination of the value of the constant R from experimental results does not require the knowledge of $N_A$; the value of R corresponds to the slope of the straight line in the diagram for the value ν = 1. This slope is 2,270/273 = 8.31. The unit of R is energy/degree. In honor of Ludwig Boltzmann, the Austrian physicist who made major contributions to statistical thermodynamics, the constant k is called the "Boltzmann constant." The universal gas constant R and the Boltzmann constant k are interrelated via the Avogadro number $N_A$, i.e., $R = k*N_A$.

The findings represented in Fig. 10.8 make it possible to measure temperatures which are outside of the interval between the freezing point and the boiling point of water. The basic idea of a so-called gas thermometer was presented to you in Fig. 10.6 where a well defined quantity of gas is enclosed in a container, and the temperature can be determined from the product of pressure and volume. Although the temperatures which can be measured by such a thermometer are not restricted to the interval between 0 °C and 100 °C, such thermometers also have a restricted range because, the gas will become fluid at very low temperatures, and the container will melt at very high temperatures. Different ranges of temperatures require different types of thermometers which are based on different physical laws.

The idea that heat is nothing but the kinetic energy of atoms or molecules being in some kind of disordered motion finally led to the law of conservation of energy. The essence of this law is the idea that energy is a physical quantity which, in the universe, cannot be generated or eliminated, but only converted. The first form of energy which I introduced in this book was mechanical work (see Fig. 7.3). Then I divided mechanical energy into the kinetic energy of moving bodies and the potential energy of bodies in the field of gravity. Two other forms of energy are chemical energy and electrical energy, both of which can be converted into heat, chemically by burning coal and electrically by current flowing through thin wires. All substances which can be burned contain chemical energy – think especially about natural gas, crude oil and coal, which play major roles in today's energy supply. In Chapter 8 on relativity theory, you learned that even the rest mass of atoms can be partially converted into energy; this occurs in nuclear power plants. Electromagnetic waves also contain energy. This can be deduced from the relationship between the units of the electrical and magnetic fields E and H. The unit of E is Volt/meter, and the unit of H is Ampere/meter. The product of these two units is Watt/square meter which corresponds to energy per second and square meter. Since we know the propagation speed c of the wave, we can determine the amount of energy which is contained in one cubic meter of the space through which the wave is passing. The formula for this computation is E*H/c.

An important question with respect to energy is how one form of energy can be converted into another form. The generators in electric power plants convert mechanical energy into electrical energy; the reverse conversion occurs in electric motors. Electricity generators are driven by steam turbines whose mechanical energy is obtained from the heat of the steam. The steam is obtained by converting chemical or nuclear energy to heat. A similar chain of conversions is needed to run a car engine: the chemical energy contained in the fuel is converted into heat, which then is transformed into mechanical energy.

It would be nice if any form of energy could be completely converted into any other form but, unfortunately, this is not the case. The various forms of energy can be ranked according to their convertibility. Energy of a higher rank can be converted completely into energy of a lower rank, while in the reverse direction, only a partial conversion is possible. Electromagnetic energy has the highest rank, and heat has the lowest. This means that any form of energy can be converted completely into heat, but heat can be only partially converted into energy of a higher rank. As an example, let's consider a gas turbine. By burning gas, its chemical energy is converted into heat, and this, according to Fig. 10.8, increases the product of pressure and volume. The resulting forces push the blades of the turbine and produce mechanical work. But this work cannot be as much as the chemical energy consumed at the input of the conversion chain, because part of the energy is used to warm up the turbine and another part leaves the turbine in form of the heat of the exhaust.

Why can't a stone which is lying in the front yard convert part of its thermal energy into kinetic energy, and then convert this into potential energy by flying onto the roof? Such a process would not violate the first law of thermodynamics, i.e., the law of energy conservation, but nobody ever observed such a process happening. This finally led scientists to believe that such chains of energy conversion are impossible, and they called this belief the *second law of thermodynamics*. In other words, this law says that certain conversions of energy are irreversible. The elliptic orbit of a planet around the sun (see Fig. 7.6) could be explained as a periodic process of converting kinetic energy into potential energy and then back into kinetic energy. Obviously, these energy conversions are reversible. But think of a container being divided into two compartments of equal size by a removable wall, with each compartment filled with the same quantity of water. The temperature in one compartment is close to the freezing point while the temperature in the other compartment is close to the boiling point. Immediately after the separating wall is removed, the water which original was cold will get warmer, and the water which originally was hot will get cooler. After a short time, all the water in the container will have a temperature of approximately 50 °C. This process of change towards a state of equilibrium is irreversible, which means that if we start with 50 °C water it cannot happen that one half of the water gets hotter and the other half gets colder. Another example of an irreversible process is the spreading of a drop of ink which falls into a bathtub full of water. Immediately after the drop hits the water, its blue color indicates the position where the drop hit on the surface. But the blue spot will then spread, and after a short time all the water will have the same extremely faint blue color. It will not happen that the ink molecules, after being equally distributed over all the water in the bathtub, will come together again at the original position of the drop. However, although the probability of such an event is extremely low, there is no reason to say that the probability is exactly zero.

The example in Fig. 10.9 should help you to understand the second law of thermodynamics concerning the distribution of particles. It illustrates the fact that the less the distribution of particles differs from the equilibrium, the higher the probability of that distribution. The figure shows a square-shaped container which appears divided into four chambers, although there are no real walls separating the chambers. We assume that eight particles are placed into the container and that neither a certain particle nor a certain chamber is favored with respect to the distribution of particles. There are $4^8 = 65,536$ different ways to distribute the eight particles among the four chambers. The probabilities are as shown in Fig. 10.9. In the figure, these cases are grouped according to how many chambers are empty. The probability that two or three chambers are empty is slightly higher than 2 percent, while the probability that no chamber is empty is approximately 62 percent. If we had not restricted the example to eight particles but had taken 100 or 1,000 particles, the probability of at least one chamber being empty would have been very close to zero. You now have to consider the fact that the number of particles in a full bathtub is not a million or a billion, and not even billions of billions, but a number which is so high that there is no name for it. Thus it becomes quite evident that the probability of the ink molecules coming together at a certain small region of the water in the bathtub actually will be extremely close to zero.

Theory of Gases

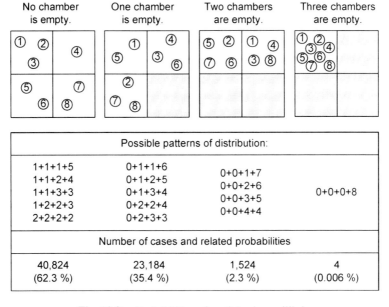

Fig. 10.9   Probabilities of particles in equilibrium

Once the idea was conceived that irreversible processes can be explained by looking at the probability distributions of certain properties such as position or kinetic energy for an extremely large number of particles, the question arose about whether an abstract physical quantity could be defined for capturing the degree of reversibility of state transitions. It was the German physicist Rudolf Clausius (1822-1888) who defined the so-called *entropy* as a property of the actual state of a closed system. It is completely irrelevant what the value of the entropy of a specific state is because it's only the difference between the entropies of two states which is interesting. This should remind you of the concept of potential which is a quantity assigned to points in space. Here, too, it is completely irrelevant what the potential of a single point is; only the difference between the potentials of two points is of interest (see Fig. 9.15). The greater the ratio $p_2/p_1$, where $p_1$ and $p_2$ are the probabilities of the two states $S_1$ and $S_2$, the higher the probability of a transition from state $S_1$ to state $S_2$. Thus, the difference $Ent(S_2) - Ent(S_1)$ between the entropies of the two states, which is a measure of the probability of a transition from state $S_1$ to state $S_2$, must be a function of $p_2/p_1$. This requirement can only be satisfied by the definition that the entropy of a state corresponds to the logarithm of its probability, since then the difference $Ent(S_2) - Ent(S_1)$ is equal to $\log(p_2) - \log(p_1) = \log(p_2/p_1)$. The difference of the entropies will be positive only if state 2 has a greater probability than state 1. Since the transition from a state with low probability to a state with high probability is more likely than the reverse transition, the entropy of a system tends to increase until it has reached its maximum value. This corresponds to the equilibrium state.

In thermodynamics, the interesting system states are determined by the distribution of the location and the kinetic energy of the particles, and these are related to temperatures and flowing thermal energies. Therefore it must be possible to compute the difference of entropy corresponding to a transition from one state to another from the values of the flowing thermal energies and the applicable temperatures. However, the deduction of the formula for this would not give you much additional insight, and therefore, I decided to omit it.

## How Particles Which Had Been Called "Indivisible" Broke Apart

As I already mentioned, the word *atom* is derived from the Greek word 'atomos' which means indivisible. Of course, the use of this name did not prevent scientists from performing experiments with the goal of dividing atoms. Such experiments had become possible by the findings of Faraday and Maxwell. They had detected a way of "shooting bullets against atoms" where the bullets are electrical charges being accelerated in an electric field. However, they could not perform such experiments themselves, since suitable bullets had not yet been found.

Michael Faraday had done experiments with electrical currents flowing through liquids. Thus, it is no wonder that other scientists tried to make current flow through gases or even through a vacuum. They produced closed glass bulbs containing two separate metal plates, called electrodes, which could be connected to an external voltage source via wires leading through the glass wall. Battery voltages were not high enough to cause a current to flow between the electrodes. But generators based on the law of induction provided much higher voltages. The experiments could be varied by choosing different gases and by varying the pressure of the gas and the distance between the electrodes. In cases where an adequate experimental setup was found, a current really did flow, and sometimes strange light phenomena were observed. Today's fluorescent tubes are successors to those experimental setups. Like children who always come up with new variations of games, physicists develop many ideas about how to vary their experiments. Thus, someone had the idea of heating one of the electrodes and making it glow. This led to the detection of electrons. The electrons which are components of the atoms close to the surface of the glowing electrode can be torn away from their atoms if the electric field caused by the voltage between the electrodes is strong enough, and if it is directed towards the glowing electrode. These electrons, which have a negative charge, will then be accelerated in a direction opposite to the direction of the electric field and will fly toward the cold electrode.

From the history of the evolution of the theory of electromagnetism, we know that sometimes an accidental observation triggers an avalanche of findings – remember Mr. Galvani observing the twitching of the frog legs or Mr. Oersted observing the turning of the compass needle. Now, again, an accidental observation stood at the beginning of a development which drastically changed human life. This development finally led to nuclear weapons and nuclear power plants. In 1896, the French physicist Antoine Henri Becquerel (1852-1908) put a piece of uranium ore into a drawer next to a pack of black-and-white film. The element uranium was discovered in 1789 and got its name in reference to the planet Uranus which had been discovered eight years earlier. When Mr. Becquerel developed the film, he detected a black spot whose shape corresponded to the shape of the piece of ore. Since both the ore and the film had been wrapped with opaque paper, the spot could not have been caused by light the ore might have emitted. Therefore, Mr. Becquerel concluded that the uranium ore must have emitted a kind of radiation which had been unknown previously. He had detected natural radioactivity. Soon, other physicists also performed experiments with radioactive materials. Best known is Maria Slodovska (1867-1934) who, in 1891, had come from Poland to Paris for a university education in physics. She is better known as Madame Curie since, in 1895, she married the French physicist Pierre Curie with whom she later shared interests in radioactivity. In 1898 she found the element radium whose radioactivity is stronger than that of uranium by a factor of more than one million. In 1903, Madame Curie, her husband and Mr. Becquerel got the Nobel Prize in physics. In 1911, Madame Curie got her second Nobel Prize, this time in chemistry.

The idea that certain materials emit an unknown kind of radiation had been concluded from the fact that these materials could produce spots on a film as if the film had been exposed to light. But what kind of radiation could this be? Though the rays could not be seen, the places where they hit the film were visible. From this it was possible to find out whether and how much the paths of rays can be changed in direction by an electric or magnetic field. The corresponding experiments showed that there is not just one type of rays, but three which were called alpha-, beta- and gamma-rays. The first two are rays of particles which can be deviated by an electric field because they have an electric charge. They differ in their masses and their charges. Once all the three components of atoms - electrons, protons and neutrons - were known, in 1932, alpha-particles were found to be nuclei of helium atoms which consist of two protons and two neutrons, and beta particles were found to be electrons. The direction of gamma-rays could not be changed by electric or magnetic fields from which follows that they are not rays of charged particles. In 1914, the final experimental confirmation found that gamma-rays are electromagnetic waves.

Looking for spots on films was not the only way to find out whether there was radiation and whether the rays could be deviated. In 1824, an interesting property of the mineral fluorite was discovered. When fluorite was exposed to ultraviolet light, it emitted light of colors between green and blue. The effect later was called *fluorescence* in reference to the mineral for which this property had been observed. Those of you, who are old enough to be familiar with television sets where the picture tube was a cathode ray tube, saw the application of fluorescence, perhaps without knowing it. For the pictures you saw on those TV monitors were produced by rays of electrons hitting fluorescent material which had been coated onto the inside surface of the picture tube.

Many of the experiments which finally provided information about the internal structure of atoms were performed in the laboratories of Ernest Rutherford (1871-1937). He was born in New Zealand, where he got his university education in physics. He then moved to Canada where he did research from 1898 until 1907. Afterwards, he worked in England, first in Manchester, then in Cambridge. In his laboratory, he found what the alpha-, beta- and gamma-rays really are. As a consequence of this, the components of atoms – electrons, protons and neutrons – could be identified. The electric charge of a proton is positive and that of an electron is negative, but the amounts of charge are the same. The mass of a proton is approximately equal to that of a hydrogen atom while the mass of an electron is less by a factor of 1836. This justifies the assumption that a hydrogen atom consists of only two components, one proton and one electron. An alpha-particle has the charge of two protons and the mass of approximately four protons. This led to the question about whether an alpha-particle consists of four protons and two electrons, in which case the negative charges of the two electrons would neutralize the positive charges of two of the protons. Another structure which also would explain the behavior of alpha-particles is that the alpha-particle consists of two protons and two other components which don't have any charge, but whose mass is close to that of protons. This second structure finally was confirmed in 1932; the components which have approximately the same mass as the protons, but don't have any charge, were called *neutrons* to indicate their electrical neutrality.

After the masses and charges of the components of atoms had been found, scientists asked themselves how they could find information about the sizes of these components. From the ratio between the masses, e.g., (electron mass):(proton mass) = 1:1836, no conclusions could be drawn with respect to the actual sizes of these components. Again, it was Ernest Rutherford who, in 1909, did experiments which provided useful information about these sizes. He exposed a very thin foil of gold to alpha-rays which were emitted by a piece of radioactive material and observed the distribution of the light spots on a fluorescent screen located behind the foil. The distribution he observed indicated that most of the alpha-particles went through the foil without being deviated, while a minor fraction of the particles was deviated more or less. Some of them were even totally reflected. This observation could be explained by assuming that the gold atoms have a very small

Components of Atoms 261

center, the so-called nucleus, where the mass of the atom and the charges of the protons are concentrated. The distance between these nuclei is so great that there is enough space in between for the alpha-particles to pass through without being deviated either by the mass or by the charge of the protons. This distance could be explained by the assumption that the electrons of an atom, which neutralize the charge of the protons, move around the nucleus at a great distance and thus determine the size of the atom.

This model of an atom, consisting of a very small nucleus in the center and a distant shell where the electrons move around, obviously conflicts with common sense, since it says that all matter, be it gold, iron or stone, which we experience to be extremely impermeable, actually consists to the greatest extent of "nothing", just empty space. Fig. 10.10 roughly illustrates the proportions of an atom, but unless you read the numbers in this figure, you would still get the wrong impression. The numbers say that the distance between two nuclei is greater than the diameter of a nucleus by a factor of ten thousand. The grey shaded overlapping shells indicate the zone which contains the electrons. Originally, Rutherford and others assumed that the electrons orbit the nucleus like the planets of the sun. But quantum theory, which shall be presented in the next chapter, showed that the assumption of electrons moving in well-defined orbits is wrong.

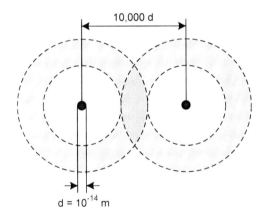

**Fig. 10.10** Proportions concerning the size of atoms

Every time I look at Fig. 10.10, my thoughts go back to a man whom, many years ago, I saw walking up and down in front of the cafeteria at our university, carrying a poster with an inscription which said, "All physicists are deceivers because they claim that solid matter consists of small atomic nuclei which sit extremely far from each other, leaving a lot of empty space in between. If this actually were the case, we could look through all solid material as if it were glass." I didn't see any way to convince him that he was wrong, since this would have meant that I must introduce him to quantum theory which tells us how matter and light interact.

When the neutron as the third component of atoms was discovered in 1932, many phenomena of chemistry including the periodic table of the elements could be explained. Nevertheless, the search for elementary particles continued, since now the question that had to be answered was whether electrons, protons and neutrons are also composed of smaller components. Many such particles have been detected since then, but until now, knowledge about them is still absolutely irrelevant for the work of engineers. The corresponding theories do not help to develop better basic materials or to design more powerful systems. But these theories are fundamental with respect to developing new hypotheses concerning the evolution of the universe. A few years ago, a physicist working in this area said, "The distance between our world of elementary particle physics and the world of common people has become breathtaking." This breathtaking distance is partially the consequence of the fact that, in this field of physics, only about 25 percent really is physics, while 70 percent is mathematics. The remaining five percent belongs to philosophy. I didn't find these numbers in any source; they merely illustrate my personal view. By reading Chapter 8 on relativity theory, it should have become obvious to you that mathematics is playing a more and more dominant role in physics. This is even more the case in the field of elementary particle physics.

A very impressive result was the detection of the so-called *antiparticles*. In 1928, the British physicist Paul Dirac (1902-1984) came up with an equation which combined quantum theory with relativity theory, and this equation has, as is the case with many equations, more than one solution. One of these solutions could be associated with the well-known electron, but the other solution seemed to correspond to an electron having negative energy. This triggered a lot of thinking about the question of how this second solution should be interpreted. Later experimental results led to the assumption that any subatomic particle has a partner which can be viewed as its mirror image and which is called its antiparticle. The theory required that some particles are identical to their antiparticles. The situation is analogous to the relation between positive and negative numbers, where the negative numbers are the mirror images of the positive numbers, and where the negative zero is identical to the positive zero. When they meet each other in a sum, the positive number and the negative number disappear, and the result is zero. In the analogy, this means that if a particle and its antiparticle meet, they both disappear and the sum of their original masses is completely converted into energy. For the electron-positron pair, this actually could be confirmed experimentally. The concept of matter and antimatter led to a completely new theory of the so-called *empty space*.

Experimental results in elementary particle physics are mainly geometric properties of the tracks which the particles leave behind after they have traveled through a detector chamber. Although the types of detector chambers have changed drastically over the years, the basic principle has not changed at all. The chamber is filled with a gas or a liquid, and the particles which enter the chamber with very high speed cause the atoms or molecules of the gas or the liquid to emit light. Then, the tracks can be saved on film or on digital memory devices. Usually, the chambers are located in strong electrical or magnetic fields which make a track of charged particles deviate from a straight line. From such a track it is possible to deduce the speed, the mass and the charge of the particle. In many experiments, there are two rays of particles entering the chamber from opposite ends, and when two particles collide, the resulting tracks provide the information about whether the particles remained complete or broke apart. In the case of a break up, the tracks provide information about the pieces.

The laws of conservation of energy and momentum apply not only to processes in the macro world, but also in the world of elementary particles. When the physicist Wolfgang Pauli (1900-1958) applied the law of conservation of momentum to the results of a certain kind of radioactive decay, he came to speculate that, in addition to each electron which is emitted, another particle which has no charge is also emitted, and if it has any mass at all, this mass must be extremely small. Later, this hypothetic particle was called the *neutrino*. Physicists for many years had no hope of ever detecting this particle experimentally because of its not being charged and its extremely small mass. Computations said that a nuclear reactor emits rays of neutrinos with a density of $10^{12}$ neutrinos, i.e., one thousand billions neutrinos per square meter and second. However, almost all of these neutrinos go unhindered through steel walls, even those with a thickness of ten meters. The probability that one neutrino gets stuck on its way is extremely small. In order to illustrate the situation, an expert in particle physics once said, "There are ten billion rain drops per second falling on our town, and one of these is pink. That's the one we have to find."

It may be that you already know that there are particles called *quarks*. The physicists who first assumed the existence of such a type of particle had to choose a name for them, and they had all the freedom of choosing any name. The proposal to call them quarks came from Murray Gell-Mann (1929- ). He took the word from the sentence, "Three quarks for Muster Mark!" which can be found in the short novel "Finnegan's Wake" by the Irish author James Joyce. Actually, it is not the word quark, but the number 3 which has a meaningful relation to this new type of particle. The hypothesis that such particles might exist was a conclusion from the observation that the track of an electron, which passes very close by a proton, does not have exactly the shape it should have if the charge of the proton were concentrated in its center. The shape of the track could be explained by the assumption that the charge of the proton is distributed asymmetrically, and this would be the case if the proton consisted of components having different charges. Today, the experts in particle physics are convinced that the proton is composed of three quarks, two of which have the same charge while the third one has a different charge. Again, names had to be chosen for these different types of quarks, and, quite arbitrarily, they were called *Up* and *Down*. The charge of an Up is 2/3 of the charge of a proton, and the charge of a Down is 1/3 of the charge of an electron which corresponds to 1/3 of the opposite charge of a proton. Thus, the charge of a proton is the sum of the charges of two Ups and one Down: 2/3 + 2/3 - 1/3 = 1. By combining two Downs with one Up, a particle with zero charge is obtained, and this is the neutron: 2/3 - 1/3 -1/3 = 0.

While two types of quarks, Up and Down, were sufficient to explain the existence and the behavior of the protons and the neutrons which are the components of the long-lived matter of our daily experiences, they could not explain the existence of certain extremely short-lived particles which occur during certain experiments. The *muon* which I mentioned in the chapter on relativity theory and which has a life span of about two microseconds, is an example of such a short-lived particle. In order to explain such particles, the existence of two additional pairs of types of quarks was assumed, and these types got the names (*Strange* and *Charm*) and (*Top* and *Bottom*). These names, again, were chosen arbitrarily – this choice has no better justification than the choice of names like Tom and Jerry or Pretty and Ugly would have had. It has not yet been possible to isolate single quarks experimentally. They have been assumed to be components of higher particles, and it seems that the forces which keep them together in certain groups are so strong that there is no way of breaking up such a group.

The question of which forces keep the components of structured particles together came up as soon as it became clear that the nucleus of an atom contains protons which have positive charges. Coulomb's law says that a force of repulsion pushes particles apart if their charges have the same sign. Thus, an explanation had to be found concerning why the protons of an atom nucleus are not pushed apart. There must be an attracting force between them which is stronger than the repulsion force according to Coulomb's law. In the field of particle physics, the concept of force has been replaced by the concept of interaction. This is the consequence of the idea that there are no static forces at all, and that forces occur only in the case of collisions. When I introduced the concept of force as one basic concept of Newton's theory of mechanics, I first introduced the concept of momentum as the original concept on which I then could establish the definition of force (see Fig. 7.3). In the view of elementary particle physics, however, the process of two particles being driven apart requires a third particle which moves back and forth between the two particles as a result of alternating collisions. Consequently, the process of two particles attracting each other would require that the third particle orbit the other two

Components of Atoms

particles and push them together from the outside, like a sheep dog running around the flock and pushing the sheep to the center. If, however, we neglect our experience from the macro world, we can also produce an attracting force by letting the third particle transmit a negative momentum at its collisions.

Since about 1945, physicists distinguish four different kinds of interactions: (1) gravity which keeps the planets in their orbits and is the reason for our being attracted towards the center of the earth; (2) electromagnetism which keeps an electron near the nucleus of its atom and is used in all kinds of technical systems such as electric motors; (3) strong nuclear interaction which makes the components of an atom nucleus stick together; (4) weak nuclear interaction which occurs in connection with radioactivity.

While Ernest Rutherford and his colleagues who clarified the structure of atoms could perform their experiments with rather simple set-ups, the experimental systems of particle physicists have become more and more complex. The biggest and most expensive laboratory in the world for research in particle physics is located near the Swiss city of Geneva. It was founded in 1952 and is called CERN, which is an abbreviation of the French name of the institution, Conseil Européen pour la Recherche Nucléaire. By the end of 2008, their biggest experimental set-up was completed, a circular tunnel with a diameter of about 10 kilometers. The equipment in this tunnel makes it possible to accelerate charged particles until they have a speed which is very close to the speed of light. Finally, particles flying in opposite directions collide and the tracks in the detector chamber provide information about the results of such collisions. The energy of a moving particle is not expressed in the usual unit Watt*seconds (Ws), but in the standard unit electronVolts (eV) which is commonly used in elementary particle physics. One eV is the kinetic energy of an electron which has been accelerated by a potential difference of 1 Volt. Modern particle accelerators make it possible for the colliding particles to have an energy in the range of $10^{12}$ eV. From the high exponent of this power of 10, you might get the impression that this is an extremely high energy, but it corresponds only to approximately one sixth of the millionth part of one Ws. This is less than the kinetic energy of a flying mosquito. But since the mass of an electron is much, much less than the mass of a mosquito, the electron would have to move faster by a factor of over one billion in order to have the same kinetic energy as the mosquito.

# Chapter 11
# How the Difference between Particles and Waves Disappeared

Before James Maxwell discovered that light is an electromagnetic wave, scientists discussed the question of whether or not light is a stream of particles, similar to those considered in the previous chapter. Such particles have a rest mass and can be accelerated. The Dutch Christiaan Huygens (1629-1695) argued that light couldn't be a stream of particles since two beams of light, in contrast to beams of water, intersect without affecting each other. Many scientists invested a lot of thought into how to experimentally determine whether light is a wave or a stream of particles.

## How Waves Can Be Forced to Show Us That They Really Are Waves

Finally, someone came up with the idea to check whether the kind of interference which had been observed in the case of intersecting water waves could also be produced with light. Fig. 11.1 illustrates this kind of interference. In this figure, the wave is coming from the top and hits a wall which has two openings. From each of these openings, a semicircular wave propagates, and these two waves interfere with each other. At points where a peak meets a peak and a trough meets a trough, the wave will be reinforced, and at the points where a peak meets a trough, the wave will be cancelled. The angles between the directions of reinforcement and cancellation are determined by the distance between the openings and the *wave length*, which is the distance between two consecutive peaks. Thus, the wave length can be determined from the distance between the openings and the angles between the directions of reinforcement and cancellation. While the wave length lies in the range of centimeters or meters in the case of water waves, it is less than a thousandth of a millimeter in the case of light waves. Although it is much easier to demonstrate wave interference in the case of water waves, it is also possible to produce it with light waves. But since interference occurs only if the wave has a well-defined wave length, no interference will be observed when sunlight is used in the experiment. Sunlight has no well-defined frequency but is made up of a combination and superposition of waves having different frequencies. The entire set of these frequencies is called the *spectrum* of sunlight. Each of these frequencies corresponds to a different color, and the colors of this spectrum become visible as rainbow colors when raindrops and sunlight come together in a certain combination, or when sunlight passes through a glass prism. The lower frequencies which correspond to longer wave lengths are at the red end of the rainbow, and the higher frequencies which correspond to shorter wave lengths are at the violet end.

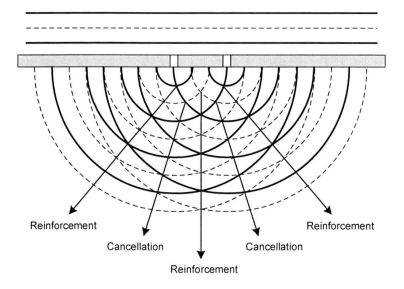

**Fig. 11.1** Wave interference following a pair of openings

When he analyzed the spectrum of the sun with very precise optical instruments, the German optician Josef von Fraunhofer (1787-1826) detected over 500 dark lines which were irregularly distributed among the rainbow colors. Later, these lines were called Fraunhofer lines. Obviously, he had detected some missing frequencies which were either never emitted by the sun or were eliminated on the way from the sun to the earth. That it's actually an elimination and how this happens will be discussed in the following paragraphs.

The frequencies in light are so extremely high that we cannot really imagine them – or can you imagine something oscillating a million of billions times per second? To understand this, we will first think of oscillators having frequencies in the range of only a few cycles per second, and afterwards we shall transfer our findings formally to the unimaginable high frequencies of light. The system which is represented at the top of Fig. 11.2 contains a block which is sitting between two

Waves and Oscillations

springs, the left one being stronger than the right one by a factor of 11. The end of the spring at the right can be moved back and forth by a stimulator. We assume that the block is sitting on an oil film and that the friction between the block and the base plane can be neglected. The form of stimulation considered is represented by the curve s(t) in the two diagrams which also contain the corresponding responses r(t). I got the stimulation function s(t) by adding five cosine functions whose frequencies have the ratio 1:2:3:4:5. Since the stimulation should begin at s(0) = 0, but the cosine function is 1 at x = 0, I had to shift the curve f(2π*t/T) to the right by 0.225 T. The lowest frequency contained in s(t) is 1/T, and this determines the cycle time T of the periodic curve. This curve doesn't have the shape of a cosine function because of the higher frequencies 2/T through 5/T. The response curves r(t) will be discussed later.

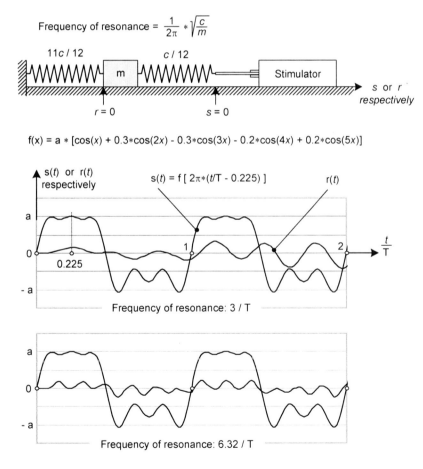

Fig. 11.2  Fourier decomposition and resonance

In his book on the theory of heat [FOU], the French scientist Jean Baptiste Joseph Fourier (1768-1830) presented a method for decomposing a given periodic function of arbitrary shape into a sum of sine and cosine functions. This method later was called the *Fourier transformation*. Without the knowledge that any periodic function can be considered as a sum of weighted sine and cosine functions having the frequencies 1/T, 2/T, 3/T ..., we could not explain the phenomenon of *resonance*. Each system which can behave as an oscillator has its *resonant frequency* which is also called its *eigenfrequency*. "Eigen" is a German word that means "own." It is universally used in mathematics and physics to indicate certain characteristics of systems. When a system which has a resonant frequency is brought into a state outside of its equilibrium state and then left alone, it will oscillate at its resonant frequency. The most illustrative occurrence of resonance I remember is related to a bus ride in my home city: each time the bus had to stop at a red traffic light, its windows began to rattle loudly. The periodic stimulation of the windows by the engine idling at a low speed obviously corresponded to the resonant frequency of the windows. Resonance effects can be so strong that they destroy a stimulated system. The system in Fig. 11.2 has a resonance frequency which is determined by the mass m of the block and the constant c which describes the strength of the springs.

The upper diagram in Fig. 11.2 represents the behavior of the system when its resonance frequency is 3/T. Since this corresponds exactly to the frequency of one of the summands of the stimulation $s(t)$, the system responds as if it were stimulated only by this frequency and therefore it shows the effects of resonance, i.e., it begins to oscillate at the resonant frequency with increasing amplitude. If there were no friction, the amplitude would continue to grow until either the system would break apart or its structure would finally limit the amplitude. In this process, the stimulator provides the energy which the oscillator absorbs. The lower diagram represents the situation where the resonance frequency is 6.32/T which does not correspond to any of the frequencies of the summands in the stimulation function $s(t)$, or to any multiple of these frequencies. Therefore, the response $r(t)$ doesn't show any resonance effects, i.e., the amplitude of $r(t)$ stays limited.

The finding that an oscillator can withdraw energy from a stimulator helps us to explain the Fraunhofer lines which I mentioned above. There is no reason to assume that the frequencies belonging to the Fraunhofer lines are not emitted by the sun. But since they are missing when sunlight reaches the earth, the corresponding energy must have been withdrawn on the way from the sun to the earth. This leads to the assumption that light passes through systems that have resonance frequencies which correspond exactly to the frequencies of the Fraunhofer lines. It took about one hundred years after Fraunhofer's discovery in 1814 before it became known what kind of systems caused these resonance effects.

## How It became Necessary to Consider Rays of Light and Heat as Flying Packets of Energy

As a child growing up in a small town, I loved to watch the blacksmith shoeing the horses of the farmers from the nearby villages of the Black Forest. Even today, I remember this blacksmith whenever I see a block of iron at red heat. Although mankind knew for thousands of years that iron at red heat emits light and heat, it was unknown what kind of rays these are until about 150 years ago. In Chapter 9, I told you the logic by which James Maxwell discovered that light is an electromagnetic wave. This led to the question of whether, perhaps, heat might also be such a kind of wave. Actually, the difference between a wave of light and a wave of heat lies only in the frequency which is higher in the case of light. Another question was how the intensity of the emitted waves depends on the temperature of the emitting body and the frequency of the wave. This intensity is defined as the energy which arrives at a square meter of an irradiated surface each second. In order to measure the dependency on the frequency, the spectrum of the rays had to be expanded beyond the spectrum we know for the rainbow. The experimental results are shown in Fig. 11.3. As long as the temperature is not high enough to make the bodies glow, the entire energy of the emitted radiation lies in the infrared section of the spectrum. Only after the bodies have been heated to higher temperatures do they also emit energy in the visible part of the spectrum. With increasing temperature, the emitted light goes from faint red to more and more intense white.

In 1900, the German physicist Max Planck (1858-1947) became interested in the curves concerning the radiation of heat shown in Fig. 11.3. Until then, no

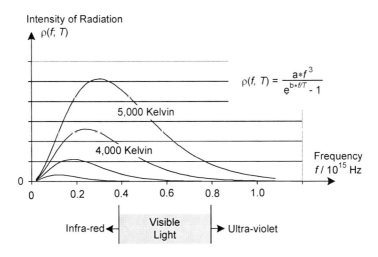

**Fig. 11.3** Radiation, heat and temperature

convincing explanation of the shape of these curves had been found. After having tried many dead ends, Mr. Planck finally brought himself to assume – "in an act of desperation" – that heat energy is not emitted continuously but as a stream of extremely small packets whose contained energy is proportional to a particular frequency. He had to choose a name for the proportional factor, and he chose the letter h as a reference to the German word "Hilfsgröße" which means "auxiliary quantity" in English. Later, h was interpreted as "action quantum," and in honor of Max Planck it is called *Planck's constant*. Its value is h = $6.626*10^{-34}$ kg*m$^2$/s; its unit corresponds to the product of an energy with a time, and thus the unit can also be written as Ws$^2$ (see Fig. 9.17). This is also the unit of an angular momentum (see Fig. 7.10). The packets of energy are called *quantums*, and in the special case of light they are called *photons*. In order to get an idea about how extremely small the energy of a quantum is, let's consider a radiator in a living room at a temperature of about 40 °C. According to Fig. 11.3, this radiator emits packets of energy in the frequency interval $10^{13}$ Hz < f < $10^{14}$ Hz, and thus the energy of a packet in the middle of this interval is approximately h*f = $3*10^{-21}$ Ws. In order to get just one Wattsecond, we would need more than one hundred billions of billions of such packets.

Planck's theory conflicted heavily with classical physics where the energy of a wave depends on its amplitude, but not on its frequency. However, even before Planck, experimental results had been found which indicated a connection between the frequency and the energy of light. In 1839, Alexandre Edmond Becquerel, the father of Henri Becquerel who discovered the radioactivity of uranium, observed a phenomenon which later was called the *photoelectric effect*. In his experiments, a current began to flow through a vacuum as soon as the negative electrode was hit by rays of light. Today, we can say that the light "hammers out" the electrons from the metal of the electrode, but when the effect was detected in 1839, the existence of electrons had not yet been discovered. Fifty years after the discovery of this effect, Wilhelm Hallwachs, an assistant of Heinrich Hertz, determined experimentally how the kinetic energy of the electrons at the moment of their leaving the electrode depends on the frequency of the light which sets them free. This dependency was not explained until 1905. For his explanation, Albert Einstein got the Nobel Prize in 1921. The light arrives at the electrode as a stream of energy packets whose energy is proportional to their frequency according to the formula h*f. When hitting the electrode, a packet cannot free an electron if its energy is below a certain threshold $E_0$ which is specific to the material of the electrode. The energy $E_0$ is absorbed in the process of detaching the electron from the electrode. Thus, the kinetic energy of the free electron will correspond to the difference between the energy brought by the packet and the energy absorbed by the electrode, i.e., $E_{kin} = h*f - E_0$.

In his book [HUN] on the history of quantum theory, Friedrich Hund (1896-1997) wrote: "Quantum theory is the description of the role which Planck's constant h plays in the universe." After Max Planck had his ingenious idea in 1900, it took almost three decades until most of the questions concerning the role of h had been answered. But the answers refer to rather mysterious phenomena which conflict strongly with our common sense. Not only laymen in the area of physics, but even professional quantum physicists cannot really understand these phenomena, but have to accept them despite how mysterious they are. The existence of these phenomena has been confirmed experimentally many, many times, and never did the experimental results contradict the formulas which describe these phenomena.

At first, the questions asked did not refer to mysterious phenomena, but to phenomena which always had been thought to be quite natural. Heated metal blocks emit rays of light and heat, and Maxwell had concluded from his equations that these rays are electromagnetic waves. Thus, an interesting question was how the atoms of the heated material could emit electromagnetic waves. Since the year 1896, when Heinrich Hertz (1857-1894) designed an experimental set-up for generating and receiving electromagnetic waves in his laboratory, it is well known how to build electromagnetic oscillators and antennas for emitting and receiving electromagnetic waves. Today, the earth is full of such devices – think of radio and television broadcasting, or cellular telephony. The wavelengths of these radio waves lie in the range of meters, centimeters or millimeters and are longer than the wave lengths of light and heat waves by a factor of 100,000. Radio waves are emitted when a current flows periodically to and from an antenna. The electrical energy which is carried away by the electromagnetic wave must be provided continuously from batteries or from a generator in a power plant. In the case of radiating atoms, the process of providing the energy for the electromagnetic waves emitted by the atoms certainly cannot be the same as in the case of radio or television stations. The energy which an atom passes to the emitted electromagnetic wave can come only from heat, and heat is nothing but the kinetic energy of moving particles in the micro world. Thus, the assumption that atoms are able to transform kinetic energy into electromagnetic energy was justified. The only way an atom can obtain energy from heat is by collisions. It seemed reasonable to assume that such a collision could cause a transition of the state of the atom from not being able to emit a wave to a state of potential radiation. An analogy to the atom would be a mechanical system containing a spring: the transition from the state where the spring is without tension to the state with tension requires mechanical work which is converted into the potential energy of the spring. The mechanical energy an atom receives in the process of a collision becomes potential energy within the atom. And later, this potential energy is emitted in form of an electromagnetic wave carrying the energy h*f. Experimental results which showed that specific atoms can emit waves only at very specific frequencies could now be explained by the assumption that an atom can be only in states where the potential energy has very specific values.

Since an atom certainly doesn't contain any springs, there must be other methods of storage for the potential energy. It seemed reasonable to assume that the different states which correspond to different values of the potential energy are determined by different locations of the electrons in the electric field caused by the positive charge of the protons in the nucleus of the atom. However, this assumption still doesn't explain the fact that there are only very specific possible values for the potential energy. Before I can introduce the theory which explains this, some other considerations must first be presented.

In the chapter on Einstein's theory of relativity, you learned that there is a relation between mass and energy which is expressed by the formula $E=m*c^2$. Since this formula is believed to be universally valid, it can be applied to a single photon which has the energy $h*f$, and from this it follows formally that a photon has a mass and a momentum (see Fig. 11.4). The speed of the photons is always equal to the speed of light c, and from this we can conclude that photons don't have a rest mass. Here, you should take the time to look back at Fig. 8.10 which represents the deductions leading to the formula $E=m*c^2$. If you try to apply this formula to the case where the rest mass $m_0$ is zero and the ratio v/c is 1 – the case that characterizes the photon -, you cannot get a resulting energy, since the formula now contains the product zero times infinity. But since the energy of the photon is known to be $h*f$, we can equate this formally with $m*c^2$. The formula for the momentum refers to the wave length $\lambda$ which is related to the frequency f and the propagation speed c by the equation $f*\lambda=c$. This equation can be easily understood as follows: f corresponds to the number of wave lengths $\lambda$ which are generated per second, and so $f*\lambda$ corresponds to the distance the wave propagates in one second, and this is equal to the speed c.

| Energy E (Planck) | Energy E (Einstein) | Mass m | Momentum p = $m*c$ |
|---|---|---|---|
| $h*f$ | $m*c^2$ | $\dfrac{h*f}{c^2}$ | $\dfrac{h*f}{c} = \dfrac{h}{\lambda}$ |

Fig. 11.4 Mass and momentum of a photon

In the area of mechanics, momentums become relevant mainly in the case of collisions. Since in Fig. 11.4 the photons have momentums assigned to them, the question arises about whether a photon can collide with another photon or with a particle which has a rest mass. As an example of such a particle, let's consider an electron which has a rest mass of $9.1*10^{-31}$ kilogram. A photon of visible light has an average frequency of about $5*10^{14}$ Hertz from which, according to Fig. 11.4, it

# Discovery of Quanta

follows that its mass is $4*10^{-36}$ kilogram. Thus, the mass of a photon is less than the rest mass of an electron by a factor of about 200,000. Therefore, a photon colliding with an electron is similar to a tennis ball colliding with a locomotive. However, there are quantums having frequencies which are much higher than the frequency of light. In 1895, the German physicist Wilhelm Conrad Roentgen (1845-1923) discovered a new type of rays which he called X-rays in reference to the variable x which is used for unknowns in equations. At that time, he could not have known that these rays are the result of two different processes. Both processes are initiated by a beam of electrons hitting a metal electrode. The electrons are slowed down and stopped, and thus lose their kinetic energy which, partially or completely, is converted directly into electromagnetic radiation. But there is also the possibility that the kinetic energy of the electrons is consumed for transitions of the energy states of the atoms of the electrode. These atoms originally are in a state of low energy and are taken into states of very high energy. They stay in the state of high energy for some time which is not defined precisely, but defined only by a probability distribution, and then they return to the state of low energy by emitting radiation at a frequency determined by the energy difference $\Delta E$ between the two states according to the equation $f=\Delta E/h$. The frequencies of X-rays lie in the range of $5*10^{17}$ Hz $< f < 5*10^{19}$ Hz which is higher than the average frequency of light by a factor of 1,000 to 100,000. These high frequencies correspond to very small wave lengths and these are the reason for the fact that X-rays can pass through bodies which are impermeable to light. In 1923, the American physicist Arthur Holly Compton (1892-1962) confirmed experimentally that collisions between X-quantums and electrons really do have the effects which correspond to their masses and momentums given in Fig. 11.4.

Clearly, our image of two colliding particles does not correspond to the idea that a photon is a wave. It is impossible to find out experimentally what a photon "looks like." Some experimental results can be explained only by the assumption that the photons are particles, while other experimental results can be explained only by the assumption that a stream of a huge number of photons is a wave. As a compromise, we can assume that a single photon is a so-called wave packet which has a shape similar to the examples in Fig. 11.5. The energy contained in such a wave packet is proportional to its length and to the square of its amplitude. From this it follows that the two wave packets in the figure have the same energy, since $8*1^2 = 32*0.5^2$.

If someone who is not familiar with the findings of Mr. Fourier looks at the functions in Fig. 11.5, he might conclude that these functions have a well-defined wave length $\lambda$ from which the exact value of the frequency can be obtained by applying the equation $f = c/\lambda$. But this conclusion would not be correct. In my comments about Fig. 11.2, I referred to the so-called Fourier decomposition. Mr. Fourier found that any periodic function can be expressed as a sum of sine and cosine functions. Later, he even showed that non-periodic functions, if they satisfy certain conditions,

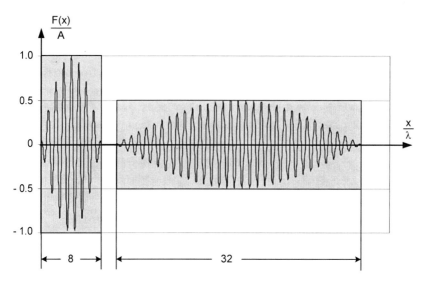

**Fig. 11.5** Two wave packets with the same wave length and energy

can be decomposed into sine and cosine functions. In this case, adding up the components must be done by integration. From this it follows that only waves which can be described by the formula sin[$2\pi*(x/\lambda - f*t)$] have an exact frequency f; all other waves are compositions of more than one sine-wave, each with a different frequency. Thus, the wave packets in Fig. 11.5 contain many frequencies.

Each function which can be drawn in a rectangular area, as it is the case for the wave packets in Fig. 11.5, satisfies the condition for being representable in the form of a Fourier integral. Thus, I could have applied the Fourier transformation to the two wave packets in Fig. 11.5. But this would have required a rather extensive computation which I wanted to avoid. Therefore I considered the simpler "wave packets" which are represented in the left half of Fig. 11.6. Each of these wave packets consists of an odd number of sine cycles. By placing the center of the coordinate system at the middle of the wave packets, I obtained functions which satisfy the condition $F(-x) = -F(x)$. The Fourier transformation of such functions requires only the computation of the sine-component $S(f)$, since the associated cosine-component $C(f)$ is identically zero. In the right section of the figure, you see the functions $S(f)$ which correspond to the wave packets on the left side. The frequency $f_0$ belongs to a wave having infinitely many sine cycles of length $\lambda$.

While for each of the functions $F(x)$ in Fig. 11.6 there exists a threshold $\Delta x$ which determines an interval $-\Delta x < x < \Delta x$ outside of which $F(x)$ is zero, a corresponding threshold $\Delta f$ for the functions $S(f)$ does not exist. Nevertheless, as a helpful approximation we can assume that for each of the functions $S(f)$ a threshold $\Delta f$

Discovery of Quanta

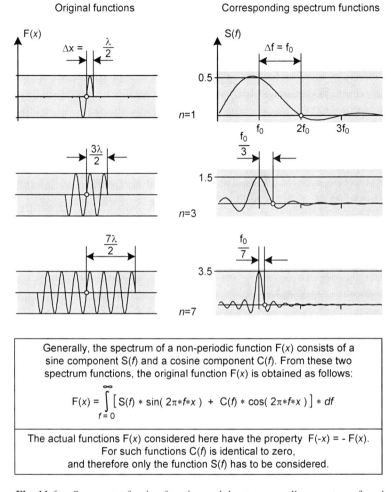

**Fig. 11.6** Segments of a sine function and the corresponding spectrum functions

exists which determines an interval $-\Delta f < f < \Delta f$ outside of which $S(f)$ is zero. The values for $\Delta f$ which we shall consider are indicated in the right half of Fig. 11.6. The ratio $f_0/\Delta f$ is equal to n which is the number of sine cycles in the corresponding wave packet. From this we can conclude that the frequency interval which must be considered is smaller when the wave packet contains more cycles. This means that considering the frequency $f_0$ as the frequency of the wave packet is more justified when the packet contains more sine cycles – which is rather trivial. But this leads immediately to the nontrivial question of how many sine cycles are contained in the wave packet of a photon. I said earlier that we cannot find out experimentally what a photon looks like. When I introduced Fig. 11.5 and said that we might assume a photon to be a wave packet as a compromise, I didn't say how many cycles of length λ the photon might contain. In cases where it behaves as a particle, it seems to be concentrated locally, which means that it does not contain many cycles in these cases. In the other cases, however, where it behaves as a wave, it seems justified to assume that the photon is a wave packet with many cycles. Since we cannot have an image of something which sometimes behaves as a particle and sometimes as a wave, it would be unreasonable to expect that there is a well-defined number of cycles which we could assign to a photon. But from Fig. 11.6, at least we can deduce that the more we know about the location $x$ of a photon, the less can we know about its momentum $p$, and vice versa.

Fig. 11.7 shows how a relation between $\Delta x$ and $\Delta p$ can be obtained from the relation between $\Delta x$ and $\Delta f$ in Fig. 11.6. This relation says that if the location of a photon is determined with an uncertainty $\Delta x$, its momentum can be determined at best with an uncertainty of $\Delta p = h/(2*\Delta x)$. This uncertainty relation was first deduced by the German physicist Werner Heisenberg (1901-1976) in 1927. The assumptions he made differ slightly from mine which lead to Fig. 11.6. Heisenberg got the value $4\pi$ for the denominator instead of our value of 2.

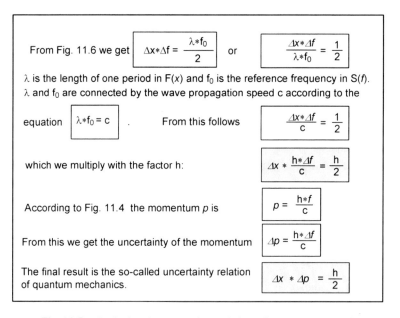

**Fig. 11.7** Deducing the uncertainty relation of quantum mechanics

In 1808, roughly one hundred years before the development of quantum theory began, the Frenchman Étienne Louis Malus (1775-1812), who had been a student of Mr. Fourier, observed a phenomenon which he could describe, but not explain. For the explanation, two concepts had to be found first, namely the concept of electromagnetic waves which was introduced by James Maxwell in 1865, and the idea of photons being indivisible energy packets which was introduced by Max Planck in 1900. In his experiments, Monsieur Malus sent light through crystals of calcite and observed that the crystals had the effect of filters. By changing the angle between the light source and the crystal, the intensity of the light which passed could be varied between almost zero and full intensity. In today's terminology, we say that Mr. Malus discovered the polarization of light and the existence of polarizing filters. In Fig. 9.10 which accompanies my discussion of electromagnetic waves, you see that the electric field E and the magnetic field H are perpendicular both to each other and to the direction of the wave propagation. A polarizing filter has a preferred direction for the electric field, which means that light passes through the filter unhindered if the electric field has the preferred direction of the filter while light cannot pass at all if the direction of the electric field is perpendicular to the preferred direction.

But instead of viewing light as an electromagnetic wave which more or less can pass through the filter, we can also view light as a stream of photons which hit the filter and can pass or not pass through the filter. In this view, for each photon it must be decided whether it is allowed to pass individually, or whether it is repulsed. In the case that the light has the preferred direction of the filter, all photons will pass, and in the opposite case, all photons will be repulsed or blocked. In other cases in between these two extremes, light passes with reduced intensity which means that a certain percentage of the photons pass while all the others are repulsed. This can be captured mathematically by a probability of passing which depends on the angle between the field direction of the light and the preferred direction of the filter. Fig. 11.8 illustrates this dependency.

You should realize that the outcome of polarizing filtering provides almost no information about the polarizing direction of a single photon which hits the filter. Even if its probability of passing is very low, it might pass, and if this possibility is very high, the photon still might be repulsed. But there is an interesting effect: each photon which succeeded in passing the filter interacted with the filter in such a way that afterwards it was polarized in the polarizing direction of the filter. This explains the results of the experiments which are illustrated in Fig. 11.9. In both the upper and lower sections of the figure, a chain of three filters is represented. The two chains differ only with respect to the ordering of the filters. In both experiments, the number of photons which pass the first filter each second is chosen as a reference which corresponds to 100 %. All these photons passed the first filter and therefore are horizontally polarized. Since the polarizing direction of the second filter in the lower chain is vertical, none of the arriving photons can pass. In the upper chain, however, the preference direction of the second filter is at 45 degrees, and therefore, half of the arriving photons will pass this filter. By interacting with

280                                                                                      11. Quantum Theory

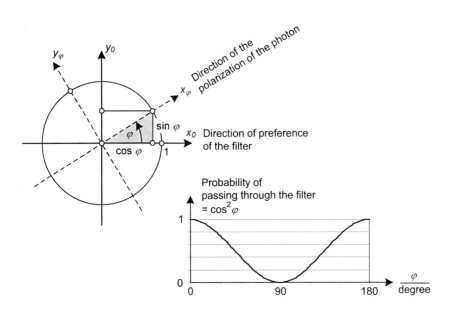

**Fig. 11.8**   Characteristic of a polarizing filter

the filter, their polarization will be changed from horizontal to 45 degrees, and this makes it possible for half of those remaining to pass the third filter.

The example in Fig. 11.9 is an example of a general law which says that all experiments in the world of quantums change the original state of the system in such a way that it is impossible to deduce from the experimental results, what the previous state was.

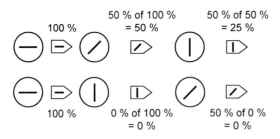

**Fig. 11.9**   Alteration of polarization by filtering

## A Theory Which Could Be Confirmed, but Stayed Inconceivable

In the table in Fig. 11.3, the mechanical quantities of mass and momentum were assigned to a photon. I am quite sure that, by looking at this table, I never would have come up with the idea to ask whether it might be meaningful to make an inverse assignment, i.e., to assign wave properties to particles having a rest mass. It was the French nobleman Louis Victor de Broglie (1892-1987) who, in 1923, proposed and wrote down such an assignment in his PhD thesis. Once he had come up with the idea, it was an absolutely formal act to make the assignments which are represented in Fig. 11.10.

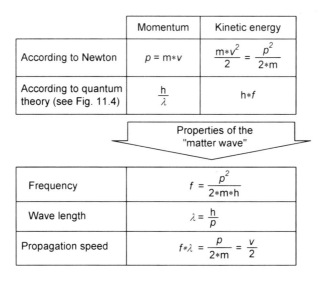

**Fig. 11.10**  Particles considered as waves (non-relativistic view: $v \ll c$)

We all know that paper accepts anything written on it, so that it can represent deep insights or pure nonsense. Therefore, de Broglie's assignments immediately provoked experimental physicists to ask how they could confirm or disprove these ideas experimentally. It didn't require a genius to come up with the idea that, if the assignments in Fig. 11.10 represented any truth concerning the physical world, it should be possible to produce interference phenomena with beams of particles with mass such as electrons. Look again at Fig. 11.1: as long as we imagine a real wave arriving at the wall with the two openings, we have no problems accepting the fact that, beyond the openings, there are directions of enhancement and others of cancellation. But it is absolutely impossible for us to imagine a process which produces the same enhancements and cancellations when, instead of a wave, a stream of particles hits the wall with the openings. Maybe you didn't think about this problem when you read about the photons in the paragraphs above. Our difficulty in understanding interference phenomena produced by streams of particles is not restricted to particles which have a rest mass such as electrons and which we think of as being "real particles," but this difficulty is the same for the case of the "particles of light" which we call photons.

This problem can be illustrated by an example which has nothing at all to do with waves or particles hitting a wall with two openings. Imagine you were sitting high up in a football stadium and were looking down on the green playing field expecting the opening ceremony of a championship game. All of a sudden, a player dressed in white comes running onto the field and stops at a position which, to you, seems arbitrarily chosen. Soon afterwards, a second player dressed in white comes and stands at another seemingly arbitrary position. One after the other, more and more players come and place themselves somewhere on the field. At first the white pattern on the green field seems to look absolutely arbitrary, but then, more and more, your conviction grows that the pattern will converge to represent a word or a text. Finally, you actually can read the word "WELCOME". Obviously, each player knew exactly where to place himself.

Now we return to Fig. 11.1. In addition to what is shown in this figure, you have to assume that there is a screen placed at some distance beyond the wall. Now we imagine a sequence of particles reaching the wall, one after the other. We assume that our view is restricted in such a way that we cannot see the moving particles, but only the positions where the particles hit the screen after they successfully passed the openings of the wall. This screen might be a film or a fluorescent plane. Some of the particles will hit the wall far away from its openings and therefore will be repulsed. But some will succeed in passing an opening and mark a position on the screen. This screen corresponds to the green field in the stadium on which we saw the players appear one after the other. Similar to the example where, at the beginning of the process, we saw only the development of a random pattern on the field, a random pattern will develop on the screen. But as time passes, this pattern will converge to the pattern which is determined by the directions of enhancements and cancellations according to the interference of a wave. Nobody can explain how this can happen; we have to accept it as a mystery, even though it can be reproduced in the laboratory as often as we want. In the case of the pattern produced by the white-dressed players on the green field, we of course know that each of these players had received clear directions about where to place himself. But in the case of the particles hitting the screen, it would be absurd to assume that they have a memory and know what they were told about their specified positions on the screen. This has remained a mystery, i.e., even the cleverest physicists haven't found an explanation. They just accept the fact that such strange phenomena can be observed in the world of quantums.

# Quantum Theory

In Fig. 11.10, you find not only the frequency f and the wave length $\lambda$ which are assigned to a particle moving at the speed v, but you also find the propagation speed which follows from these assignments. Surprisingly, this speed is only half the speed v at which the particle moves. But since we have already learned that, when dealing with the quantum world, strange phenomena must be accepted without explanation, we shall not ask how this difference between the speed of the particle and the speed of its associated wave might be interpreted.

The fact that interference patterns can be produced by letting particles flow against a wall with two openings, quite naturally leads to the question about what kind of wave it might be which is assigned to the moving particles. From the interference pattern, the wave length $\lambda$ can be deduced, but this does not provide any information about the kind of wave to which this wave length belongs. The final interpretation was introduced by the German physicist Max Born (1882-1970) who suggested that such a wave be viewed as a moving probability density distribution with respect to the location of the particle. In the coming paragraphs, more details will be presented concerning the connection between particles and associated probability densities.

Now we shall come back to the question of how it could be explained that specific atoms can emit radiation only at specific frequencies. The explanation will be based on certain probability densities. It was not a single person who found the solution to this problem, but four scientists contributed equally to this solution: the Dane Niels Bohr (1885-1962), the Austrian Erwin Schroedinger (1887-1961) and the two Germans Werner Heisenberg and Max Born. When I was looking through many different textbooks in order to see how the authors introduce quantum theory to their students, I was very surprised to note that often this theory is introduced as if it fell from heaven. It is true that the central formalism of quantum theory differs a lot from all the methods and laws which characterize classical physics. Therefore, it is no wonder that anyone who sees this formalism for the first time asks himself how the development of such a strange formalism could happen. It was postulated in the year 1927, and since then no experimental results have been found which contradict the predictions according to this formalism. It is impossible to deduce this formalism as the result of a sequence of logical steps. It could be found only because its originators were willing to leave all conventional paths and move like sleepwalkers in directions which they could not justify by logical reasoning. This is illustrated by the words which Heisenberg [HEI] wrote about Bohr: "We all could sense immediately that Bohr had obtained his results not from computations and proofs, but by dreaming and guessing, and that he now finds it difficult to defend them in front of the critical faculty of mathematics in Goettingen." And Bohr himself once said: "We must be aware of the fact that in our field the natural language can be used only as it is similarly used in poetry, where the purpose is not the accurate description of facts, but the generation of mental images and connections between ideas." And in addition, Heisenberg said: "We got used to accepting that concepts and images which had been transferred from early physics to the area of atoms in this new context, were only half right and half wrong. Therefore, we could apply them without being guided by strict rules. This gave us the freedom to guess the correct mathematical relationships if they could not be deduced logically. Thus, we concentrated our efforts on guessing adequate formulas which were similar to the formulas of the classical theory, and yet satisfied the new requirements." I cited these statements from the originators of quantum theory in order to convince you that there is no simple way of introducing this theory. But while those originators were searching for an unknown goal, I am in a better position since I already know the results they found. And therefore, I now can present a sequence of more or less intuitive steps which lead to this result.

If a mathematician were asked if he could suggest a formalism by which a set of discrete values could be singled out from a continuum, he might refer to the concept of *eigenvalues* of matrices. In my comments concerning Fig. 11.2, I mentioned that the German word "eigen" indicates a kind of ownership. In the case of Fig. 11.2, it was a mechanical system which had a resonant frequency. By analyzing the formalism of matrix multiplication (see Fig. 3.4), mathematicians discovered a new kind of ownership. Until now, we used matrices here to capture the relation between two different coordinate systems, and we had no reason to consider a matrix as an owner of something. But with the help of Fig. 11.11, I shall now show you that matrices actually own certain properties in the form of characterizing numbers. Fig. 11.11 illustrates the fact that a matrix can be interpreted in two different ways. Either it is interpreted as a description of the relation between two different coordinate systems, or it is considered as an interpretation of a reversible mapping between the points of two spaces which have the same coordinate system.

In the upper left corner of Fig. 11.11, you see a plane the points of which are described by the coordinates $(x_1, y_1)$. Six points of this plane have been selected to become the corners of the grey-shaded hexagon. The matrix underneath defines a mapping between the pairs of coordinates $(x_1, y_1)$ and their partners $(x_2, y_2)$. The figure shows that the pairs $(x_2, y_2)$ can be interpreted in two different ways. They can be interpreted as alternative descriptions of the points of the original plane. In this case, only the coordinate system has been changed, but not the shape of the original hexagon as shown in the upper right corner of the figure. But the pairs $(x_2, y_2)$ can also be interpreted as the coordinates of the points of a second plane which are mapping partners of the points of the original plane. In this case, the two planes have the same coordinate system assigned, but shapes such as the hexagon no longer remain the same as shown in the lower right corner of the figure.

Quantum Theory

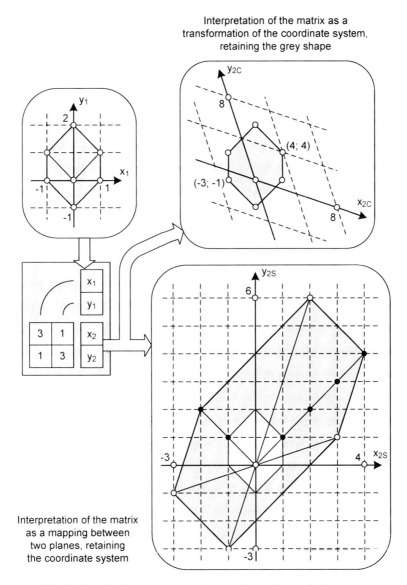

**Fig. 11.11** Background for introducing *eigenvalues* and *eigenvectors*

Now you should focus on the rays which run from the center of the coordinate system to the corners of the hexagon. When the hexagon is transformed from its original symmetric shape to the asymmetrical shape, only two of the six rays in the new version still have the same directions as before, while the other four rays don't point in the original directions any more. The directions which were not changed by the application of the matrix are called *eigendirections* of the matrix, and the factors by which the lengths of the rays which point in these eigendirections differ from their original lengths, are called *eigenfactors* or *eigenvalues*. In the example in Fig. 11.11, the eigenvalues are 2 and 4 which is illustrated by the black dots on the corresponding rays. Each square-shaped matrix with n∗n entries has exactly n eigenvalues. Since the matrix in Fig. 11.11 has 2∗2=4 entries, it has 2 eigenvalues.

It is still too early for an explanation about the role of these eigenvalues in quantum theory. Therefore, I kindly ask you to be patient and trust me that I shall guide you to a useful application. Just wait a little bit until we have finished all the preliminaries. The next step on our way is represented in Fig. 11.12. Once we have found the eigendirections and eigenvalues, we can define a coordinate system whose axes point in the eigendirections. It is always possible to find a matrix which describes the relation between the original coordinate system ($x_1$, $y_1$) and this new one ($x_E$, $y_E$). Since this is only a transformation of the coordinate systems, the corner points of the hexagon in the original plane will not be mapped to new points, and therefore the original shape of the hexagon will not be changed. But as in Fig. 11.11, we now want to transform the shape of the hexagon. Once the shape has been described in the coordinate system whose axes point in the eigendirections, the transformation requires only the multiplication of the coordinates $x_E$ and $y_E$ by the corresponding eigenvalues. This corresponds to a multiplication of the vector ($x_E$, $y_E$) by a so-called diagonal matrix which has the eigenvalues in the positions on the main diagonal while the other positions contain zeros. The diagonal matrix which corresponds to our example is shown in the lower left corner of Fig. 11.12.

Although I used an example from geometry in my introduction of the concept of eigendirections and eigenvalues, the application of these concepts in quantum theory is absolutely formal and does not refer to geometric shapes. The possibility of formalizing these concepts is based on the structure which is represented in Fig. 11.13. Here, the definition of the concepts is given in the form of formulas only, without any reference to diagrams. The formulas refer to so-called *eigenvectors* which are vectors pointing in the eigendirections.

I shall now apply the formalism to a case where there are no rays having eigendirections and no eigenfactors describing ratios of real lengths. In the left part of Fig. 3.7, a matrix is presented which describes the relation between the two coordinate systems in Fig. 3.6. Since, in this case, the coordinate system ($x_r$, $y_r$) is obtained by rotating the coordinate system ($x_o$, $y_o$) counter-clockwise by the angle φ, there can't be any ray which keeps its original direction. But even in this case, eigenvalues and eigendirections exist; only they can no longer be real (see Fig. 11.14). No one can associate more meaning to these complex numbers beyond their being solutions of the equation in Fig. 11.13 with respect to the matrix from Fig. 3.7. It is an interesting experience that people such as me, after having applied mathematics intensively over a long period of time, get used to applying formulas

# Quantum Theory

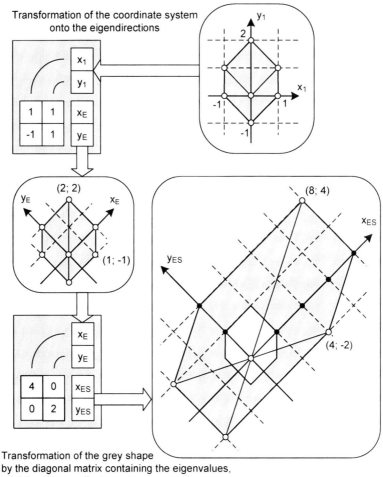

**Fig. 11.12** Application of eigenvalues and eigenvectors

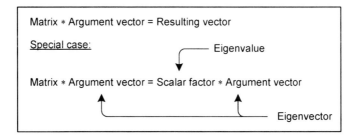

**Fig. 11.13** Roles of eigenvalues and eigenvectors in matrix multiplication

in a strictly formal way without asking what their meaning might be beyond their arithmetic correctness.

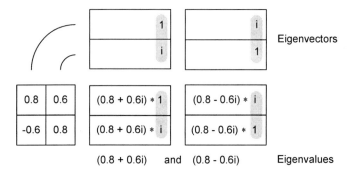

**Fig. 11.14** A matrix with complex eigenvalues

Now we have reached the point where we can apply the concepts of eigenvalues and eigenvectors to a real problem of quantum physics. The example which is well suited for demonstrating the application of this formalism is the polarizing filtering of Fig. 11.8. In quantum theory, the possible outcomes of an experiment are taken as eigenvalues. In my introduction to the concept of eigenvalues, I first had to choose a matrix, and then I could show that it has eigendirections and eigenfactors. Now it is the other way around, i.e., we are now given the eigenvalues and we ask for the matrix to which they belong.

In the case of polarizing filtering, there are only two possible outcomes with respect to a photon which hits the filter: either it passes or it is blocked. We are used to expressing these possible outcomes in natural language, but since this is not suitable for computations, we have to encode the outcomes by assigning numbers to them. Because of the symmetry of the situation where neither of the two outcomes {passage, blockage} is preferred, it is reasonable to choose the numbers +1 for passage and -1 for blockage (see Fig. 11.15). Once the eigenvalues are given, the diagonal matrix is defined and its eigendirections can be determined. Because of the fact that only the entries on the main diagonal have non-zero entries in a diagonal matrix, each eigenvector will also have only one non-zero component. The value of this non-zero component is not determined by the matrix, but can be chosen arbitrarily. The reason for choosing the value 1 comes from the requirement that each eigenvector can be interpreted as the representation of a probability distribution, and this requires that its length be 1.

Fig. 11.15 shows that the corresponding matrix for a given set of eigenvalues is not yet completely determined. The entries of the matrix in the upper right corner depend on the value of the angle $\varphi$ which can be chosen arbitrarily. The diagonal matrix in the upper left corner is obtained for $\varphi=0$. If two matrices differ, but have

# Quantum Theory

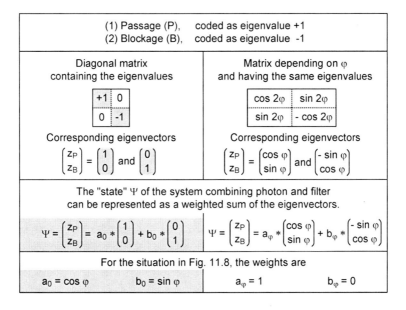

**Fig. 11.15** Application of eigenvalues to polarizing filtering

the same eigenvalues, they must necessarily have different eigendirections. Thus, by choosing different values for the angle φ, we get different eigenvectors.

By interpreting the eigenvalues as names for the potential outcomes of an experiment in quantum physics, we decided not to be interested in the properties of an isolated particle, but only in the behavior of the particle in a specific experimental set-up. In the example of polarizing filtering, the matrix and its eigenvalues and eigenvectors do not belong to the isolated photon, but to the system containing the photon and the filter. When the theory which says that it is nonsense to talk about the state of a particle as long as it is not observed was introduced, it caused a lot of discussions and objections within philosophical associations. But today, this theory with its strange consequences is as well-accepted as relativity theory.

In quantum theory formulas, the state of a system of interest is represented by the Greek letter $\Psi$. Using a mathematical view, this state corresponds to a point in space to which the describing matrix refers. In the case of polarizing filtering, this space has two dimensions, and therefore, the state $\Psi$ can be captured by two numbers $a_0$ and $b_0$. In Fig. 11.8, the dependency of the experimental outcome was given by the function $(\cos \varphi)^2$ which corresponds to $a_0^2$ in Fig. 11.15. As you may remember, sine and cosine are related by the equation $(\sin \varphi)^2 + (\cos \varphi)^2 = 1$ from which it follows that $a_0^2 + b_0^2 = 1$. As I mentioned above, the length of an eigenvector is chosen to be 1 because it is to be interpreted as a representation of a probability distribution. Now you see that the squares of the components of these eigenvectors and the squares of the weight factors $a_0$ and $b_0$ represent the probabilities of specific outcomes of the experiment in the set-up which is represented by the state $\Psi$.

The grey-shaded structure in the left half of Fig. 11.15 can be transferred to any kind of quantum system. The diagonal matrix which contains the eigenvalues determines a set of eigenvectors, each of which has only one component with a non-zero value which is 1. The state $\Psi$ of the system can always be represented as a sum of the weighted eigenvectors where the squares of the weights – in Fig. 11.15 these are $a_0^2$ and $b_0^2$ – are the probabilities of the outcome which belongs to the weighted eigenvector.

Fig. 11.16 illustrates the fact that this kind of state description, where the potential outcomes are subject of a probability distribution, can also be applied to systems which are not quantum systems at all. The system to which the state description in Fig. 11.16 belongs consists of a person holding a die which he will soon throw. The set of possible outcomes contains six elements, each of which will occur with a probability of 1/6. Since the probability is equal to the square of the weight of the corresponding eigenvector, the weights are $1/\sqrt{6}$.

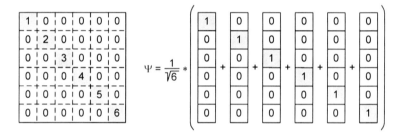

**Fig. 11.16** Diagonal matrix and state representation for the throwing of dice

Quantum physicists like to say that the state $\Psi$ of a quantum system is formed as a superposition of different "pure states." By this, they refer to the fact that $\Psi$ can be represented as a sum of weighted eigenvectors, each of which corresponds to a potential outcome of the actual experiment. Sometimes, they even say that the system is "in different states at the same time." But this is an inadequate description of a mathematical structure. As far as the system described in Fig. 11.16 is concerned, no one would say that the person who is prepared to throw dice is in six different states at the same time. An appropriate description would be to say that the system is in a state which will lead to one out of a given set of potential outcomes, each of which has a known probability.

While in both the case of polarizing filtering and in the case of throwing dice the probability distribution does not change over time, such a change characterizes the system which Erwin Schroedinger dreamed up and which later became known as

"Schroedinger's cat." The idea is that a cat is enclosed in a box which also contains a mechanism for poisoning the cat. This poisoning mechanism is triggered by the potential decay of a radioactive atom. The probability that the decay which has not yet occurred until the point in time $t_0$ will occur within the interval $t_0 \leq t < t_0+\Delta t$ is given by the function $p(\Delta t) = 1 - e^{-\Delta t/T}$. If the box is closed with the cat alive at the moment $t_0$, and opened again after time $\Delta t$ has passed, the probability for finding the cat still alive is $1 - p(\Delta t) = e^{-\Delta t/T}$; as time $\Delta t$ grows, this probability gets closer and closer to zero. Schroedinger invented this set-up for making clear that it would be nonsense to say the cat were in a superposition of two states. Either the cat is still alive or it is dead, but it can't be in a state which is the sum of 40% being alive and 60% being dead. It would be wrong to consider the cat as the quantum system whose state cannot be observed, but it is the indicator of the actual outcome of the experiment. The kernel of the quantum system is the radioactive atom whose state cannot be observed, but can be described only by a probability distribution. There are two potential outcomes of the experiment, and their probabilities change over time.

We are getting closer and closer to our goal of being able to explain why atoms can emit radiation only with frequencies which are specific for the material. The last preliminary step leads from eigenvectors to eigenfunctions. In Chapter 5 on probability theory, you learned the difference between probabilities and probability densities, and this is also relevant in quantum theory. As long as the set of potential outcomes is finite, the system can be described by a square matrix whose dimension corresponds to the number of different potential outcomes, each of which has its specific probability. In the case of polarizing filtering, the set contained only two elements, and therefore we had to consider only two discrete probabilities. Probability densities must be considered in all cases where the set of potential outcomes is infinite. Since there is no reason to assume that the number of energy states of the electrons in an atom is finite, we can no longer use square matrices to describe the set of possible states. Fortunately, the great German mathematician David Hilbert (1862-1943) found a way to transfer eigenconcepts from the world of matrices to the world of functions. In this transfer, vectors become functions and matrices become so-called *operators*.

The left part of Fig. 11.17 illustrates a way of considering a vector as a discrete function. Each component $F(j)$ of the six dimensional vector is the result of a function whose argument is the position index $j$ of the component. In the example given, the result of $F(j)$ is 13 for $j=5$. In the case of continuous functions, the continuous variable $x$ replaces the position index $j$. While the set of values for $j$ is finite and therefore the function $F(j)$ can be defined by listing its resulting values, the domain of $x$ is an infinite set which makes it impossible to list all the resulting values of $f(x)$. In this case, the function must be defined by an expression for computing its values.

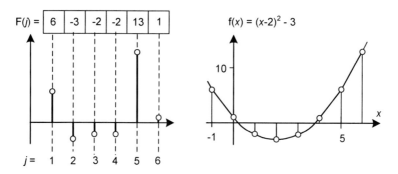

**Fig. 11.17** Possible way of relating a vector to a function

Fig. 11.18 shows that it is possible to transfer the eigenconcepts from the world of matrices to the world of functions. Now the owners of the eigenvalues are no longer square matrices, but operators. The number of eigenvalues for a matrix is equal to the dimension of the matrix. In the case of polarizing filtering, the matrix is a 2x2-matrix which has two eigenvalues. In comparison, the set of eigenvalues for an operator is infinite. Such an operator must define a way for getting from a given argument function to a resulting function. These operators belong to the world of differential equations, and therefore it might be helpful for you to take another look at Fig. 3.17 which presents some examples of differential equations. Whenever a differential equation is of a particular type, it can be written in the so-called operator representation. This is shown in Fig. 11.19. From the infinite set of eigenvalues, I selected two samples which are represented in Fig. 11.19 together with their corresponding eigenfunctions.

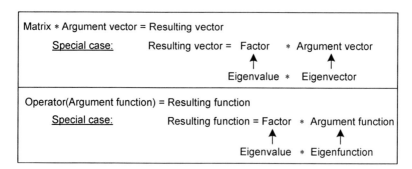

**Fig. 11.18** Transferring eigenconcepts from vectors to functions

You are now well-equipped to follow me on the path which will lead us to the energy operator of quantum mechanics. At the beginning of this path is the sine wave function which propagates with constant speed u in the direction $x$. Look again at Fig. 9.12 which represents a rope along which the drawn shape is running

Quantum Theory

---

The following differential equation determines a set of functions $\Psi(x)$:

Standard representation: $\dfrac{d^4\Psi}{dx^4} + 13 * \dfrac{d^2\Psi}{dx^2} - a*\Psi = 0$

This differential equation can alternatively be represented in the so-called

Operator representation: $\left(\dfrac{d^4}{dx^4} + 13*\dfrac{d^2}{dx^2}\right)\Psi = a*\Psi$

The functions $\Psi(x)$ are the eigenfunctions of this operator.

| | | | | | |
|---|---|---|---|---|---|
| $\left(\dfrac{d^4}{dx^4} + 13*\dfrac{d^2}{dx^2}\right)$ | $\sin(2x) + \cos(3x)$ | = | -36 | * | $\sin(2x) + \cos(3x)$ |
| | $e^{5x}$ | = | 950 | * | $e^{5x}$ |
| Operator | Argument function = Eigenfunction $\Psi$ | = | Eigen-value | * | $\Psi$ |

**Fig. 11.19** Example of an operator with eigenvalues and eigenfunctions

to the right with the constant speed v. If the wave, instead of having the shape shown in the figure, had the shape of a sine function, it could be described by the expression in the upper left corner of Fig. 11.20. The reason for choosing the letter u instead of the letter v for the propagation speed of the wave is that, in the case of de Broglie's matter waves according to Fig. 11.10, the speed of a particle and the speed of its corresponding wave differ by a factor of two. In Fig. 11.20, we are not considering a particle, but a wave and its speed.

$$A*\sin\left(\dfrac{2\pi}{\lambda}*(x-u*t)\right) = A*\sin\left(\dfrac{2\pi}{h}*\left(\overset{=p}{\dfrac{h}{\lambda}}*x - h*\overset{=f}{\dfrac{u}{\lambda}}*t\right)\right)$$

$$= A*\sin\left(\dfrac{2\pi}{h}*\left(p*x - E_{kin}*t\right)\right)$$

$$= \operatorname{Im}\left(A*e^{i*\left(\dfrac{2\pi}{h}*(p*x - E_{kin}*t)\right)}\right)$$

$$= \operatorname{Im}\left(\underbrace{A*e^{i*\left(\dfrac{2\pi}{h}*p*x\right)}}_{\psi(x)} * \underbrace{e^{-i*\left(\dfrac{2\pi}{h}*E_{kin}*t\right)}}_{\Theta(t)}\right)$$

**Fig. 11.20** Part of the path from an arbitrary sine wave to the energy operator of quantum mechanics

The mathematical approach illustrated in Fig. 11.20 leads from the wave function in the upper left corner via four steps to the expression in the fourth row. The first step leads from the left side to the right side of the first row. In this step, the denominator λ is moved from outside of the brackets to the inside and, additionally, Planck's constant h is introduced as a factor inside the brackets and compensated by the denominator outside. The second row is obtained by substitutions according to Fig. 11.10. The third row is reached by applying Euler's equation which is represented in the bottom line of Fig. 3.22. Euler's equation says that a sine function is equal to the imaginary part of an exponential function having an imaginary exponent. Since the exponent in the third row is a sum of two summands, the exponential function can be written as a product of two exponential functions, each having one of the original summands as its exponent. This is well known from the world of natural numbers: $2*2*2*2*2=2^5=2^{3+2}=2^3*2^2$.

The expression Im[$\Psi(x)*\Theta(t)$] requires that the complex results of the functions $\Psi$ and $\Theta$ must be first computed and then multiplied, and the imaginary part must be taken from their product. When I told you the story of the creation numbers, I illustrated the numbers as points in a plane (see Fig. 2.6). From this it followed that there are four different ways of assigning real numbers to a complex number: we can take its real part, its imaginary part, its angle or its radius. Each time when mathematicians leave the world of real numbers and functions and enter the world of complex numbers and functions – as we did in Fig. 11.20 – the computations with complex numbers can be performed without constantly indicating which path will be chosen for returning to the world of real numbers. Although computations in the world of complex numbers make sense only if the final results are interpreted in the world of real numbers, this does not require that the mapping between the two worlds is explicitly expressed in all equations. Therefore, we now omit the term "Im" in subsequent equations. And since we are interested only in systems whose energy does not change over time, we can also omit the function $\Theta(t)$ and restrict our consideration to the function $\Psi(x)$ which is given in the grey shaded area in Fig. 11.20. This expression contains the momentum p as a factor of the exponent, but it does not contain any term which could be interpreted as an energy. According to Fig. 11.10, the kinetic energy is equal to $p^2/2m$. Therefore, we now look for a way which leads from $\Psi(x)$ to an expression which contains $p^2$.

Fig. 11.21 illustrates the fact that the derivatives of an exponential function are also exponential functions which have the same exponent as the original function. The derivatives differ only by a constant factor which comes from the exponent. Therefore, the second derivative of $\Psi(x)$ which is given in the top row of Fig. 11.22 has $p^2$ as a factor, and this makes it possible to rearrange the equation in such a way that instead of $p^2$ it contains the variable $E_{kin}$ for the kinetic energy. This corresponds to the path which leads from the first row to the third row in Fig. 11.22. Since it is our goal to get an operator representation with respect to the

| f(x) | $\frac{df}{dx}$ | $\frac{d^2f}{dx^2}$ | $\frac{d^3f}{dx^3}$ | $\frac{d^4f}{dx^4}$ |
|---|---|---|---|---|
| $e^{a*x}$ | $a*e^{a*x}$ | $a^2*e^{a*x}$ | $a^3*e^{a*x}$ | $a^4*e^{a*x}$ |

**Fig. 11.21** Derivatives of an exponential function

function $\Psi$, we isolate the term $E_{kin}*\Psi$ and then substitute $E_{kin}$ for the difference between the total energy $E_{total}$ and the potential energy $E_{pot}$. The resulting equation is given in the middle part of the figure. From here, two simple rearrangements lead to the final representation in the bottom row which is an operator representation of the differential equation defining the function $\Psi$ and where the total energy $E_{total}$ is the variable of the eigenvalues. The operator in the grey shaded area is called the energy operator of quantum mechanics. It was first developed by Erwin Schroedinger in 1926 and, in his honor, the equation in the bottom row of Fig. 11.22 is called "Schroedinger's equation." This equation contains the variable $E_{pot}$ for the potential energy. When the equation is applied to an actual system, this variable must be substituted using a specific function $E_{pot}(x)$ which captures the specific properties of the actual system. It is no wonder that the resulting eigenvalues depend strongly on this function.

The computations for solving Schroedinger's equation are rather simple if the potential energy is proportional to $x^2$, and therefore both Schroedinger and

$$\frac{d^2\psi}{dx^2} = \left(i*\frac{2\pi}{h}*p\right)^2*\psi$$

$$= -\left(\frac{2\pi}{h}\right)^2*p^2*\psi$$

$$= -\left(\frac{2\pi}{h}\right)^2*2m*E_{kin}*\psi$$

$$E_{kin}*\psi = -\left(\frac{h}{2\pi}\right)^2*\frac{1}{2m}*\frac{d^2\psi}{dx^2} = (E_{total} - E_{pot})*\psi$$

$$-\left(\frac{h}{2\pi}\right)^2*\frac{1}{2m}*\frac{d^2\psi}{dx^2} + E_{pot}*\psi = E_{total}*\psi$$

$$\left(-\left(\frac{h}{2\pi}\right)^2*\frac{1}{2m}*\frac{d^2}{dx^2} + E_{pot}*\right)\psi = E_{total}*\psi$$

**Fig. 11.22** Energy operator of quantum mechanics ("Schroedinger's equation")

Heisenberg chose this case when they first checked whether their theories provided reasonable results. Therefore, we now follow the steps of these great physicists and ask what the function $\Psi(x)$ is in the case of $E_{pot}/E_0 = (x/x_0)^2$. By selecting convenient values for the reference constants $E_0$ and $x_0$, the operator from Fig. 11.22 can be simplified to the form in Fig. 11.23. Fortunately, the eigenvalues and eigenfunctions of this operator have already been found by the French mathematician Charles Hermite (1822-1901), and thus they could be used by Schroedinger and Heisenberg. Each positive odd number is an eigenvalue of this operator. The eigenfunctions for the first three eigenvalues from the infinite set of eigenvalues are given in Fig. 11.23.

| Operator | Eigenfunction F(x) | = | Eigenvalue | * | F(x) |
|---|---|---|---|---|---|
| | $e^{-\frac{x^2}{2}}$ | = | 1 | * | see second column |
| $-\frac{d^2}{dx^2} + x^2 *$ | $2x * e^{-\frac{x^2}{2}}$ | = | 3 | * | see second column |
| | $(4x^2 - 2) * e^{-\frac{x^2}{2}}$ | = | 5 | * | see second column |

**Fig. 11.23**  Eigenvalues and eigenfunctions of the energy operator for a particular case

How these eigenvalues and eigenfunctions are interpreted with respect to quantum physics shall now be explained by reference to Fig. 11.24. The left part of this figure applies to the simple mechanical oscillator which was introduced in Fig. 11.2. We assume that no stimulation occurs and the mass block oscillates with the eigenfrequency of the system. The oscillation is characterized by a periodic exchange between the kinetic energy of the moving block and the potential energy stored in the springs. The value of the potential energy is $0.5*c*x^2$, i.e., it is a function of the distance $x$ between the actual location of the block and its equilibrium position. The sine-curve in the lower left corner of the figure illustrates how the location $x$ changes over time. In its extreme left and right positions, the speed of the block is zero because the direction of the motion of the block changes at these positions. Here the kinetic energy is zero and the potential energy is equal to the total energy. The diagram above the sine-curve shows a parabola which represents the function $E_{total}(x_{extreme})$, i.e., it describes how the total energy depends upon the amplitude of the oscillation. The total energy which corresponds to the actual amplitude is marked by the horizontal line which connects two points of the parabola. The white area just above this horizontal line represents the probability density of the time dependent location of the oscillating block. Each point on the horizontal line corresponds to a possible location, and so the set of possible locations is infinite. It is therefore impossible to assign non-zero probabilities to specific points. Probabilities may be assigned only to intervals $x_{left} \leq x \leq x_{right}$, and the probability of the block actually being at a position in such an interval corresponds to the white area above the interval. Thus, the whole area must have the value 1, because the block must always be somewhere on the horizontal line. The probability of the block being in the middle zone is the lowest since there it has its maximum speed and doesn't stay long. The zones with the highest probability are those in the neighborhood of the extreme positions because there the speed of the block is low and the block spends more time.

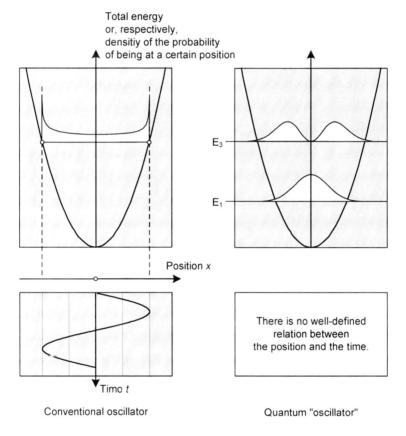

**Fig. 11.24** Conventional oscillator and its counterpart in the quantum world

Now we leave the left side of Fig. 11.24 and look at its right side. The diagram represents the results which are given in Fig. 11.23. The energies considered here can be thought to belong to an electron which moves within an electric field. The potential energy is assumed to have the same quadratic dependency from the location $x$ as in the case of the mechanical oscillator on the left side of the figure. The question of where this electric field might come from is not considered. It cannot be the field which is caused by the positive charge of the protons in the nucleus of an atom since, in this case, the potential energy is not proportional to $x^2$ but depends on (-1/x). This dependency had already been given in the case of a planet which orbits the sun (see Fig. 7.6).

The most conspicuous difference between the left and the right sides of Fig. 11.24 is in the fact that the sine function on the left side, which describes how the position of the block changes over time, has no corresponding partner function on the right side. On the left side, we could begin with the sine function, and from this we could deduce the probability density function above. On the right side, we have only the probability density functions which we obtained from Fig. 11.23. The two functions belong to the first two eigenvalues of the given energy operator; however, they are not the original eigenfunctions from Fig. 11.23. In my comment about Fig. 11.15 which represents the eigenvalues and eigenvectors for the case of polarizing filtering, I pointed out that the probabilities are not equal to the weights and components of the eigenvectors, but to the squares of these weights, and that therefore the components had to be chosen such that the sum of their squares is equal to one. These conditions must now be applied correspondingly to the eigenfunctions which means that a probability density function is obtained by squaring the corresponding eigenfunction and multiplying it by a factor such that the white area under the curve has the value 1.

On the left side, there are no restrictions concerning the amplitude of the oscillator; from this it follows that the total energy can have any value including zero. In the case of the quantum system on the right side, however, the total energy can have only values which correspond to the eigenvalues of the energy operator. In particular, it cannot have the value zero, a fact which could also have been deduced from the uncertainty relation. If the total energy were zero, the momentum would also be zero, and consequently, the product $\Delta x * \Delta p$ of the uncertainties of the location and the momentum would be zero. This would contradict the uncertainty relation $\Delta x * \Delta p \geq h/2$ which was deduced in Fig. 11.7.

The probability density functions on the right side of Fig. 11.24 have a property which we have to accept as a formal result but which is opposite of common sense. In the case of the mechanical oscillator, it would be nonsense to assume that the block's position could be beyond the limit which is given by the parabola. This parabola represents the maximum potential energy which corresponds to the amplitudes where the block changes the direction of its motion. Therefore, the probability density function on the left side does not reach beyond the parabola. In the case of the quantum system, however, the probability density functions have non-zero values outside of the limiting parabola. This forces us to assume that, although it is with a rather low probability, an electron can have a potential energy which is greater than that given by the parabola. Since the total energy, the sum of the kinetic and the potential energy, corresponds to the eigenvalue of the energy operator, its value must be the same in all situations on the horizontal line inside and outside of the parabola. The kinetic energy is zero in the situations which correspond to the two points on the parabola, is positive in the situations between these points, and it must be negative outside of the parabola. We have no idea how to imagine a negative kinetic energy – but we have to accept it as one more of those mysterious results of quantum theory.

Although the curves in the diagram on the right side of Fig. 11.24 do not apply to an electron in the electric field of the protons of an atom, two of the results can nevertheless be transferred to the situations of the electrons of an atom. Each electron can have only a total energy which corresponds to an eigenvalue of Schroedinger's energy operator. An electron can have a negative kinetic energy for an extremely short time duration and with extremely low probability.

Whenever an electron "jumps" from its initial energy level to a different one, this corresponds to a change of the energy state of the atom. If the electron falls from a higher level to a lower one, the atom emits a quantum of radiation; in the opposite case, the atom must receive the corresponding amount of energy from its environment. The distance of the energy jump of an electron is not restricted to the distance between two neighboring levels, i.e., a jump may go from the initial level to any other level.

My deduction and interpretation of Schroedinger's energy operator were restricted to the one-dimensional case, since I considered only the location coordinate x and omitted the other two coordinates y and z. But physics always occurs in the three-dimensional space, and therefore two more eigenvalues will exist. However, a Cartesian coordinate system (x, y, z) with three axes perpendicular to each other is not the best way to capture the space around the nucleus of an atom. A more appropriate coordinate system for this case is the so-called *spherical coordinate system* which specifies a point by its distance from the center of a sphere and by two angles. Think of the earth where each point on the surface has approximately the same distance from the center and is determined by the two angles of longitude and latitude. The distribution of the probability density of the location of an electron corresponding to the three coordinates must be imagined as a three-dimensional cloud whose position and shape is determined by corresponding eigenfunctions. The shapes of the probability density curves in Fig. 11.24 justify the assumption that the probability density cloud will be like a spherical shell around the nucleus of the atom only in the case of the lowest energy level. This shell depends only on the distance from the center, but not on the two angles. In the case of higher energy levels, there will be angles where the probability density has high values, and other angles where the value is low or even zero.

Certain experimental results indicated that even three eigenvalues were not enough for completely capturing the state of an electron in an atom. There is an analogy between an electron moving around the nucleus of an atom, and the earth orbiting the sun. The actual astronomical state of the earth is not completely described by the knowledge of where the center of the earth is located in relation to the sun, since we also need to know the value of the angle of rotation of the earth about its axis as additional information. Physicists introduced a property of the electron which they called *spin* which corresponds to the rotation of the earth about its axis. In classical mechanics, the term spin stands for the angular momentum which was introduced in Fig. 7.8 in order to explain the phenomena shown in Fig. 7.7. In the case of a wheel of a bicycle or of a car, we can easily see how fast and in which direction it is spinning, but there are no direct means for detecting whether an electron is spinning. An electron carries an electrical charge and, according to the findings of Michael Faraday and James Maxwell, a spinning charge must behave like a bar magnet. If a bar magnet is moved through a magnetic field, certain forces, whose directions depend on the direction of the bar, will act on it. Using very sophisticated equipment, today's physicists can actually detect that there is something which might be called the spin of an electron. From mechanics we know that ordinary angular momentum is a directed quantity which, in a Cartesian coordinate system, has three components, one for each of the three coordinates x, y and z. Rather mysteriously, the results of measuring the components of the spin of an electron always have values from the binary set $\{-h/4\pi, +h/4\pi\}$, regardless of the directions of the coordinate axes. I mentioned previously that the unit of Planck's constant h is that of an angular momentum. The spin of other particles such as protons and neutrons was detected and measured later, after the assumption of electrons having a spin was confirmed. The fact that the neutron, which has no electrical charge, also has a spin proves that spin does not require an electrical charge. Systems which are composed of such particles, i.e., atoms and molecules, also have spins. The possible values for the results of measuring a component of such a spin are always integer multiples of the two possible values for the spin of a single electron.

After the discovery that particles like electrons and atoms have a spin, the hypothesis that a photon also has a spin had to be checked. In experiments where light was absorbed by free atoms, this hypothesis could be confirmed: each time a photon was absorbed by an atom, the spin of the atom changed, and the value of the difference was $h/2\pi$. It is a basic law of mechanics that a collision between two bodies or particles does not change the sum of the angular momentums. Consequently, when a collision changes an atom's spin by $h/2\pi$, the spin of the partner in the collision must change by the opposite amount, $-h/2\pi$. If the spin of this partner can have only one of the two values from the set $\{-h/4\pi, +h/4\pi\}$, a change of $h/2\pi$ corresponds to a reversal of the spin's direction.

The simplest case for applying Schroedinger's energy operator is when the atom contains only one electron. This applies only to the hydrogen atom, since the atoms of all other chemical elements have more than one electron. These multiple electrons have interactions that make it very difficult if not impossible to solve Schroedinger's equation. Each of these electrons has a specific state within the atom, and this state is described by four eigenvalues, the so-called *quantum numbers*. No two electrons can have the same state, i.e., they must differ with respect to the value of at least one of the four quantum numbers.

An isolated atom is not nearly as relevant as materials which consist of many atoms. In the simplest case, all atoms of the material are of the same kind. But much more interesting are materials or composed structures which contain different atoms. Once the theory had been developed which described the configurations of the electrons in a single atom, it became possible to explain experimental results which could not be understood previously. Think of Mr. Volta's battery which he invented more than one hundred years before Heisenberg and Schroedinger thought to use eigenvalues for explaining the energy states of electrons in atoms. Now such batteries could be explained by the fact that the energy levels of electrons in atoms of different elements are not the same. Even the simple question about why metals are good conductors of electric current while other materials are bad conductors or insulators cannot be answered without reference to quantum theory. This also applies to so-called *superconductivity* which was discovered by the Dutch Heike Kamerlingh Onnes in 1911. He observed that the electric resistance of mercury dropped to zero or an extremely small value which could not be measured whenever the temperature of the metal was decreased below 4.2 degrees Kelvin. Since then, many other materials were found whose electrical resistance gets extremely small when the temperature drops below a certain threshold value which is specific to the material.

Knowledge about the probability densities which describe the local distribution of electrons with respect to their actual energy levels provided the basis for an explanation of valencies which were introduced by chemists for explaining the specific combinations of atoms in molecules. Now the molecules could be considered as three-dimensional structures where the atoms are connected by bridges of different strengths. Steel is a material which consists mainly of iron and a small fraction of carbon, where the atoms are not bound in molecules, but nevertheless are located in a regular spatial structure. The specific properties of steel could now be explained as a consequence of the bridges between neighboring atoms. It also became possible to explain how these properties could be modified by adding low percentages of manganese, chromium or nickel.

The new insight into the structure of atoms motivated physicists to experiment with structures similar to those of Alessandro Volta. He built a sandwich consisting of three layers of different materials, and found that this structure behaved as a voltage source. Alternative sandwiches can be obtained not only by modifying the materials of the layers, but also by modifying their thicknesses. Today, it is possible to implement layers with thicknesses less than one-thousandth of a millimeter. It is primarily the electrical resistance of such sandwiches which is measured. In general, it is quite difficult to predict the properties of such sandwiches by deduction from theory, but once experimental results are known, the properties can be explained by reference to the theory. Two types of sandwiches consisting of so-called semiconductor material are of extreme importance, both in information electronics and in power electronics. Semiconductors are materials with a low conductivity which depends strongly on temperature and whose conductivity can be increased by inserting atoms of appropriate other elements. Today's semiconductor technology is based almost exclusively on silicon. The two semiconductor sandwiches which are fundamental to modern electronics are the diode which is a two-layer sandwich, and the transistor which is a three-layer sandwich. The diode is an "electric valve" which is open only for a current flowing in one direction and is closed for the opposite direction. A transistor can be considered as a pair of diodes sharing one electrode. The current flowing through one of these two diodes controls the current through the other diode; the ratio between the controlling current and the controlled current is a constant amplification factor. While electron tubes were used to amplify the audio signals in the sound systems of your grandparents, no electron tubes can be found anymore in the corresponding systems of today – all amplification is now done using transistors.

Even today, experiments with sandwiches can provide surprising results. This can be seen from the fact that in 2007 the Nobel Prize in physics was awarded on the basis of a sandwich-effect. The German physicist Peter Gruenberg (1928--) and his French colleague Albert Fert (1938--) had done experiments with sandwiches which consisted of many extremely thin layers of different metals. Each layer had a thickness of only approximately one-millionth of a millimeter. They found that the electrical resistance of such sandwiches could be varied by extreme amounts using rather small magnetic fields. This effect made it possible to significantly increase the density of the digital information stored on storage devices such as disks and memory cards.

## Phenomena Which Even Einstein Thought to Be Impossible

Everything I told you in the previous sections about the granularity of matter and energy had a significant influence on the progress of technological developments, especially in the areas of materials and microelectronics. Now, however, before closing this chapter, I shall describe certain theories and corresponding experimental results which have not yet affected today's world of technology. Nevertheless, I include them here because they are said to bring revolutionary technological

Entanglement of Quanta 303

progress once this field has matured. Quantum physicists claim that the findings in this field will make it possible to build failure-free communication systems and computers whose speed and storage capacity will exceed the capabilities of today's computers by factors of one million or more. Computational results of the underlying theories had already been found when Albert Einstein was alive, and to him these results seemed absurd. Therefore he was convinced that they never could be confirmed experimentally. I shall now try to represent these absurdities.

The central concept in this area is the so-called *entanglement* of quantum states which can best be introduced and explained by presenting an example of a so-called *quantum bit*. The word "bit" was created by information technologists and is the result of combining the two words binary and digit. A bit characterizes a situation which can result in only one out of two possible outcomes. A simple example of a bit is a light switch which can be only in the states ON or OFF. When the term bit is used in the quantum world, the examples we may think of are the spin of an electron or the relation between a photon and a polarizing filter. In these examples, the corresponding experiments can have only one out of two possible results: in the case of the electron it is the sign of the spin; and in the case of the photon it is the decision about its passing or not passing through the filter.

Although the same word "bit" is used in both areas, there is a significant difference between a bit in conventional information technology and in the world of quantum systems. In conventional information technology, the state of a system can be observed and the result of the observation is equal to the state. Think of the light switch whose state is either ON or OFF. Which of the two it actually is in can be seen just by looking at the switch. In contrast to this, the state of a system in the quantum world cannot be observed at all since the system interacts with the observation equipment. The only information which can be obtained is the result of this interaction which changes the state. The relation between the state and the result of an observation can be described only by a probability distribution. This distribution can be imagined as a point on the surface of a sphere in a space whose dimensions correspond to the different possible outcomes of an observation (see Fig.11.25). This sphere has radius one because then the sum of the squares of the coordinates will have the value 1 according to the generalized law of Pythagoras (see Fig. 4.14). Consequently, the square $x_i^2$ of a coordinate will correspond to the probability that the observation will provide the result which is associated with this coordinate.

Now let's look again at figures 11.8 and 11.15 which I introduced in order to explain the phenomenon of polarizing filtering. As an example of entangled states, I now use two such systems, each consisting of a photon and an associated filter. In this case, there are four possible results of an observation, since the two possible results from each of the two photons must be combined. In my comment about Fig. 11.15, I pointed out that the possible results of an observation must be

| System composed of | |
|---|---|
| n conventional binary memory cells | n quantum bits |
| Interpretation of the state as an observable situation | Interpretation of the state as a probability distribution of observable situations |
| Representation of the state as a corner of a cube having the edge length 1 in a space with n dimensions: $Z = (x_1, x_2, x_3, \ldots x_k)$ with k=n, where the domain of the values $x_i$ is the binary set $\{0, 1\}$. | Representation of the state as a point on the surface of a sphere having the radius 1 in a space with $2^n$ dimensions: $Z = (x_1, x_2, x_3, \ldots x_k)$ with $k=2^n$, where the domain of the values $x_i$ may contain all kinds of numbers. |
| There are $2^n$ different states. | There are infinitely many different states. |
| The number of different observable situations is $2^n$ (equal to the number of states). | The number of different observable situations is $2^n$ (less than the number of states). |

**Fig. 11.25** Differences between conventional binary memory cells and quantum bits

encoded as numbers in order to perform matrix computations. The observation that the photon passed the filter was encoded with the number +1, and the opposite case that the photon was blocked by the filter was encoded with the number -1. Now, we have two photons and four possible results of an observation to which we must assign four numbers. In principle, we have great freedom in choosing an assignment, but not all assignments are equally suitable. The interpretation of the numbers should be as easy as possible, i.e., it should be possible to come from the given number to the corresponding observation result by simple logical reasoning. The right side of the table in Fig. 11.26 represents a method for encoding the $2^n$ observation results of a system of n quantum bits; the left part shows the actual code numbers for n=3, where the letters P and B stand for passage and blockade.

In the case of a system of two polarized photons, the set of observation results is {BB, BP, PB, PP} which is encoded as {-3, -1, +1, +3}. The actual state of the system is considered to be a point on the surface of a four-dimensional sphere having radius 1. The values of its coordinates determine whether a point represents an entangled or a non-entangled state. In Fig. 11.27, two state vectors are represented whose coordinates are determined by the values of the two angles $\varphi_1$ and $\varphi_2$. Recognizing the significant difference between these two vectors requires that you remember the relation $\sin^2(\varphi) + \cos^2(\varphi) = 1$ which follows from the definitions of the sine and cosine functions in Fig. 2.13, together with the law of

Entanglement of Quanta                                                      305

| | Component code | | | Combination code |
|---|---|---|---|---|
| | $z_1$ | $z_2$ | $z_3$ | $= 4*z_1 + 2*z_2 + 1*z_3$ |
| B B B | -1 | -1 | -1 | - 7 |
| B B P | -1 | -1 | +1 | - 5 |
| B P B | -1 | +1 | -1 | - 3 |
| B P P | -1 | +1 | +1 | - 1 |
| P B B | +1 | -1 | -1 | + 1 |
| P B P | +1 | -1 | +1 | + 3 |
| P P B | +1 | +1 | -1 | + 5 |
| P P P | +1 | +1 | +1 | + 7 |

Formula which combines n component values $z_i$ from the set $\{-1, +1\}$ to one combination code:

$$N = 2^{n-1} * z_1 \\ + 2^{n-2} * z_2 \\ + 2^{n-3} * z_3 \\ \vdots \\ + 2^1 * z_{n-1} \\ + 2^0 * z_n$$

**Fig. 11.26** Coding possible observable situations of a system

Pythagoras. Because of the fact that the squares of the vector components represent probabilities, their sum must be 1. This requirement leads to the common factor $1/\sqrt{2}$ in front of the brackets of the lower vector, since here the sum of the squared components within the brackets is 2. In the case of the non-entangled state which is described by the upper vector, the probabilities for photon 1 passing or not passing its filter are $\cos^2(\varphi_1)$ and $\sin^2(\varphi_1)$, and the corresponding probabilities for

State vector of the non-entangled pair =

$$\sin(\varphi_1)*\sin(\varphi_2)*\begin{pmatrix}1\\0\\0\\0\end{pmatrix} + \sin(\varphi_1)*\cos(\varphi_2)*\begin{pmatrix}0\\1\\0\\0\end{pmatrix} + \cos(\varphi_1)*\sin(\varphi_2)*\begin{pmatrix}0\\0\\1\\0\end{pmatrix} + \cos(\varphi_1)*\cos(\varphi_2)*\begin{pmatrix}0\\0\\0\\1\end{pmatrix}$$

State vector of the entangled pair =

$$\frac{1}{\sqrt{2}} * \left( \cos(\varphi_1-\varphi_2)*\begin{pmatrix}1\\0\\0\\0\end{pmatrix} - \sin(\varphi_1-\varphi_2)*\begin{pmatrix}0\\1\\0\\0\end{pmatrix} + \sin(\varphi_1-\varphi_2)*\begin{pmatrix}0\\0\\1\\0\end{pmatrix} + \cos(\varphi_1-\varphi_2)*\begin{pmatrix}0\\0\\0\\1\end{pmatrix} \right)$$

The sequence of the summands corresponds to the order of the eigenvalues and the corresponding observable situations:
(-3, -1, +1, +3) and (BB, BP, PB, PP)

**Fig. 11.27** State vectors of an entangled and a non-entangled pair of photons

photon 2 are $\cos^2(\varphi_2)$ and $\sin^2(\varphi_2)$. This means that the decision about whether photon 1 passes its filter is completely independent from the corresponding decision concerning photon 2, and vice-versa. Such independence does not exist in the case of the entangled state, since here all sine and cosine functions have the same argument, namely $(\varphi_1-\varphi_2)$. Consequently, changing the value of one of the angles will always have an effect on the probabilities concerning both photons.

The difference between non-entanglement and entanglement becomes clear by comparing the second row with the third row of the table in Fig. 11.28. All entries in the cells of this table are products of two factors. Each factor of the products in the cells of the second row represents a probability concerning one of the two photons, and this reflects the fact that the decisions concerning the two photons are independent of each other. In the third row, however, one of the two factors of the products in the cells always is either $p_E$ or $p_U$, with E standing for "equal" and U standing for "unequal." This reflects the fact that in this state the two decisions concerning the two photons are not independent of each other. The two decisions are equal, i.e., both photons pass or both are blocked, or they are different, i.e., one photon passes and the other one is blocked. Therefore, the sum of $p_E$ and $p_U$ must be one. The entries in the third row refer to the probabilities of photon 1, and the entries in the fourth row refer to the probabilities of photon 2. But in each column, the entries in the two grey shaded cells represent the same value, and from this it follows that each of the two photons passes its filter with a probability of 50 %.

We now consider the case where the two angles $\varphi_1$ and $\varphi_2$ are equal. From $(\varphi_1-\varphi_2) = 0$, it follows that $p_E=1$ and $p_U=0$ which means the decisions concerning the two photons will always be the same. Quantum physicists say that this does apply even to the case where the two photons have been flying in opposite directions for some time before reaching their filters. You should remember the fact that the decision about whether a photon passes its filter is still completely undetermined,

| Alternative observable situations | BB | BP | PB | PP |
|---|---|---|---|---|
| Probabilities for the case of non-entanglement | $p_{B1}*p_{B2}$ | $p_{B1}*p_{P2}$ | $p_{P1}*p_{B2}$ | $p_{P1}*p_{P2}$ |
| Probabilities for the case of entanglement | $p_{B1}*p_E$ | $p_{B1}*p_U$ | $p_{P1}*p_U$ | $p_{P1}*p_E$ |
|  | $p_{B2}*p_E$ | $p_{P2}*p_U$ | $p_{B2}*p_U$ | $p_{P2}*p_E$ |
| This is only possible for $p_{B1} = p_{B2} = p_{P1} = p_{P2} = 0.5$. |||||

**Fig. 11.28** Restrictions on probability for an entangled pair

and occurs only with a probability of 50 %. But in contrast to the non-entangled case, the random decisions occurring at the two filters are now strongly coupled in such a way that either both photons pass or both are blocked. This immediately leads to the question of how one of the partner systems, consisting of a photon and its filter, could know what happens at the same time at the site of its partner. Wouldn't this constitute a case of information transmission with a speed higher than the speed of light, maybe even with infinite speed? Although the phenomenon is rather mysterious, it should not be considered a case of information transmission since this would require that the information to be transmitted could be chosen arbitrarily by the sender. In the case of the entangled photons, no one on either side has any way to prescribe which information should be transmitted, i.e., there is no way to "tell the photons" what they should do.

Ideas have been developed about how entanglement could be used to build safe communication channels where it is possible to recognize any reading or tampering of the transmitted information by a third party. While two entangled quantum bits are sufficient for building such communication channels, many more bits are required if entanglement is to be used for implementing new concepts of information processing. The vision of being able to build computers for solving problems which are much too complex for today's systems is based on the difference between conventional bits and quantum bits (see Fig. 11.25). The states of a system containing n conventional bits correspond to the corners of an n-dimensional cube. Thus, a transition from one state to another corresponds to a jump from one corner to another. In contrast, the states of a system containing n quantum bits correspond to points on the surface of a sphere in a $2^n$-dimensional space. Here, the state can be changed continuously.

I shall not continue this subject any further since it is still a research area today. Optimists predict that in a few years, results will be provided which can be applied for building marketable systems. Pessimists, however, say that it isn't at all certain whether this field will ever provide useful applications. At the very least, entanglement is no longer a purely theoretical concept. Since 1995, quantum physicists successfully performed experiments entangling more than two bits. But these entangled states are extremely sensitive and the slightest interaction with the environment destroys the entanglement. At any rate, you now have an idea of what is meant when you hear or read the terms quantum bit or quantum computing.

# Chapter 12
# How "Recipes" in the Cells of Living Organisms Were Found and Can Be Rewritten

There is such a plethora of questions concerning life that, of course, the subset I shall consider in this chapter must be relatively small. In particular, I shall not consider Darwin's theory of evolution because it has not contributed to the progress of technology, i.e., there is no technological product or process whose designers or implementers have used any findings from the theory of evolution. In his book *"On the Living"* [CH 2] the biochemist Erwin Chargaff (1905-2002) wrote that it is impossible to give a satisfying and final definition for the term "life." He was convinced that whatever definition is given, some facts will exist which do not fit it. But this didn't bother him, since he didn't see any need for such a definition. However, he got upset whenever someone behaved as if he knew exactly what life is. In Chargaff's texts, I also found the statement that "life can never be the subject of any research, because scientists can look only at this or that living object." Since I share his opinion, the following paragraphs will not be about life, but only about the living.

## How Organization and Life Are Connected

Although living can certainly be the subject of research, this research must be rather restricted by narrow and insurmountable borders. This is a consequence of the fact that any living object is an extremely complex system which allows only very restricted experimental interventions. Therefore, for a long time, no really essential experimental findings could be obtained. The findings obtained referred to objects which once had been living, but could be analyzed only after death. It is quite obvious that all of the methods of physics and chemistry can be applied in order to obtain scientific insights about something that once had been living. But it is doubtful whether these findings tell us anything about life. At least there are opposing opinions concerning this question, and no representative of either side has sufficiently strong arguments which could convince the other side of being wrong.

Another reason for the restriction of research concerning living creatures is the difference between subjective experiences and objective results of observations. In particular, I am considering the difference between the results of observing chemical and physical processes in the brain and the nervous system on one side, and the subjective experience of the person who is subject to those observations on the other side. Today, there are many methods for observing the processes taking place in the brains of living animals and persons. A primitive method consists of inserting electrodes into the brain and recording the voltages between them. More advanced methods which have been applied to build so-called tomographs are based on quantum theory.

The term *"tomography"* (from the Greek words *tomos* for layer and *graphein* for writing) stands for generating images by evaluating the intensity of radiation which has passed through an inhomogeneous body. The inhomogeneity refers to the spatial distribution of a physical quantity such as permeability for X-rays or the magnetic spin resonance of atoms. What is actually measured is the intensity of a signal whose actual value results from an integration along a straight line leading through the body. By stepwise changing the direction of the observation within a constant plane, a set of many measured values is obtained from which the spatial distribution of the quantity of interest can be computed, and finally made visible as an image. The reconstruction of the spatial distribution from the measured values requires a lot of computing power, and this is the reason why tomographs always contain rather powerful computers. Think of a ball made of colored acrylic glass, and assume that the intensity of the color is inhomogeneously distributed within the ball. By looking at this ball from different perspectives, the intensity of the color you actually see will vary with the direction in which you are looking. Unless you apply very complex mathematical methods, you shall not be able to reconstruct the spatial intensity distribution merely from what you have seen.

There are only a few philosophers interested in science or scientists interested in philosophy who strongly defend the opinion that there are only neuronal states and no mental states. They believe, or at least they pretend to believe, that all words referring to subjective experience such as consciousness, pain, fear, conviction, etc. should be replaced by words referring to objective neuronal states. Here again there are two opposing positions without the likelihood of really proving convincingly that one position is wrong. Nevertheless, I myself am convinced that both worlds must be taken into account: the world of objective scientific facts on one hand and the world of subjective experiences on the other hand. Frankly speaking, and in my opinion, the opposite position is just nonsense.

I don't know when the question was asked for the first time about whether there is life beyond our earth, on planets in our neighborhood or far away in other galaxies. When people ask this question, they should have spent some time considering the problem of choosing which phenomena they would accept as proof of the existence of such life. Couldn't it be that the living creatures found in such places look and behave quite differently from what we are used to seeing here on earth? Since living extraterrestrial objects presented in movies always look like modified humans, animals or plants from our daily experience, we may conclude that, in this respect, human imagination is rather restricted.

From what we know today, we are convinced that human individuals and bacteria both are living objects, although they don't have much in common except for their capability to reproduce and the inevitability of their death. As an electrical engineer who knows how to build computers and robots, I asked myself why I wouldn't consider a robot which could build copies of itself to be a living object. Certainly, it cannot exist forever, i.e., it will "die" which means that it will break down or be destroyed, and thus get into a state of not being able to repair itself anymore. If we repair such a broken robot, this must not be considered "bringing this robot back to life," but rather using the material of the former robot to build a new one. Although I would never consider such a robot as a living object, I am not able to clearly state the criteria on which my decision is based. If I really could list these criteria, I would know the exact definition of life. The capability of reproducing itself and being subject to death are, at least for me, only necessary but not sufficient attributes for living objects. The philosopher René Descartes is known for his statement, "Cogito, ergo sum." which is Latin and has the English meaning, "I think, and therefore I am." My personal position with respect to separating the living from the not living can be expressed by a modified version of Descartes' statement: "I experience myself, and thus, I live." I can talk about other living objects only as their being analogous to myself.

Up until the present time, nobody knows how life first appeared on earth. Nice stories have been invented such as the one in the Bible (1. Mose 2, Verse 7) where Adam is made from a lump of clay by a divine breath being blown into it, but these stories have had no consequence for the life of mankind. Although the question is asked, "what came first, the chicken or the egg?," everybody knows that this question should not really be taken seriously. The German physician and bacteriologist Rudolf Virchow (1821-1902) was convinced that all we can know is that life comes from life. His belief was the result of his dealing with living cells where, by segmentation, two cells are created from one. The question of where the first cell might have come from was considered irrelevant and not being answerable by humans.

The knowledge that all living objects are either single cells or compositions of cells was obtained only after high resolution microscopes became available. By looking through such a microscope, a single-cell creature can be seen as a whole, while the cells of multicellular creatures can be seen only by placing very thin layers of tissue under a microscope. The German botanist Matthias Schleiden

(1804-1881) and the German physiologist Theodor Schwann (1810-1882) can be considered the fathers of the so-called cell theory. Schleiden discovered the cells in plants, while Schwann discovered the cells in animals. It was an essential finding that a cell is not just a simple component used for building living objects. Instead, a cell is itself a unit of living, i.e., an individual containing all the requirements for living and which, together with other cells, can constitute higher forms of living organisms.

It is useful to compare organisms and organizations. The essence of organisms and organizations lies in the fact that living objects are connected and constitute a higher unit by coordinated interaction. Presumably, the largest organization on earth is the United Nations Organization which has its headquarter in New York City. Other examples of human organizations are industrial companies, cities and churches. In these cases, a group of people constitute the organization by coordinated interaction. But a human being who consists of lower interacting living units is a higher unit himself. These interacting living units are the organs such as the heart, the lung, the liver, etc. Certainly, you may doubt whether such an organ really is a living unit since it cannot survive on its own once it has been taken out of the human body. But it is not really too extreme to consider an organ as a living individual whose survival is guaranteed only as long as the higher organization exists. We can consider an organ in analogy to a department of a company or an administration. When the company or the administration is closed down, its departments loose their rights to exist and will also be closed. The division of labor between the departments or the organs, respectively, led to a specialization which made it impossible for the departments or the organs to exist on their own. Nevertheless, is it reasonable to consider a department or an organ as a living unit. The adequacy of this view was confirmed in the last decades by the development of organ transplantation. When a person dies, his organs don't all die at the same time. Therefore, a living organ, e.g., a kidney, can be removed from the dead body and placed in a special container where it can be kept alive for a certain time. It can be transplanted only into the body of a patient if it has not died in the meantime. In analogy, a finance department might be extracted from a company which went bankrupt and be "transplanted" into a different company. This, too, requires that the department doesn't die in the meantime which means that its employees - analogous to the cells - still keep their ability to function cooperatively as before.

From this it follows that understanding the living requires that the structure of cells, the processes inside and their interactions with their environment, is understood. We can talk only about these processes and interactions by using the same words we use when talking about organizations whose cells are human individuals. But we must not forget that within a cell or an organ there are no agents having a free will which determines their goals. The processes within cells and between them are strictly determined by the laws of physics and chemistry.

# Organization and Life

We know that the trajectory of a ball which is flying along, after having been kicked by a soccer player, is not the result of the ball's eagerness to reach the goal. In the same way, molecules and particles have no idea what their moving through the cell or taking part in chemical reactions might be good for. But considering the fact that there are hundreds of thousands of different kinds of molecules in a cell, and millions of reactions between them each second, we have no chance at all to obtain the least understanding of this complex system unless we assume that the concepts which apply to human organizations like companies and administrations also apply to cells.

We now consider systems which are clearly separated from their environment and where matter, energy and information is flowing both inside and across their borders. To these systems we apply abstractions which have proven to be very useful in system theory. Although we can observe only flowing matter and energy, we also talk about the flow of information. Matter and energy are physical quantities, while information is the result of an interpretation of a form and, of course, this form can exist only as formed matter or formed energy. Think of written text on paper as formed matter, or of radio waves as formed energy. If we see matter or energy flowing, the purpose of this flow could be communication, but this is not necessarily so. When I put a leaf of green lettuce into my mouth, I certainly don't want to interpret the form of this leaf, but I eat it because it is a piece of matter which is conducive to the processes of metabolism. The concept of information shall be discussed in more detail in Chapter 14. Fig. 12.1 presents examples of the different purposes to which the two kinds of flowing quantities, matter and energy, might serve.

| Purpose of the flow: Fulfilling a demand for | Type of the flow |          |
|---|---|---|
|  | Matter | Energy |
| Matter | Bricks |  ×  |
| Energy | Coal | Heat rays |
| Information | Newspaper | Radio waves |

**Fig. 12.1** Purposes of flows in systems

Research concerning the structure of cells and their related processes belong to specific areas of physics and chemistry. Biophysicists are interested in the forces between the molecules which cause their motion and deformation. Biochemists look for the different kinds of molecules which can be found in cells, and they are interested in the reactions which consume or produce such molecules. In the early years of chemistry, the field was not yet subdivided into specific areas. But today, there are specific branches of chemistry from which three are of interest here: inorganic chemistry, organic chemistry and biochemistry. The substances which inorganic chemistry deals with are all elements from the periodic table (Fig. 10.4) and all compounds which existed before the appearance of any kind of life. Examples of such compounds are water, sulfuric acid, common salt and carbon dioxide. The substances which organic chemistry deals with are those which occur as components of, or are produced by, living objects. For a long time, chemists were convinced that they could only decompose these substances, but could not synthesize them using only elements as components. From this they concluded that there must exist something within living objects which they called "life power" and which they thought to be required for certain chemical reactions. When I read about the concept of life power for the first time, it reminded me of the concept of a "fire substance" which had been assumed by early chemists when the essence of heat had not yet been understood. Of course, when the first organic substance had been successfully synthesized in the laboratory, it became obvious that such a thing as "life power" is not a scientific concept.

Major contributions for establishing organic chemistry as a discipline of its own were made by the Swede Joens Jacob Berzelius (1779-1848) and the German Friedrich Woehler (1800-1882). The latter was a student of Berzelius. In 1828, Woehler successfully synthesized the organic substance urea in his laboratory. The structure of the urea molecule is shown in Fig. 12.2. This molecule contains the four elements, carbon, oxygen, nitrogen and hydrogen, which had previously been found as components of most organic substances that had been decomposed. Carbon is the only element which had been found in organic substances in all cases. The important role the carbon atom plays is a consequence of its having

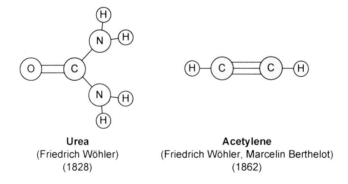

**Urea**
(Friedrich Wöhler)
(1828)

**Acetylene**
(Friedrich Wöhler, Marcelin Berthelot)
(1862)

**Fig. 12.2** First organic substances synthesized in a laboratory

four electrons on its outermost shell. This enables it to have connections to at most four other carbon atoms. Thus, rings and linear or branching chains can be built which can become the backbones of huge molecules. This explains the enormous plentifulness of organic substances. The valencies which are not used for connecting further carbon atoms are mostly used as bridges to atoms of hydrogen, oxygen, nitrogen, sulfur and phosphor.

Besides inorganic substances, organic chemists use substances which once were produced by living plants or animals, although these plants and animals may have died long ago. Think of coal and crude oil coming from plants and animals which died 350 to 400 million years ago. In contrast to this, biochemists explore processes in which living cells play a part. Besides the word "biochemistry," you may also have heard the term "molecular biology." Both disciplines look at processes which occur in cells and communities of cells. The biochemist asks what the structures of the participating molecules are, and how they can be synthesized. The molecular biologist asks what the purposes of the processes are, and how they can be influenced. Of course, we can easily find many ways of influencing such complex interactions between hundreds of thousands of different kinds of molecules. But unless we have at least a rough understanding of the system, our intervention will in most cases have the effect that the system collapses and the cells die.

A person who swallows a chemical substance certainly intervenes in the processes which take place within his body. A certain chemical in an appropriate dose can have the effect of healing the sickness of a patient or of improving the performance of a sportsman; the latter is called "doping." But in most cases, swallowing chemical substances will often cause the death of the person. Even thousands of years ago, people had knowledge about the positive effects of swallowing certain substances, although they certainly had no knowledge at all about the underlying processes. Unlike their predecessors, today's biochemists and molecular biologists know a lot about these processes. For example, they know exactly which substance it is which causes a person to die after eating a piece of a poisonous mushroom. And they can tell us how the process goes which finally leads to death. But we don't have much reason to assume that complex living systems, with their huge large number of different interacting molecules, will one day be understood well enough so that we could control the processes in any way we want. We shall always depend on doing experiments in order to find out whether a certain substance or a mixture of substances helps against a certain disease or has sufficient negative side effects to prohibit its use.

I find it interesting that certain substances which proved to be very beneficial drugs against certain bad diseases have been called antibiotics; antibiotic is a Latin word meaning "against life." They terminate the life process of bacteria without having bad effects on the life process of humans and animals. It seems to be a general law that different kinds of living creatures are competing against each other in the sense that an advantage for one kind is compensated for by a disadvantage for another kind.

## How the Living became "Technological Matter"

When you read in the previous chapters about the theories of mechanics, electromagnetism, relativity, atoms and quantums, you probably had no doubts that these theories are fundamental findings which had an effect on the progress of technology, and therefore must be included in this book. Therefore, I could refrain from starting those chapters with a justification explaining how the theories presented became relevant for technology. But here it seems necessary for me to put such a justification at the beginning. About twenty years ago, most people would not have expected a chapter on biology in a book on the fundamentals of technology. The only reason for including such a chapter then would have been an interest in the technology of fermentation or sewage treatment. Technology, as I understand it, has always had the purpose of providing something useful to us humans, something which nature does not provide at all, or not enough to satisfy our demands. Steel is an example of a technological product which is used in manifold ways in mechanical engineering and which we must produce because nature does not provide it. An example of an energetic service, a service based on the use of energy, which nature does not provide is the elevator which takes us from the first to the tenth floor in a very short time. In addition, we cannot overlook the examples of technological systems which help us to transport and process information, since television sets, telephones and computers are almost everywhere.

When we consider the area of living things, we first ask what kind of benefit we can obtain from living creatures, and we begin our consideration of them without insisting upon a connection to any technology. There are two kinds of benefits, matter and service. Matter can be food, drugs or materials for making products like clothing, jewelry, tools or housing. Matter can be provided by a living creature, e.g., milk, apples or maple syrup, or in form of the material of the creature itself, e.g., bones, fur or wood. Service can be energetic or informational: think of an ox pulling a plow, a horse carrying a person or a carrier pigeon transporting a written message. All these benefits are provided by nature without requiring the use of technology. But since people were never satisfied with what they actually had, they began to look for ways to increase the quantity and the quality of what they got from nature. They wanted oxen which were stronger than those they had, and they wanted cherry trees providing bigger and sweeter cherries than those from the trees they had. This was the beginning of breeding plants and animals which dates back thousands of years. Someone first realized that, with a high probability, a strong bull will have stronger offspring than a weak bull, and that a better tree will grow from the seeds of a tree which produces first class fruit than from one that grows from the seeds of a tree which produces puny fruit. This was all known before the Augustinian monk Gregor Mendel (1822-1884) began to search for the laws of inheritance. The only way to experiment was to select plants whose pollen and blossoms were brought together for pollination, or to pair the animals brought together for mating. Although this might be considered as an intervention into the processes of nature, it certainly would be inappropriate to call it technology.

Once Charles Darwin (1809-1882) came up with the idea of the evolution of genetic information, it was only a question of time before someone would try to modify genetic information by physical or chemical means. The American Hermann Joseph Muller (1890-1967) had the idea that X-rays might have an effect on the genetic information in semen or eggs, the consequences of which would become visible as unusual properties of the resulting creatures. Of course, nobody expected that such experiments would provide "improved" creatures, but they would confirm the assumption that eggs contain genetic information which can be influenced. The first object chosen for such experiments was the fruit fly (*drosophila melanogaster*) which could not object about being used for such a procedure. People always assume that the more an animal differs from a human being, the less it can suffer. As expected, the exposure of the eggs to the X-rays did not lead to "improved" flies, but to a variety of crippled flies. This was considered great scientific progress, and therefore Mr. Muller was awarded the Nobel Prize.

Breeding of plants and animals had always involved a procedure for exploiting arbitrary changes of genetic information. Later, such changes were called "mutations," and for a long time, the causes of these mutations were unknown. But as soon as the essence of genetic information and the biochemical background of mutations were discovered, changing genetic information became a technological issue.

## Like the Mother, Like the Father - How Inheritance Works

It didn't require any research to discover that inheritance exists. Everyone who attentively observes generations of humans around him becomes aware of amazing similarities between parents and children. I still remember quite clearly how amazed I was when, many years ago, I was walking behind my brother-in-law and his five year old son. The little boy, holding the hand of his father, was walking exactly with the same characteristic gait as his father. I had noticed long ago that their faces strongly resembled each other, but that their ways of walking were the same came as a great surprise. It is quite natural for someone who detects such a similarity to ask himself how properties or traits transfer from parents to their children.

After high resolution microscopes became available, it was discovered that all higher animals and humans always evolve from a single cell which results from the merger of a female cell with a male cell. Thus, all genetic information transferred from the mother to the child must be contained in the egg, and correspondingly, all similarities between the father and the child must be the result of genetic information contained in the sperm cell. Research, of course, was not restricted to higher animals, but to all kinds of living objects such as plants and bacteria. In the case of bacteria, there is no merger of an egg with a male cell, but a splitting of the bacteria, i.e., one bacterium gives up its individual existence in order to produce two new bacteria. While the process of a cell splitting has the purpose of reproduction in the case of bacteria, such processes also occur in multicellular animals and plants where they are the basis for growth and regeneration of tissue. At first, after an animal has been conceived, it consists of only a single cell. Then it grows by successive cell divisions: the original cell splits into two cells, these split again and generate four cells, etc. The number of cells is very large and these cells of multicellular creatures are not all of the same kind. They are specialized according to the role they must play within the whole system. Since the creature consists of a single cell only at the beginning, this specialization must take place in the course of the successive cell divisions. The fact that all of these highly developed creatures which populate our earth have finally evolved is certainly a great miracle. It could be assumed that cell divisions and the related cell differentiations are somehow controlled by genetic information. But until the middle of the 20th century, no one had the slightest idea about how the information about the genetic properties of mother and father determine these processes.

When quantum physics reached a mature state in about 1935, it was possible to build so-called electron microscopes which have a much higher resolution than conventional microscopes which operate using light. This is a consequence of the fact that the wave length determines the resolution. The wave length of the de Broglie wave (see Fig. 11.10) associated with a beam of electrons is much shorter than that of a beam of light, and this explains the difference of resolution between electron microscopes and conventional light microscopes. However, the principles of building light microscopes could not be applied to building electron microscopes because light waves are electromagnetic waves while de Broglie waves are probability density waves. Therefore, electron microscopes cannot be explained without referring to the particle aspect of the electrons. Imagine that you are directing a jet of water at a marble sculpture and measuring how much water is splashing away in different directions. By varying the direction of the water jet, the reflections of the water provide enough information to deduce the shape of the sculpture. In an electron microscope, the direction of a beam of electrons is controlled by electrical and magnetic fields, and the electrons are reflected and scattered by the atoms of the "observed" object. The number of the scattered electrons which arrive at a certain location in a given time interval determines an electric current which can be measured. By measuring such currents at different locations around the object of interest, enough information is obtained to deduce an image of the object.

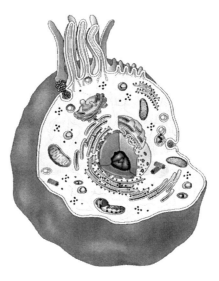

**Fig. 12.3** Cutaway view of a cell (from [BR])

Once the electron microscopes became available, they could be used to "look" at cells and produce images. Fig. 12.3 represents an example of such an image. Although the figure represents a drawing, the proportions and shapes of its components nevertheless correspond to the image which originally was produced by an electron microscope. But even such excellent images don't provide much help for finding the answer to the question of how inheritance works. Here again, we have to remember the old truth: "if you don't know what it looks like, you won't be able to find it."

The path which finally led to the knowledge about what should be looked for began in the middle of the 19th century in a city which today is called Brno, a city located in the Czech Republic. At that time, it was the city of Brünn and a part of the Austrian monarchy. A young man whose original name was Johann Mendel joined the local Augustinian monastery where he got the new first name, Gregor. He spent much time in the monastery garden and performed a lot of experiments with peas. He was very familiar with the methods of artificial pollination where the pollen of one plant is placed on the stigma in a blossom of another plant which was selected systematically. This plant then produces seeds from which new plants are obtained, and then these plants can again be subject to artificial pollination. In the years between 1854 and 1868, he used and observed over 25,000 plants of peas and recorded every single step of his experiments very accurately. From his notes he derived certain general laws about the inheritance of specific properties. In 1868, he became the abbot of the monks, and from then on he did not have time to act as a gardener and breeder.

Although Mendel's experiments extended over a time interval of fourteen years, the essence of his findings can be condensed into three figures, 12.4 through 12.6. Figure 12.4 represents Mendel's results concerning the transfer of one specific property from the parents to their children. If one parent provides the genetic information that the child's property or trait should be an A, while the other parent provides a B, there are three possibilities concerning the property which the child will finally have. Either A or B could be a so-called *"dominant property"* which means that the child will have this property and the information from the other parent will be ignored. If neither A nor B are dominant, the child will have a property C which is equal to neither A nor B. The property considered in Fig. 12.4 is the color of the blossoms of plants of peas; A is red, B is white and C might be pink.

| Genetic information (e. g. color of the blossoms) | | |
|---|---|---|
| From one parent | From the other parent | Property of the child |
| white | red | white (White is dominating.) |
| | | red (Red is dominating.) |
| | | combination color (no dominance) |

**Fig. 12.4**  Consequences of genes from parents

Fig. 12.5 shows how the genetic information from grandparents can influence a property of a grandchild. The assumption is made that both the mother and the father of the grandchild are children according to Fig. 12.4, i.e., that they both obtained different genetic properties from their parents; their actual properties, however, are not of interest. From the four genetic properties provided by the grandparents, only two can reach the grandchild, one coming from its mother and the other coming from its father. The selection is the result of a random process.

As a result, the probability is the same for each of the four combinations which correspond to the four fields in Fig. 12.5. The probability that a property of the grandchild is determined by two A's or two B's is 25 % for each case, while the probability that the property is determined by one A and one B is 50 %. Whether

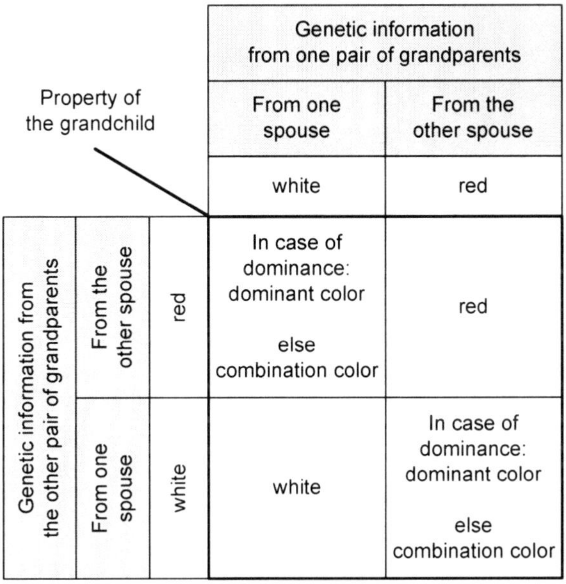

Inheritance and DNA

| Property of the grandchild | Genetic information from one pair of grandparents ||
|---|---|---|
|  | From one spouse | From the other spouse |
|  | white | red |
| Genetic information from the other pair of grandparents — From the other spouse — red | In case of dominance: dominant color  else combination color | red |
| Genetic information from the other pair of grandparents — From one spouse — white | white | In case of dominance: dominant color  else combination color |

Fig. 12.5  Consequences of genes from grandparents

these considerations correspond to reality can be checked by breeding many grandchildren from a given pair of grandparents and counting how many of these grandchildren have the property according to AA, BB or AB, respectively. The ratio of these numbers depends on whether or not one of the properties A or B is dominant. In the case of dominance, the property arising from AB is equal to either AA or BB. Let's assume that A is dominant; then AB generates the same property as AA. In this case, the number of grandchildren having this property will be, by a factor of three, greater than the number of grandchildren having the property according to BB. If, however, neither A nor B is dominant, there will be not only two, but three different properties, and the ratio of the numbers will be AA·BB·AB=1·1·2. All of these ratios were confirmed by Mendel's experiments.

Mendel's findings can be illustrated by a single summary diagram which is given in Fig. 12.6. There are properties whose actual appearance in an individual is determined by two genetic factors, one coming from the mother and the other from the father. When a child is conceived, only one of the two factors which the father had obtained from his parents can be transferred to the child, and the same applies to the mother. The actual selection occurs at random. Fig. 12.6 shows the flow of the two genetic factors which determine a particular property of the individual at the bottom. One of these two factors is represented by a white rectangle and the other one by a shaded rectangle. The diagram shows that these two factors must already have occurred in the set of the 16 factors associated with the eight great-grandparents. The remaining 14 factors became lost in dead ends and cannot appear again in later generations unless the couples considered do not have just one child each, but have larger numbers of children.

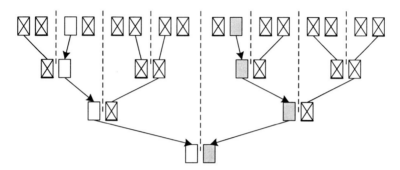

**Fig. 12.6**   Flow of genetic information from generation to generation

Though Gregor Mendel published all his results, no one showed any interest in them for many years and his findings had no effect for a long time. It took over thirty years, i.e., until about 1900, when his results became the starting point for further investigations. Mendel had used the term "factor" for genetic information concerning a specific property or trait. In 1909, the Dane Wilhelm Johannsen (1857-1927) replaced the word factor by the term "gene." The Brit William Bateson (1861-1926) had introduced the name "genetics" for the science of inheritance three years earlier; this term refers to the Greek word *genetikos* (creation or production) which is also the origin of the word generation. At that time, the chemical nature of genes was still completely unknown. But it became rather clear, only a few years later, where to search for the genes. The most probable locations were in the so-called *chromosomes* which had been discovered in 1843; but their involvement in inheritance was not confirmed until 1910. Using appropriate chemicals for dying, chromosomes can be made visible under a normal light microscope, but only in a certain phase of the cell division process. The name "*chromosome*" refers to the two Greek words *chroma* for color and *soma* for body. The chemical structure of chromosomes was not found until 1940, but this knowledge was not required for the conclusion that genes must be associated with chromosomes. Again the corresponding experiments were done using the fruit fly, the most favored experimental object of geneticists which I mentioned earlier in connection with exposing eggs to X-rays. By performing extensive breeding experiments accompanied by careful observation of the chromosomes, it became clear that properties and traits are inherited via the chromosomes.

The fact that genes must be contained in the molecules of desoxyribonucleic acid (DNA) was concluded from experiments which were performed by the

Canadian physician Oswald Avery (1877-1955) in 1944. However, he was not the discoverer of DNA, since it was discovered in 1869 by the Swiss biologist Friedrich Miescher (1844-1895) who called it "nucleic acid." Avery wanted to find out by his experiments whether the genetic information is in the DNA or in the proteins which are contained in a great variety in the cells. In his experiments, he used two types of pneumococcal bacteria which were known to be the bacteria which cause pneumonia. One kind of pneumococcal bacteria is covered by a protective surface of slime which gives them a smooth appearance. This cover is missing in the other kind, making their appearance much rougher. From the bacteria having the smooth cover, Avery extracted all the substances which were candidates for carrying the genetic information and, one after the other, he inserted these substances in cultures of purely rough bacteria. He observed that only with the insertion of DNA and after a certain time, the culture which originally had contained only bacteria of the rough kind now also contained more and more bacteria of the smooth kind. This effect, however, could not be observed when an enzyme which was known to decompose the DNA molecules was inserted with the DNA.

Once it had become clear that the genetic information is located within the DNA molecules, chemists began to analyze DNA in order to find its structure. It was not at all sufficient to determine which elements are contained and the quantitative ratios between these elements. Instead, it was necessary to find the chemical structure showing which elements are connected via their valencies. Such structures can be represented as two-dimensional graphs (Fig. 12.2). DNA, however, is not a single substance but a huge variety of substances built according to a common structure principle. The forces which hold a molecule together depend on the relative locations of the atoms in three-dimensional space. The final clarification of the three dimensional structure of DNA was provided in 1953 by the Englishman Francis Crick (1916-2004) and the American James Watson (1928- ). The three-dimensional model they proposed was the result of their theoretical considerations which were based on experimental results found by other scientists. I am convinced that the most important experimental contributions came from the Austrian biochemist Erwin Chargaff (1905-2002). From 1935 until he retired, he was a professor at Columbia University in New York City. After Avery succeeded in confirming that genetic information is contained in the DNA molecules in 1944, Chargaff asked himself what kind of chemical analysis would provide more insight into the chemical structure of this information. If the genetic information is contained in the DNA molecules of an individual, the molecules must differ both between species and between individuals within a species. And furthermore, the differences between two different species most probably would be greater than the difference between two individuals of the same species, since a dog differs from a human being much more than one human being differs from another.

Here, we should stop and think about the tremendous task Chargaff had decided to tackle. At first, a sufficient quantity of DNA had to be extracted from cells which contain hundreds of thousands of different kinds of molecules. In contrast to Avery, Chargaff could not restrict his analysis to bacteria, but had to analyze the DNA of a great variety of living creatures, and this required completely new methods. It is much easier to extract the DNA from bacteria than from cells of higher creatures where the DNA molecules are entangled with protein molecules, forming clusters which are located in the chromosomes in the cell nucleus. After having obtained a sufficient quantity of a certain DNA, he had to solve the even more difficult problem of finding the pattern of the chemical structure. Depending upon the actual species the DNA comes from, the number of atoms within one DNA molecule can lie in the range of millions or even billions. Such huge molecules are always composed systematically of small structured building blocks. If you look again at Fig. 10.2 where you find the molecular structures of two kinds of alcohol, you'll easily see the principle on which these molecules are built. The backbone of such a molecule is a chain of links, each consisting of one atom of carbon and two atoms of hydrogen. By adding more such links to the chain, new types of alcohol can be obtained.

The essence of Chargaff's findings, which he published in the years 1949 through 1952, consists of the ratios between the four substances adenine, thymine, guanine and cytosine which he had found as building blocks of the DNA molecules. Although such molecules contain more than these four building blocks, the others are not relevant with respect to genetic information. Perhaps you have read or heard that the four relevant substances are called *bases*. In chemistry, the complementary concept to a base is an acid, and an aqueous solution of a substance can be basic or acidic. You also may have learned in high school that the test for finding out which of the two cases actually applies consists of dipping a piece of litmus paper into the liquid. In the case of an acid, the paper turns red; otherwise it turns blue. If, however, you ask a professional chemist for a definition of the base/acid pair, he presumably will not refer to the litmus test, but provide the definition which is illustrated in Fig. 12.7. When talking about atoms or molecules, we quite naturally assume that they are electrically neutral particles, i.e., that the electrical charge of their electrons is compensated for by the positive charge of the same number of protons. But in chemistry, atoms and molecules are also of interest when they carry an electrical charge, i.e., when they contain more electrons than protons or vice versa. Atoms and molecules which are not electrically neutral are called "ions." There are four different ways a particle which originally is neutral can become an ion: we can either add or take away an electron or a proton. But which of the four possibilities can be realized depends on the structure of the original particle. For example, a proton can be taken away only from a molecule which originally contains at least one atom of hydrogen.

Inheritance and DNA

| Acid<br>can dispense protons,<br>i. e. hydrogen nuclei | Base<br>can absorb protons,<br>i. e. hydrogen nuclei |
|---|---|
| Sulfuric acid<br>$H_2SO_4$ | $HSO_4^-$ |
| $NH_4^+$ | Ammoniac<br>$NH_3$ |
| $C_5H_6N_5^+$ | Adenine<br>$C_5H_5N_5$ |

**Fig. 12.7**    Examples of acid/base pairs

The definition given in Fig. 12.7 says that the base/acid pair is characterized by adding or taking away a proton.

In reference to the four bases adenine, thymine, guanine and cytosine, Chargaff's findings can be condensed to the following four statements which today are called "Chargaff's rules":

1. The ratios between the bases differ from species to species.
2. DNA samples from different tissues of an individual have the same base ratios.
3. The base ratios of an individual are independent of his age, his alimentation and his environment.
4. In all DNA molecules, the number of adenine molecules is the same as the number of thymine molecules and the number of guanine molecules is the same as the number of cytosine molecules.

The first three rules are well suited to the assumption that the base ratios of a DNA molecule determine the genetic information of an individual. But at first glance, the fourth rule seems to contradict the idea that the DNA molecule could be viewed as a text, where the letters correspond to the four bases. It doesn't make much sense to assume a text written with an alphabet of four letters {a, t, g, c}, where each letter has a partner in the text which occurs exactly as often as the letter itself. Someone who is used to thinking in formal systems - I remind you of Chapter 4 - will nevertheless soon find ways to consider the DNA molecules as texts with letters from a given alphabet. Instead of the four bases, the pairs adenine/thymine and guanine/cytosine could be taken as letters, and in this case, the ordering of the partners in a pair could be either relevant or irrelevant. If the order is irrelevant, the letters would be the two unordered sets {a, t} = {t, a} and {g, c} = {c, g}. If, however, the order is relevant, the letters would be the four tuples (a, t), (t, a), (g, c) and (c, g).

I had the privilege of getting to know Erwin Chargaff personally when he was quite old and, on the basis of our discussions and correspondence, I can say that he was an ingenious, wise and far-sighted person. I would have liked to have asked him why he himself did not consider the relationships among the four bases and the possible alphabets for genetic texts as I presented them above. Unfortunately, Chargaff died before I began thinking about this subject.

As far as the solution of the alphabet problem is concerned, whether the DNA molecules are viewed as three-dimensional structures or as two-dimensional graphs in a plane is completely irrelevant. The problem is a logical and not a physical one which makes it possible to present the solution in the form of a graph. The need for adding a third dimension comes up only when considering how the logical structure is implemented physically. Therefore, I now shall present the logical solution. Based on this, it is rather easy to describe the corresponding three-dimensional physical solution.

It follows from the fact that the four bases occur in pairs that there must be forces which keep the partners within each pair together. In my design of Fig. 12.8, I chose a layout where the bridges between the partners of the pairs are represented by horizontal dashed lines. These bridges are called hydrogen bridges since, on one side, they end at hydrogen atoms. The dashed lines don't correspond to chemical valencies. You can easily check this by counting the continuous lines which end at the atoms. The sums you get always correspond to the number of valencies of the atom considered, e.g., 1 for a hydrogen atom, 2 for an oxygen atom and 4 for a carbon atom. Both the forces which belong to the valency connections and those which belong to the hydrogen bridges can be explained by referring to the distribution of the electrons at different energy levels according to quantum theory. All forces, both those within a molecule and those between molecules, are either attracting or repulsing electrical forces. The bigger the molecules, i.e., the more atoms they contain, the more the molecules affect each others shapes. Think of a molecule as being a group of people where each person corresponds to an atom. Then, each valency connection can be viewed as an elastic rope between two persons. Furthermore, you should assume that each person has applied an individual body spray with a fragrance which some of the neighbors like and others don't. If two or more of such "molecules" get close to each other, they will try to optimize their shape in such a way that the distances between persons who like each others fragrance will be minimized, while in the opposite case the distances will be maximized. Such deformations also occur in the case of huge molecules, although in that case there are no fragrances of body sprays, but only electrical forces.

From each of the four bases in Fig. 12.8, a continuous line leaves the molecule and it is not shown where such lines end. If such a base is an isolated molecule, the corresponding line is connected to a hydrogen atom, but if the base is a building

Inheritance and DNA

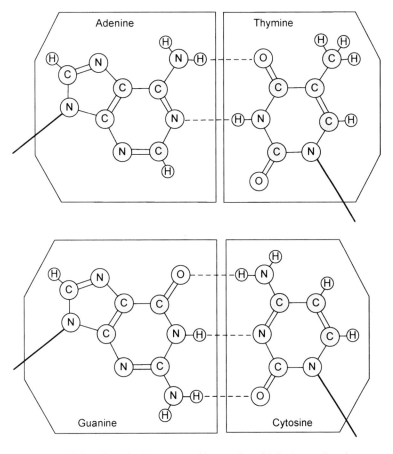

**Fig. 12.8** Chemical structure of base pairs with hydrogen bonds

block of DNA, the line leads to the DNA backbone as shown in Fig. 12.9. The DNA molecules are chains which can be considered as ladders where each pair of complementary bases corresponds to a step whose ends are connected to its corresponding side rail. In Fig. 12.9, I represented the steps as pairs of electrical connectors such as you have seen at the ends of electrical extension cords. At each pair, the pins of one connector are plugged into their corresponding sockets. According to the hydrogen bridges shown in Fig. 12.8, the pairs G/C have three pins while the pairs A/T have only two pins.

In contrast to a normal ladder, the two side rails in Fig. 12.9 have directions which oppose each other. These directions result from the unsymmetrical chemical structure of the building blocks of these side rails. The labels 3' and 5' result from the rules which organic chemists apply for identifying the atoms which are components of a cycle. The inverted comma indicates that the building blocks of

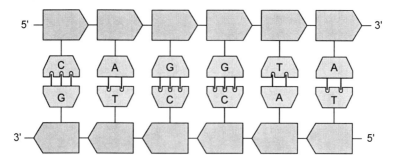

**Fig. 12..9** Section of DNA corresponding to a piece of genetic information in text form

the side rails contain two cycles, and that the numbers refer to the second cycle. The actual chemical structure of these building blocks is of no interest in the given context, but it might be interesting for you to know that they contain phosphor.

For someone who knows Chargaff's rules and adds the assumption that genetic information is encoded as a linear text, the structure in Fig. 12.9 seems to be a straight forward conclusion. But I suppose that in 1950 the hypothesis of genetic information being encoded as a linear text was not at all trivial. Nowadays, the idea that each cell contains a "cook book" containing recipes which can be read by the cooks, who also are located in the cell, is common not only to scientists, but to most educated people. We can immediately see the analogy to computers and their stored programs. But between 1949 and 1952, when Chargaff was working at the analysis of DNA molecules, computer science as an academic discipline did not yet exist, although the computer had been invented. In those years, mathematicians and engineers first began to realize that computers are good not only for crunching numbers, but for processing any kind of information. The specific way of thinking which has been introduced by computer scientists and which today is considered as basic as the elementary multiplication table, was not yet a part of the curriculum for educated people. While the concept of energy already had matured and it had the same meaning as today, the concept of information was still maturing. The history of science shows us again and again that concepts and ideas which for us are quite evident were outside of any human thinking process in previous years. Think of the hurdle René Descartes had to overcome when he developed the concept of a coordinate system (see Fig. 3.2).

While Chargaff and his assistants needed many years to find the results which could be condensed into the four rules, the two scientists Crick and Watson, who became famous as the discoverers of the DNA structure, needed only one year for "knocking together" their three-dimensional model of a DNA molecule on the basis of the experimental findings of others. By looking at their DNA model in reality or in a photo, you can assure yourself that the term "knocking together" is not far fetched at all, since it is a structure built out of wooden balls, wires, cardboard and sheet metal. The two scientists not only knew Chargaff's rules, but they also had gained access to photos which showed the results of X-ray analysis experiments with DNA molecules. These photos were made by Rosalind Franklin (1920-1958), an English biochemist who successfully had crystallized DNA. It is the characteristic of crystals that the atoms or molecules they contain are regularly positioned in space. Since the distances between the particles in a crystal are in the same range as the wave lengths of X-rays, interference patterns can be obtained by sending X-rays through thin layers of crystals. From such interference patterns, it was possible to determine the lattice structure of the crystal. Franklin's images led to the conclusion that a DNA molecule looks like a twisted rope ladder. While the property of being twisted is not necessary for understanding the logical structure of DNA molecules (Fig. 12.9), the twisting is required in the physical structure, because otherwise the complementary partners of the base pairs would not face each other adequately and the hydrogen bridges would not result.

Because the side rails have opposite directions, a question arises concerning the direction in which the text whose letters are the base pairs should be read. In reference to Fig. 12.9, we could arbitrarily determine one of the side rails, the upper or the lower one, but in reality, the two rails cannot be distinguished by referring to their positions. Thus, the only way to determine the direction for reading the sequence of base pairs is by looking at the text itself, since texts may contain sections which can be recognized as patterns, and these patterns determine a direction. Think of a sequence of digits which might be a phone number containing only the digits 0, 1 and 8, e.g., 0811801. If someone gave you this phone number on a slip of paper where he used straight vertical lines for the ones, the sequence would still look like a phone number even if you held the slip upside down. You could not know whether you should read the correct number 0811801 or the reverse number 1081180, unless you have additional information. This could be the restriction that a certain sequence will always be contained, while the corresponding reverse sequence will always be prohibited. Assume that, in our example, the sequence 01 is required and its reverse 10 is prohibited. This will help you to decide whether you hold the slip in the right direction or upside down. At the time when the chemical structure of DNA molecules was discovered, it was only a hypothesis that each DNA molecule contains sequences which determine the direction for reading its base sequence. This hypothesis could not be confirmed until successful methods were found for deciphering the actual base sequences. These methods were found about half a century after Crick and Watson published their paper on the three-dimensional DNA structure. Then the hypothesis concerning the determination of the reading direction could actually be confirmed.

The structure proposed by Crick and Watson not only answered the question about how genetic information is encoded, but it also solved the problem of how exact copies of DNA molecules could be made in the process of cell division. Fig. 12.10 shows the scheme for doubling a DNA molecule. The chain in Fig. 12.9 can be viewed as a zipper which can be opened from either side. "Opening" in this case means that the hydrogen bridges are cut. If this opening process occurs while the molecule is floating in a liquid which contains DNA building blocks of all kinds and in great abundance, the bases of the opened zipper (grey, shaded blocks in Fig. 12.10) can replace their lost complementary partners by acquiring appropriate components (white blocks in Fig. 12.10). Obviously, the possibility of such a straight-forward procedure for doubling a DNA chain is the consequence of the very tricky structure of these molecules. Nevertheless, you should not forget that this tricky structure is only a helpful initial condition for this doubling process which would not occur unless a great variety of "clever agents" cooperate with perfect coordination.

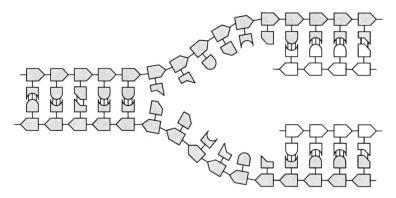

**Fig. 12.10**   Doubling a DNA molecule

Try to imagine that the structures shown in Fig. 12.9 and 12.10 were not chemical structures of molecules, but mechanical structures being composed of blocks having dimensions in the range of meters. With this view, Fig. 12.10 could belong to a process which occurs in an assembly shop. From the left, an assembly line bringing the undivided grey structure enters the shop. This assembly line ends at the center of the shop, and from here two separate lines take over and lead to the right. At this point, there must be a workman who opens the connections to make sure that two separate halves can flow to the right. Additional workers have to stand at each of the two continuing lines, for they must connect the white blocks to the grey structure. In order to make sure that these workers always have a sufficient supply of white blocks, other workers have to work in the background to produce such blocks and bring them to the lines. If this system were not only an analogy but a real assembly line, we would not wonder how it works because, of course, we could assume that all the workers had been very well trained for their specific jobs. But since the process occurs in a biological cell and not in an assembly application, it appears to me and presumably also to you to be like a miracle that all the necessary steps are performed accurately by "biochemical agents." If we consider the fact that the DNA molecules to be doubled can consist of chains of many millions of base pairs, the whole phenomenon becomes even more miraculous.

In our analogy, we assigned all difficult tasks to well-trained workers, and in biochemistry these workers correspond to the so-called *enzymes*. In Renneberg's book [RB], I found the following short introductory paragraph on the subject of enzymes:

> "Enzymes control almost all chemical reactions in living cells. Until now, over 3,000 different enzymes have been described in detail. It is suspected that up to 10,000 enzymes exist in nature. There are kinds of enzymes which occur in cells with only a few molecules, and others which occur in great numbers up to 100,000. All enzymes act as biological catalysts: they transform substances into other products without being changed themselves, often in fractions of a second. The cells have diameters between a tenth and a thousandth of one millimeter, and within them, thousands of coordinated enzymatic reactions occur each second. This can work only if each enzyme is able to recognize, among thousands of different substances contained in the cell, exactly the one specific substance which it has to transform to its product."

When I read this text, I immediately was reminded of the analogy between a cell and an assembly line, since this text addresses enzymes as if they were workmen or machine tools.

The path which led to today's knowledge about enzymes began with the work of chemists who were interested in understanding the processes which transform fruit juices, flour paste and milk into alcohol, leaven or cheese. It had been known for a long time that something has to be added to the original substances in order to start the corresponding transforming processes. If the final product is to be alcohol or leaven, the substance to be added is yeast. Certain other organic substances are required for getting cheese from milk, for example a liquid from the stomach of calves. Regional particularities led to the well-known plenitude of different kinds of cheese. The Latin word for fermentation is "fermentum," and this is the reason why the word "Ferment" was used in the German language since the 15th century for all the substances which are required for triggering those transforming processes. In 1878, the German Wilhelm Friedrich Kühne (1837-1900), who was a professor of physiology at the University of Heidelberg, proposed the word "Enzym" as a substitute for the former Ferment. Its origin is the Greek word "enzymon" which means yeast or leaven. Before the year 1897, it was an open question whether fermentation required living cells or whether appropriate non-living organic substances would be sufficient. This question was answered by the German chemist Eduard Buchner (1860-1917) who succeeded in triggering a fermentation process using an extract of yeast which didn't contain any cells and undoubtedly was not living. Today, more than one enzyme involved in fermentation is known. They have been isolated from different species and can be characterized biochemically.

In the text about enzymes which I cited above, enzymes are said to be biological *catalysts*. Today, most people associate the word catalyst with a technical component in the exhaust emission system of automobiles. This component is required by law because the original exhaust gases are harmful both for human health and for the environment. The catalyst in the car's catalytic converter transforms the harmful gases into harmless products as they pass from the engine to the exhaust pipe. The term "catalyst" was introduced in 1835 by the Swede Berzelius whom I already mentioned as one of the fathers of organic chemistry. He discovered that many reactions occur only if certain substances are present, although these substances are not consumed in the reaction. The Baltic chemist Wilhelm Ostwald (1853-1932) provided major contributions to a deeper understanding of the thermodynamic fundamentals of catalysis. He gave the following definition: "A catalyst is a substance which increases the speed of a chemical reaction without being consumed or changed, and without changing the final thermodynamic equilibrium of the reaction." This means that all such reactions would still occur without the presence of the catalyst, but that their speed would be much lower. The factor between the speeds with and without a catalyst can be in the range of billions.

There are many "jobs for enzymes" within a cell, but there is also a need for substances which transport "information" from one location of an organism to another. If the organism is animal or human, these "messenger substances" are called "*hormones*"; they are produced within certain cells and then transported by the blood stream to other specific cells where they cause certain effects. Since hormones can have effects at many different locations in an organism, they are always involved in processes which result in basic changes of the whole organism. For example, think of puberty which is the transition of a human being from a child to a sexually mature adult.

When reading a newspaper or a journal, you might sometimes be confronted with a foreign word from the area of biochemistry. For many years, I didn't know why producers of detergents put hormones into their products, or why athletes take hormones to improve their performance. During my search for the answer to these questions, I also encountered the need to distinguish vitamins from enzymes and hormones. "*Vitamins*" is the summary name for a group of substances which chemically do not have much in common. They are defined to be indispensable for

Inheritance and DNA

animal and human organisms, but they cannot be synthesized by the organisms themselves, and therefore must be provided as components of food.

Why do I tell you about enzymes, hormones and vitamins? One reason is that I want you to know what these words mean when you encounter them in printed texts or in television reports or discussions. But mainly, I want you to see quite clearly the complexity of a living organism which can live only if thousands or even billions of internal chemical reactions which occur every second are well coordinated. For me, the existence of life will stay a miracle, even though researchers are still discovering more and more details of the underlying biochemical processes.

You learned about the location and form of genetic information from my explanation of the structure of DNA molecules. The next question we shall consider is how the genetic information provided by the mother and the father are combined and become the genetic information of the child. The set of chromosomes contained in the nucleus of a cell of a human male is shown on the left side of Fig. 12.11. The cell considered might be a cell from any organ, e.g., a cell from the brain or the liver. Such a cell contains 46 chromosomes, half of which come from the mother and half of which come from the father. In the figure, two of the 46 chromosomes are shaded grey, while the others are white. The white chromosomes are labeled with a number in the range 1 through 22, and since each of these numbers occurs twice, a letter is added to indicate the source of the chromosome: M stands for mother and F stands for father. Two chromosomes having the same number always have the same structure and contain information about the same properties. In Fig. 12.4, we considered a child which got the genetic information "white blossoms" from one parent and "red blossoms" from the other parent. These two contradictory pieces of information must be located on two chromosomes having the same number which means that such pairs of chromosomes are contained not only in the cells of animal and human individuals, but also in plants. The grey shaded chromosomes in Fig. 12.11 are labeled not with a number, but with the letters X and Y. These two chromosomes determine the sex of the individual. The mother will always provide an X-chromosome while the father will provide either an X- or a Y-chromosome. If the chromosome which came from the father is an X, the child will be female; otherwise it will be male.

Each of the 46 chromosomes on the left side of Fig. 12.11 contains a DNA molecule consisting of many base pairs. When we think of such molecules, we usually imagine them as straight lines; this leads to the views shown in Fig. 12.9 and 12.10. But in reality, the DNA molecules are never stretched out, because otherwise they would not fit into the nucleus of a cell. They are extremely thin threads with lengths in the range between centimeters and meters. Since the diameters of cells are less than 1 millimeter, the DNA molecules must be wound in coils - think of many meters of sewing thread which don't need much space if they

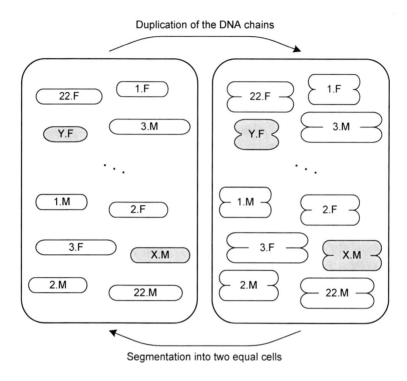

**Fig. 12.11**  Duplicating a cell of a male person

are wound on a small spool. In the case of DNA molecules, the spools are protein molecules. Thus, a chromosome not only contains a DNA chain, but also the protein molecules around which the DNA chain is wound. The whole ball is kept together by electrical forces because the DNA chain carries a negative charge while the proteins carry a positive charge. During the doubling process shown in Fig. 12.10, at least that section of the chain which is actually affected by the enzymes must stick out from the ball. At the end of the doubling process, the two identical copies of the chain still stick together at a certain point as if a drop of glue had been applied there. This situation is shown on the right side of Fig. 12.11. When the cell is divided, these connections are cut and each of the two partners goes to a different one of the two new cells. At the end of the cell division, there are two new cells, where each has 46 chromosomes and can be viewed as the cell on the left side of Fig. 12.11.

The purpose of the cell divisions shown in Fig. 12.11 is growth of an organism or regeneration of tissue. In contrast, we now consider cell divisions associated with sexual reproduction. The top of Fig. 12.12 shows a cell which contains the same pairs of identical DNA chains as that on the right side of Fig. 12.11, but now the

Inheritance and DNA 335

pairs, which have the same number and differ only in their origin indicator M or F, are clustered. While the cell on the right side of Fig. 12.11 contains two grey and 44 white pairs, the cell on top of Fig. 12.12 contains two grey pairs and 22 quartets. For these quartets to form, the original pairs must move to their corresponding partners having the same number, and once they reach each other, they must adjust themselves in order to form a well-structured package. This process can be observed

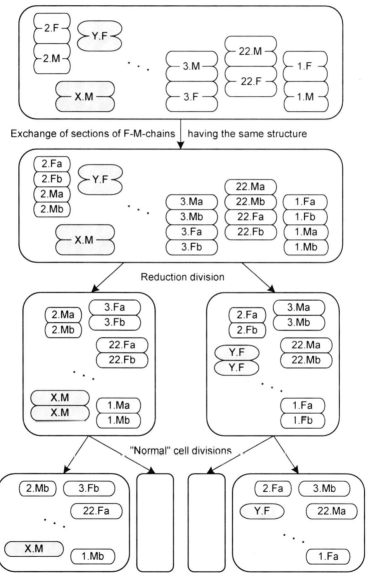

**Fig. 12.12** Building a germ cell - a human sperm cell

under a microscope, but a physical or chemical explanation of this process has not yet been found. It seems as though the original pairs know exactly what they have to do. Since the chains which came from the same parent don't differ, each quartet initially contains two pairs of identical DNA chains. But gradually, the four chains cross each other and this leads to an exchange of chain sections. At the end of this exchange process, the four DNA chains in a quartet are different, with a very high probability. We can conclude that the process which starts with the cell structure shown on the right side of Fig. 12.11 and ends with the structure shown under the top of Fig. 12.12 is very complex, because it takes about two weeks. Subsequently, the cell splits into two cells, each of which gets one grey pair and one half of each quartet. Thus, the two cells which result from this so-called *reduction division* differ significantly from the cell on the right side of Fig. 12.11 which contains two grey pairs and 44 white pairs, and where the partners in a pair are identical. In contrast to this, a cell which results from reduction division contains only one grey pair and 22 white pairs, where the partners in a white pair differ with high probability. Each of these two cells splits again, i.e., its pairs are divided and their halves go to two new cells. Thus, each of the four cells at the bottom of Fig. 12.12 has 23 chromosomes, each of which contains one DNA chain.

Each of these four cells is a male sperm cell which contains a selection of the genetic information which the man had obtained from his parents. While the cell at the top of Fig. 12.12 contains all the genetic information which he obtained from his parents and this information is clearly separated, the cells at the bottom contain only some of this information. This selected information is the result of the consecutive steps leading from the top to the bottom of the figure. This random process of selection is so sophisticated that it is extremely unlikely that two cells might be found having two completely identical sets of genetic information within the huge set of sperm cells which are generated during the lifetime of this man. The process of producing eggs has essentially the same structure and differs only with respect to the "grey pairs." If I had chosen a cell at the top of Fig.12.12 which belongs to a women, it would contain two X-pairs instead of one X- and one Y-pair. There is, however, still another significant difference between men and women with respect to the production of germ cells. The production of male germ cells, i.e. sperm cells, begins with the end of puberty and lasts almost a life time, while the production of female germ cells, i.e., eggs, occurs only during the growth of the embryo and is finished when a girl is born.

While almost nothing was known about genetic information in the past, many details have been discovered in the meantime. We not only know where genetic information is located, but we also know that this information is encoded in a text using an alphabet of four letters and how it is transferred from one generation to the next. Physicists and chemists have developed methods for analyzing DNA molecules in such depth that the sequences of base pairs can be deciphered even if

# Inheritance and DNA

their length is in the range of hundreds of millions of base pairs. In this context, you should note that there is an important difference between the two verbs "to decipher" and "to decode." If someone gives us his address in written form, it could be that we cannot decipher his handwriting which means that we cannot decide what sequence of letters he wrote. An absolutely different case is when someone gives us a clearly written sequence of letters which doesn't contain any words we know. In this case, the text seems to be an arbitrary sequence of letters, although it might well be that there is a meaning behind it. We can decode it only if we know the rules for interpreting this sequence. The successful deciphering of long DNA chains is a great achievement, that's for sure; but now, the scientists were confronted with the problem of decoding these sequences. What about the DNA chains in the cells of my brother-in-law? What do they have to do with his very special way of walking which he passed on to his son?

If we get text which we cannot decode, it's reasonable to ask someone who knows the decoding rules. However, the DNA text was not written for communicating with scientists, but for transferring information within a cell. And who reads it there? The readers are specific enzymes! The DNA text should not be compared to a novel, but rather to a cookbook. Many processes within a cell, except for the process of doubling the whole text, refer only to specific sections of the text. This is analogous to a cook who doesn't read the entire cookbook, but reads only those recipes which are relevant to the actual meal he is preparing. In cells, the recipes which are encoded in DNA are used for the production of proteins. Proteins are composed of so-called *amino acids*. There are 20 different amino acids and, with these as components, many thousands of different kinds of proteins can be built. The DNA text specifies which amino acids must be combined in order to form a molecule of a certain protein, but the text does not say how the amino acids shall be produced. The text refers to the amino acids only by their name which requires that the "cooks," i.e., the enzymes which build the protein, know how to interpret these names. Since there are only 20 amino acids which must have names, such a name requires only three letters from the alphabet of the four base pairs. Using this alphabet, $4^3 = 64$ different words of length three, called triplets, can be formed, but from these 64 different triplets only 20 are required as the names of amino acids. The analysis of the processes in the cell and especially the synthesis of proteins has shown that some amino acids have more than one triplet name assigned to them, and that other triplets are not names at all, but are used to divide the DNA text into "phrases" and "paragraphs," i.e., they correspond to punctuation marks such as commas, semicolons and periods.

Soon after the connection between the DNA text and the production of proteins had been discovered, it became clear that this text does not consist only of recipes for producing proteins. In the case of humans, only approximately ten percent of the entire DNA information in a cell describes the composition of proteins.

A rather high percentage of the remaining sequences have been called, perhaps prematurely, "junk DNA," i.e., litter or trash. We should never forget the complicated processes which lead from a single cell to a grown up living creature, and which afterwards guarantee that this creature stays alive. Compared to the knowledge which would allow us to say that we really have understood these processes, the knowledge we actually have, even though it is very impressive, is almost nothing. My knowledge that triplets in a DNA chain are used as names of amino acids doesn't help me at all to understand the fact that my nephew, although he was only five years old, already walked in the same distinctive manner as his father.

## How New Recipes Can Be Smuggled into Living Cells

There is no definitive answer to the question of whether an English text which we appreciate as a beautiful poem should still be called a poem after nobody is around who can speak or read English. Thus, we should not call a DNA chain a description of a living creature without adding that a living cell is required for the DNA chain to make sense. Genetic engineering quite naturally starts from the assumption that living cells do exist, and the question about where the first cells came from is not asked. At present, genetic engineering consists of changing the genetic information in cells by smuggling in additional sections of DNA chains. Let's consider the analogy between a musician acquiring a new set of notes and a cell getting new DNA sections. It certainly is much easier to put a sheet of new music on a music stand than to insert a new section of DNA into a living cell. And while the composer who writes the new music knows exactly how the musician will interpret what he has written, a genetic engineer cannot know exactly how the cell and the organism to which it belongs will interpret the sequence of base pairs which he composed. The comparison becomes more realistic if we assume that the person who writes the new music knows only the formal look of notes, but has no idea how they are interpreted. This person could "compose" acceptable music only by copying sections of music from competent composers. Thus, genetic engineers actually don't completely compose new sequences of base pairs, but copy and arrange "useful" sequences which they found in other cells. In the following paragraphs, I shall explain the solutions to the two basic problems of genetic engineering: how sequences can be found which are worth "transplanting," and how such sequences can be cut from one chain and inserted into another.

At present, the DNA chains of a living object cannot be interpreted as a whole, and we cannot know whether this will ever be possible. Looking at the DNA chain of a single living object doesn't tell us which sections of the chain determine particular properties. In order to try to find such associations, DNA chains of a great number of individuals of the same species were compared. Such comparisons showed that long sections of the chains were absolutely identical while others differed strongly. By neglecting the so-called junk DNA and restricting the comparison to the recipes for protein production, causal relationships between some recipes and certain properties of the creatures could be detected. Because of the extreme lengths of the DNA chains, which lie in the range of millions of base pairs or even more, these comparisons can be done only by using computers. This led to the new profession of a bio-information scientist. Among other results, the connections between certain typical modifications of human genetic information and specific diseases could be confirmed. An example is a bleeding disorder disease which is characterized by a total or a partial absence of certain substances in the blood which cause clotting in order to close wounds. Persons having this disease are always in the danger of bleeding to death. This disease is caused by a deficiency on an X-chromosome unless it is dominated by a second healthy X-chromosome. It is quite unlikely that girls or women have this disease, because they have two X-chromosomes, each coming from a different parent. They will have the disease only if both the mother and the father provided deficient X-chromosomes. Thus, it is no wonder that the disease is almost exclusively found in male members of a family.

Another example of a disease which could be related to a typical disorder within the DNA with absolute certainty, is the so-called *Down's syndrome*. It is named after the English physician John Langdon-Down (1828-1896) who, in 1866, was the first to describe this disease or disability. He called it "mongolism" because of the specific shape of the face and the almond-shaped eyes of the affected persons which gives them a look that somehow reminds us of people from Mongolia. A comparison of the DNA of those having this syndrome with the DNA of other people showed that chromosome 21, or at least significant parts of it, occur three times in the case of Down-syndrome persons. Such a triplication can happen if the process which is represented in Fig. 12.12 doesn't go as it should. There are many possible failures which might occur during the production of an egg or a sperm cell. Most of these will not cause the birth of a disabled or sick baby, but will cause the early death of the embryo.

Quite naturally, the question of whether or not there is a connection between genetic information and the different kinds of cancer has been asked. Cancer is characterized by an uncontrolled cell division process. When I explained the basics of cell division for growth (Fig. 12.11), I didn't mention that a cell cannot decide on its own when it begins its division. There must be a flow of information which triggers the division processes of all cells of the living object in order to guarantee that everything is well coordinated with respect to the life of the whole organism,. If this information flow is disturbed, the result might be cancer. Actually, specific DNA sections whose disorder could cause certain failures of the information flow have been identified. These DNA sections normally still have their regular structure at the time of birth, but failures during the process of cell divisions could produce such dangerous deficiencies.

Mankind does not yet know enough details to make all kinds of purposeful changes in human DNA chains without taking the risk of causing bad side effects, but we never know what the future might bring. A simple case - at least that's what many biochemists believe - is the process of inserting just one recipe for the production of a protein into the cell of a living organism. At present, the species which are "enriched" by such insertions are not higher animals or humans, but are bacteria and plants. In most cases of inserting an additional section of a DNA chain into a bacterium, the purpose is to transform this bacterium into a "machine" which produces a certain kind of protein, e.g., a useful enzyme. Once such a machine has been "built," this machine will immediately start to produce more machines of the same kind, and after a rather short time, the cell division process will have produced many millions of such machines. The species which has been used in this way as a machine for the production of a large variety of proteins is the bacterium *escherichia coli*. It occurs in high numbers in the intestines of higher animals and humans, and can be kept alive in culture mediums in the laboratory. The substances produced by such machines are not really new, but were produced previously by some living organisms; otherwise, the corresponding recipe would not have been known. Today, more and more substances which are produced in this way, but previously could be produced only by applying very complicated and expensive processes, are helpful in medicine. A typical example of such a substance is *insulin* which is required for our metabolism, but cannot be produced at all or not in the required quantities by diabetic patients.

Genetic engineers were not the first who smuggled additional DNA chains into bacteria, for this was done by viruses from the very beginning. About a hundred years ago it was discovered that there are organisms which can cause diseases, but are much smaller than the known bacteria. These organisms are called viruses, and they differ significantly from bacteria, not only in size. They contain DNA chains, but don't have the enzymes which are required for reproduction. Thus, viruses can be considered as "cookbooks" having a nice cover, but which must be brought into a "kitchen," i.e., into a cell where enzymes can read the recipes and act correspondingly.

Bacteria don't have a nucleus containing chromosomes. Their DNA molecules are closed rings which float in the cytoplasm. Scientists, who focused their research on enzymes which deal with DNA chains, once discovered some special enzymes which don't do anything except cutting DNA chains at specific sequences of base pairs, and other enzymes which just do the opposite, namely "gluing" pieces of DNA chains together if their ends fit. Fig. 12.13 should remind you of someone who wants to glue two boards together in order to get a larger board. To do this, he will cut the profiles of the sides of the boards in such a way that

# Genetic Fingerprints

they fit together exactly. The enzymes cutting the DNA chains act accordingly, i.e., they don't make straight cuts, but produce "profiles" which later can be detected by the gluing enzymes.

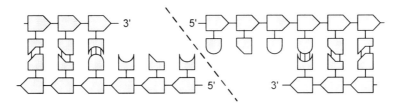

**Fig. 12.13**   Oblique cut through a DNA-chain

Almost all known modifications of the genetic information of plants were sought for the purpose of increasing the harvest, and not of improving the quality of appearance or taste. An increase of the harvest could be reached by making the plants immune to certain agents. Such agents are not only bacteria or insects, but also certain chemicals which the farmer wants to spray over his fields in order to kill weeds or pests. Creating such an immunity requires that an organism be found which already has this immunity. Then, the DNA section causing this immunity can be searched and transplanted into the cells of the plant to be "improved." It is also possible to enable the plant to produce certain proteins whose odor is disliked by specific insects and makes them lay their eggs elsewhere. The reason tomatoes begin to decay from their inside relatively early is the existence of a certain putrefactive agent. Therefore, the genetic information of tomatoes could be expanded by the recipe for the production of an enzyme which neutralizes this agent. Of course, this was possible only because this recipe had been found previously in another organism.

## How to Provide Evidence Confirming "Who It Was"

At the end of my short introduction to the basics of genetic engineering, I shall now briefly explain a genetic fingerprint and how it is obtained. No two human individuals have identical genetic information except for monozygotic twins. Of course, certain sections of the DNA chains must be absolutely identical, since a large variety of proteins have to be produced in each human organism in exactly the same way. There are two questions which can be answered with the help of genetic fingerprints. One question of interest arises when human cells are found somewhere, and we would like to know which person was the source of these cells. The other question arises when there is some indication that two persons may be related, and we would like to get clear evidence about this. This is especially the case when there are doubts concerning fatherhood. Theoretically, the answer to the first question could be obtained by complete decipherment of the DNA, both from the cells found and from cells of all persons which come into question. But this really is only a purely theoretical possibility, since the DNA chains in a human cell consist of over three billion base pairs, and their decipherment would be much too laborious and expensive with respect to the actual purpose. Therefore, only certain characteristic properties of the DNA chains are compared which differ from individual to individual, not with absolute certainty but with an extremely high probability. This is similar to the case of actual fingerprints where the possibility that two persons have the same fingerprints cannot be excluded with absolute certainty. When a certain blood relationship is to be confirmed or excluded, it is not the identity of the genetic information which must be checked, but the existence or absence of certain characteristic similarities that must be detected.

Interestingly, only those sections of DNA which at present cannot be interpreted at all are used for genetic fingerprints. This means that the recipes for the production of proteins are not taken into account here, because only minor differences between individuals would be found anyway. Thus, the genetic fingerprint is based on an analysis of certain sections of the so-called junk DNA. Each of these sections is a multiple of a typical sequence of base pairs, e.g., four times the sequence GGACTAG. The numbers of multiples may differ between the corresponding sections which the person got from his mother and his father. It is extremely unlikely that two individuals will have exactly the same combination of multiples of the different base pair sequences in all periodic sections considered.

Before the analysis can be started, a great quantity of DNA chains must be provided. In many cases, only a few cells are available originally, and only one DNA chain is contained in each chromosome. Therefore the process of generating a genetic fingerprint always begins with a multiplication of DNA chains. This is done by successive doubling of the chains according to the scheme in Fig. 12.10, although now the process takes place in a laboratory and not in the cells of an organism. The enzymes which are required for this process can be extracted from specific bacteria. The chains which are multiplied and then analyzed are not the original molecules contained in the chromosomes, but only the specific periodic sections which I described above. These sections can be selected by the specific enzymes used; I cannot tell you any details about this because I don't know them either. Thus, the multiplication process finally provides a mix of DNA chains, and the distribution of the lengths of these chains characterizes the individual person from whom the original cells came. This distribution of chain lengths can be made visible by forcing the chains to "race through a thicket" where their speeds depend upon their lengths. The shorter a chain is, the higher its speed will be; at the same time, the longer chains will stay behind. Some time after the race has been started, a snapshot can be taken which shows the actual positions of the different chains along the race track. This picture is the genetic fingerprint which is specific for the individual considered. How can such a race be established? DNA chains, of course, are not sportsmen who can race through a thicket, but nevertheless they can be placed on a certain race track and forced to move forward. If the chains carry an electrical charge, which is the case for the DNA sections considered here, they can be placed in an electric field which lies between two electrodes and which pulls the chains from one electrode to the other. The thicket corresponds to a piece of moistened felt. The race can be stopped at any time by turning off the voltage between the two electrodes. By adding certain substances to the liquid in the felt, the actual positions of the DNA chains can be made visible and photographs can be taken. Since the purpose of this procedure is always the comparison between two or more genetic fingerprints, two or more parallel race tracks, each of which is assigned to the chain mix of a specific individual, are used. Then the patterns on the different tracks can easily be compared in order to see if they are identical or by how much they differ.

Ordering huge molecules according to their lengths or their weights occurs not only in the process of generating a genetic fingerprint, but it is also required in organic and biochemistry for different purposes. If the molecules are floating in an appropriate liquid and are electrically neutral, i.e., if they don't carry an electrical charge, then a field of gravity will cause a distribution where the heavier molecules are closer to the bottom. But in many cases, the field of gravity of the earth is much too weak to cause the desired effect. Much stronger gravitational fields can be generated by using so-called ultracentrifuges at speeds up to 80,000 revolutions per minute. Having a radius of 20 cm, they cause centrifugal forces which are greater than the forces of natural gravitation by a factor of one million.

Besides nuclear power engineering, genetic engineering is the most controversial technology today. Erwin Chargaff once said, "By abusing two kinds of nuclei, the nucleus of the atom and the nucleus of the living cell, mankind has crossed borders from which they should have shrunk back." I hope that you, dear readers, can accept my abstinence concerning any questions about ethics. In this book, I want only to present and explain the findings which enabled us to develop today's technologies, and there is no space remaining for any serious evaluation of all the activities which became possible by applying these technologies.

# Part III
Fundamentals of Engineering

# Chapter 13
# Why Engineers Are "Playing with Models"

Before you began reading this book, you certainly did not doubt that mathematics, physics, chemistry and biology are different kinds of sciences. But I would not be surprised if you had doubts concerning the question of whether engineering itself is a science. Couldn't it be that all the knowledge which is required for designing, producing and repairing technical products belongs either to the sciences mentioned above, or is a kind of "know-how" which lacks the characteristics of scientific findings? If technology really amounted to no more than many tricky applications of the findings presented in the previous chapters, I would have finished my manuscript at this point, because I already would have reached my goal. In the first chapter of this book, I said, "The purpose of this entire book is to provide the knowledge which leads the reader to believe that these products can be conceived and built." If you are interested in the details of specific products, you still must ask a competent specialist. But he will enjoy giving you the necessary explanations only if he is not required to begin "at Adam and Eve," i.e., if he need not first teach you all the fundamentals which are presented in this book. Now you should ask yourself whether or not there are technical products for which it is not obvious how the fundamental findings presented in the previous chapters could be used to conceive and build them.

## What Engineers Are Needed for

Certainly there are many technical products which can be explained easily on the basis of the knowledge previously presented here. Imagine that Socrates had been transferred into our time and now, for the first time in his life, he sees a bicycle or a steam engine. There would be no need to explain to him what the bicycle is good for and how it works, since he would immediately understand the purpose and the construction of this system by just watching someone riding on it. In the case of the steam engine, he would need a brief explanation because the control of the flow of the steam is hidden. The fact that steam generates forces was already known to the ancient Greeks, for steam lifted the lid of the pot of boiling water cooking a hen in their kitchens in exactly the same way as it does today. The function of an engine running on gasoline or diesel can also be explained to him since, at this point, he has already become familiar with the concept of energy, and he knows the fundamentals of mechanics and electro-magnetism. Thus, even an entire car would not seem mysterious to him, and he would have no problems understanding the ignition, the fuel injection pump, the headlights and the windshield wipers. But what he learned from the previous chapters would not give him the faintest idea about how an antilock braking system (ABS) or a global positioning system (GPS) could be built. Certainly, Socrates would be very confused when, all of a sudden while he is riding in the car with his friends, he hears the enjoyable voice of an invisible lady announcing that an intersection is coming in 1/4 of a mile, and adding the recommendation that the driver should prepare to turn right. Obviously, there are hidden technical concepts which cannot be explained just by referring to the fundamentals of physics and chemistry. These hidden concepts are abstract system models which constitute the main results of engineering science. Teaching these models requires an effort which is comparable to that of teaching the basic findings of mathematics and natural sciences. There are some system models which are applied in almost all areas of technology, while the use of other models is restricted to only one or a few specific areas. The models which I shall present in the following sections are so universal that they can be said to be part of "the native language of engineers." Therefore, these models should be known to everyone who wants to understand how engineers see the world.

When talking about the world of engineers, it seems reasonable to draw a line which marks the difference between engineers on one side and all other people being involved in creating technical products on the other side, e.g., architects, designers, craftsmen and artisans. Architects and designers sit in the middle between engineers and artists, since the main criterion for the quality of an architect's or a designer's product is esthetics, while this is of minor importance for the work of engineers. This is a consequence of the fact that most of the things engineers design and construct are invisible to the users of their products. A car engine is hidden under the hood, electronic circuits are hidden in the cabinet of a TV set and a concrete foundation is hidden under a building. Today, car bodies which are designed to please the tastes of the buyers are not designed by engineers, but by designers having a specific education. It is an interesting fact that architects have no objection at all against transferring the responsibility for structural engineering to civil engineers. Engineering science is required whenever craftsmen or amateurs can no longer achieve acceptable results. At first thought, it seems that engineering is nothing but applied physics and mathematics. Certainly, engineers in the traditional fields often must solve problems of applied physics and applied mathematics. But there is a definition of the role of the engineering professional which does not apply to the so-called "applied physicists" or "applied mathematicians:"

Need for Engineers 349

> It is the mission of the engineer to design and construct technical works on the basis of fundamental findings in natural sciences and engineering sciences, to consider economic and social implications, and to direct and supervise the implementation of the designed work.

I found this definition in Meyer's Encyclopedia [ME], and I cite it here because it exactly expresses my opinion of what engineering is all about. In this definition, natural sciences and engineering sciences are referred to as two sciences of comparable relevance. The activities of engineers can be classified into two completely different areas to be optimized. The first area concerns the products or systems and the processes of their implementation which must be optimized by applying special mathematics methods. The second area concerns the communications between the people involved in the processes, communications which must be optimized by applying appropriate concepts of abstraction and the corresponding forms of standardized representations. Optimal communication is required since many of the products and systems are so complex that their planning, design and implementation require an extremely high degree of division of labor. The people involved cannot make optimal contributions unless each individual gets the complete information he needs in time for the specific performance he is expected to provide. The concepts of abstraction which are the basis for optimal communication are provided in the form of system models. Please note that the word "model" has at least three different meanings. In connection with figures 4.11 and 4.12, a model meant a specific case of application of an abstract theory. A second meaning, however, is just the opposite, i.e., here a model is an abstract structure which can be applied to specific real structures. And a third meaning of the word model is used in the case of children playing with a model railroad; here, the word model means that real trains and tracks have been scaled down to one hundredth or less of their original size. It was the last meaning which caused me to speak of "engineers playing with models" in the title of this chapter although, of course, most of the system models engineers use are not scaled down copies of the real world.

No engineers are needed for the design and implementation of systems if they are sufficiently simple; it may even be that people who have no special education in technology can get a system to work. But with the growing complexity of systems, engineers become indispensable. The complexity of a system is not determined merely by the number of its components, but by the heterogeneity of the interacting components. Once you understand the interaction of two gear wheels, you also understand the interaction of a thousand gear wheels. However, the systems considered here are composed not only of gear wheels, but of a huge variety of interacting components. A transistor for an electronics engineer is just a simple component like a gear wheel in mechanics. Of course, physicists still can put a lot of effort into optimizing transistors for different purposes, and the electronics engineer greatly appreciates their efforts. But the engineer has to deal with systems containing many millions of transistors which must interact in a well-coordinated way, and mastering such systems is not a problem of physics. Examples of "complex systems" are an automobile assembly plant, an aircraft for over 500 passengers and a telephone switching system.

In contrast to mathematical problems where a proposed solution must be either correct or incorrect, solutions of technical problems usually don't allow such a definite judgment. In the field of technology, a proposed solution is only more or less appropriate. Engineers who plan and design technical systems have to move through a set of trade-offs which is characterized by the fact that the improvement of one property can be obtained only by degrading another property.

When systems have very high complexity, it is generally quite unlikely for the first straight-forward design to constitute an acceptable solution. The situation of the engineer is similar to that of a pharmaceutical chemist who is developing a new medication. Both have to submit their new product to a large variety of experiments in order to find out whether there are unacceptable side effects. Consider unplanned resonance effects which might cause unacceptable noises in a car or a bus, or of unplanned electromagnetic radiation which might cause failures of nearby electronic devices. Certainly, engineers know a lot of tricks for reducing the probability of such bad effects, but before a system has been put into operation and intensively tested, they can never be sure whether all unacceptable effects have been eliminated.

When a product with a completely new functionality is brought to market, the users are often expected to acquire an extensive amount of user know-how. On the other hand, the complexity the engineers had to cope with in designing the product was rather low. A good example which clearly shows this fact is the evolution of the automobile and particularly the necessity for shifting gears. In the early years of this technology, the gear shift was not synchronized, and this forced drivers to learn how to shift using a technique called double-clutching. The gear box had a rather simple structure which didn't challenge the engineers. Today, cars are equipped with completely automatic transmission systems so that, in the drivers' view, the problem of shifting gears has disappeared. The engineers, however, now have to deal with a complex system which has almost nothing in common with the simple gear box of the early years.

In general, today's engineers are confronted with a level of complexity which is far above that of a few decades ago. Therefore, the criterion which characterizes the difference between engineers on one side and applied mathematicians and applied physicists on the other side has become much more relevant than in the past. This criterion is determined by the problem of mastering the complexity which results from the system requirements and the subsequent high degree of division of labor in the process of designing and building the system. Of course, there is still the need to provide engineering students with a deep knowledge of mathematics and physics, but among the additional subjects needed are methods for mastering complexity, methods which must have a greater share of the engineering curriculum than they did in the old days. Engineering students soon become aware that their later contributions to the design and implementation of complex systems will always be restricted to an extremely small fraction of the total effort required. Therefore, they can easily see how important it is that they learn, at the beginning of their education, how to communicate effectively with all other people involved in the technological process in order to understand their problems and decisions. Engineering professors should repeatedly emphasize that it is more important to have learned how to acquire an overall view of the whole system and how to let others participate with their own individual knowledge, than to be familiar with a huge number of methods for optimizing details. An engineer, after having obtained a good professional education, can be confident that he will always succeed in finding and understanding the appropriate optimizing methods whenever they are needed.

In recent discussions concerning the updating of engineering curricula, I often heard it said that more emphasis should be put on making engineers better team players. A team is a group of people having a common task which requires that they communicate very closely. Usually, such a team consists of five to ten persons who live together most of the day, and thus have the ability to talk to each other whenever they want. Soon after they join a team, they become very familiar with each other's strengths and weaknesses. In contrast to this, the division of labor I mentioned above is quite different. The idea of having a group of five to ten people is absolutely unrealistic in the design of complex engineering systems. Here, sometimes many thousands of people with different backgrounds must work together in planned cooperation, although most of them will never meet and talk to each other. The "glue" which connects all these individuals to a single body must be their clear and unambiguous understanding of a common goal, and this understanding can be established only on the basis of appropriate documentation. Therefore, the definition of representation standards must be considered one of the most relevant activities of engineering communities. The following statement, although a rough exaggeration, highlights the kind of engineering I am considering: "Only when he deals with technical drawings does an engineer really work as an engineer; as long as he operates only with formulas, he is an applied mathematician."

Sometimes a grandma wonders why her grandson who is studying electrical engineering cannot repair her broken TV set. She doesn't know that there is a great difference between having learned the essential concepts of television systems and the specific knowledge about how the system functions are distributed among the electronic components of the circuits of a specific brand of TV. It's for the same reason that the president of an institute of technology should not expect a professor from the Department of Mechanical Engineering to be able to repair the president's official car.

At the end of my general considerations about engineering, I now cite a short paragraph from the essay "Considerations on Technology" by the Spanish philosopher José Ortega y Gasset [OYG 2]:

> "You should realize the rather strange fact that technology is mostly anonymous, and that its creators don't enjoy the personal reputations which are connected with other ingenious individuals. By definition, the engineer cannot reign or rule as the final authority. His role is gorgeous and adorable, but inevitably second rate."

In the writings of Ortega y Gasset, I found many surprising conclusions with which I could agree completely, but here I think the philosopher got on the wrong track. After all, as far as I am concerned, I can remember many occasions where my role was not second rate!

## A Look into the Toy Box of Engineers

There is such a great variety of technical products and systems surrounding us that I could easily fill many thick books with descriptions of the functions and structural details of all the devices which we encounter regularly and everywhere. Surely you are realistic enough not to expect me to present that kind of information – especially, since you see that the major part of the book already lies behind you and that there are not many more pages left for technological subjects. I shall use these remaining pages to provide a virtual pair of glasses for seeing the world through the eyes of an engineer. In some texts, the Italian artist and scientist Leonardo da Vinci (1452-1519) is said to be the first engineer. The authors base their conclusion on the many drawings of technical devices and systems left by da Vinci. Among those drawings is one which shows a system which can be considered a helicopter. In my view, da Vinci was more an inventor than an engineer. It is true that he had many innovative ideas, but he did not have the means to actually implement his designs. We recognize something as an engineering achievement only if it has been created and stands the test of reality in its implementation.

Although it is impossible by just writing a few sentences to make you see the world through the eyes of an engineer, nevertheless, I can introduce you to some essentials of this kind of world view by presenting a single typical example. Some time ago, I was traveling with a group of friends through the Italian part of the Alps which is called Dolomites. When we reached the Passo Pordoj, a mountain pass, the bus stopped for a rest break of about one hour. The pass is located at an altitude of 2,239 meters, and from there a cableway leads up to a summit of approximately 3,000 meters. Standing in the cabin and hovering upwards with almost 60 other people, I heard some of my fellow passengers say that this really was an impressive engineering achievement. The cableway bridges the entire distance between the valley station and the mountain station without a single intermediate support. Later I got the information that the cable has a length of almost 1.5 kilometers. There are a tremendous number of problems which must be solved during the process of designing and constructing such a cableway, and these problems require the cooperation of engineers from different disciplines – civil, mechanical, and electrical engineering, and others. All of the people involved have the same characteristic picture in their mind's eye, a picture showing a rope suspended at both ends.

An engineer who thinks about a technical system always sees many aspects which could guide his interest. The choice of a particular aspect determines the set of physical quantities which must be considered. We now choose the most illustrative of all aspects concerning the cableway, the location of the components in space. That means that we now restrict our interest to the curve of a rope between its two fixed ends. There are only a few quantities which determine this curve, namely the coordinates of the two suspension points and the length of the rope. The considerations which finally lead to a formula describing the curve must start with the constraint that the forces at the ends of the rope must be directed

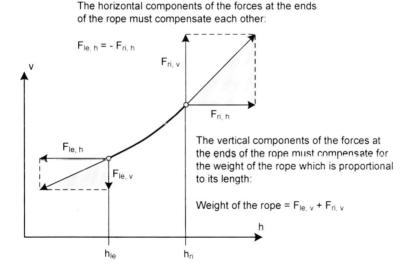

**Fig. 13.1** Forces at the ends of the wire-rope of a cableway

tangentially in the direction of the rope at these points. This means that the slope of the rope at each point determines the direction of the corresponding force. This is illustrated in Fig. 13.1. The forces caused by the weight of the rope must be vertical, since gravity acts downward. From this it follows that the sum of the vertical components of the forces at the two suspension points must point upward and its total value must be equal to the weight of the rope. From the fact that the rope does not introduce any horizontal forces, it follows that the horizontal components of the forces at the two suspension points must compensate for each other, i.e., they must have the same values, but opposite directions. These considerations are sufficient for the deduction of the formula which describes the curve of the rope in the (h, v) coordinate system where h stands for horizontal and v for vertical.

When I had the idea that the example of the cableway would be perfect to introduce you to the way engineers think, I immediately remembered the formula which had been given to me by my math teacher in senior high school. Actually, I hadn't encountered any problem which required the application of this formula during the entire 40 years of my professional life. You find this formula in the upper left corner of the diagram in Fig. 13.2. The curve which is described by this formula is called a *"catenary."* I won't guide you through the formal steps which lead to this formula because you wouldn't learn much from its derivation. The

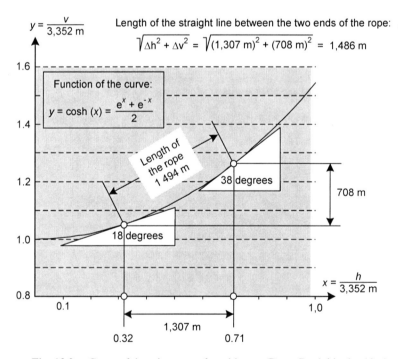

**Fig. 13.2** Curve of the wire-rope of a cableway (Passo Pordoi in the Alps)

System Models 355

mathematical methods required for this derivation, namely differentiation and integration, were developed about 300 years ago and were discussed in Chapter 3. Before these methods became available, mathematicians assumed that the curve of a suspended rope is a parabola like that in Fig. 3.2.

Once the correct formula of a catenary had been found, mathematicians began to explore this result in all directions. During this exploration, they realized an amazing structural similarity which is shown in Fig. 13.3. In the right column of the table you find equations describing properties of the two trigonometric functions $\sin(x)$ and $\cos(x)$ which I introduced in connection with the graphic representation of complex numbers in Fig. 2.15, and which occurred again in Euler's formula in Fig. 3.22. To the left of the trigonometric functions, you find the so-called *hyperbolic* functions, one of which describes the catenary. Because of the striking similarity, the hyperbolic functions were called $\sinh(x)$ and $\cosh(x)$ where the letter h refers to the Latin word *hyperbolicus*. This word should remind you of one of the conic sections shown in Fig. 7.2. Both the trigonometric and the hyperbolic functions can be used to describe specific types of conic sections. In an (x, y) coordinate system, the two functions [x=cos(p), y=sin(p)] describe a circle with radius one, and the two functions [x=cosh(p), y=sinh(p)] describe one branch of the so-called unit-hyperbola. The letter p stands for "parameter." The value of p is restricted to the interval $0 \leq p < 2\pi$ in the case of the circle, while all values from the interval $-\infty < p < +\infty$ are required for the definition of the hyperbola.

| | Formulas associated with the catenary (hyperbolic) | Formulas associated with the triangle (trigonometric) |
|---|---|---|
| $f(x)$ | $\frac{1}{2}*(e^x + e^{-x}) = \cosh(x)$ | $\cos(x) = \frac{1}{2}*(e^{i*x} + e^{-i*x})$ |
| $\frac{df}{dx}$ | $\frac{1}{2}*(e^x - e^{-x}) = \sinh(x)$ | $\sin(x) = \frac{1}{2i}*(e^{i*x} - e^{-i*x})$ |
| Relationship between $(f(x))^2$ and $\left(\frac{df}{dx}\right)^2$ | $\cosh^2(x) - \sinh^2(x) = 1$ | $\cos^2(x) + \sin^2(x) = 1$ |

**Fig. 13.3** Structural similarity between the hyperbolic and trigonometric formulas

Are you wondering why I take you with me on such excursions through the world of mathematics? I want you to see the world through the eyes of engineers, and a part of this view concerns how observed phenomena can be described by mathematical functions. We have to be aware that mathematical descriptions apply only under certain idealistic assumptions. Thus, the formula which describes the curve of the suspended rope applies only if the rope is not elastic and its weight is equally distributed over its whole length. The formula given does not describe the curve of a rope carrying a cable car; the mathematical description of this situation is much more difficult. In reality, the rope is elastic and therefore its length depends on the distribution of the load. And the fact that the length varies with temperature must also be taken into account.

The cableway is not only an appropriate example for showing you that engineers always look for mathematical descriptions, but it also provides an illustrative example for showing that engineers always decompose the world into two parts. One part contains the object of their actual interest, and the other part is the rest of the world. The union of the two parts constitutes a "system" which consists of the *system kernel* and the *environment*. This demarcation characterizes the *aspect* under which the system is actually considered. The aspect determines the physical quantities which actually are of interest. Their time-dependent values result from the interaction between the system kernel and its environment.

In the example of the cableway, the system kernel consists of only the rope, and this kernel is connected at two points with the environment. The environment not only introduces the coordinates of the two suspension points, but also the gravitational field. The physical quantities which characterize this aspect of our cableway are represented in Fig. 13.4. Some quantities are determined by the kernel or the environment alone, while others are determined from the interaction between the two. The locations where the physical quantities can be observed are symbolized by rounded nodes, and the agents which provide or require these quantities are symbolized by rectangular nodes. By using a dashed borderline for the coupling agent, I wanted to indicate the fact that this agent is virtual and cannot be considered a real object which could be cut out and taken away as could be done with the rope and the mountains. Although the coupling node is virtual, it is still a required agent, because there is nothing else which could provide the curve of the rope and the forces at its ends.

The diagram in Fig. 13.4 is an example of a so-called *directed bipartite graph* which you will encounter more often in the following sections. Such graphs are characterized by two types of nodes which can be connected by arrows, but only if the two connected nodes are of different types. Usually the two types of nodes are symbolized by rounded and rectangular shapes. Bipartite graphs are used in different interpretations. The graph in Fig. 13.4 must be interpreted as a so-called *composition model*, and it is virtual because not all of its components are real objects. In such a composition model, the two types of nodes represent agents which must provide something, and containers into which the providers can put their results. An arrow leading from a container to an agent means that the connected agent requires the contents of the container for producing its output, but it does not mean that the contents are necessarily consumed and the container left empty after the operation of the agent. It could be that the contents of a container are not matter or energy, but information which can be used just by reading it, and therefore not taking it away from the container. Fig. 13.4 shows that the kernel and the environment provide something which is used by the coupling agent to produce the contents of the container at its output at the bottom of the figure.

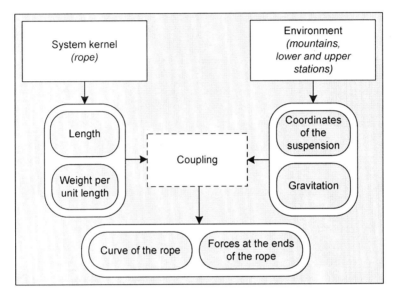

**Fig. 13.4**  Virtual composition model of a cableway

Real or virtual composition models are of great importance for an engineer doing a good job. I want to emphasize this with Fig. 13.5. The only thing which is actually given is reality; everything else is models being thought about or represented on paper. The mathematical model is assigned to the composition model, i.e., a mathematical model can be conceived only after reality has been idealized and appropriately abstracted. Sometimes we forget that modeling reality always requires some kind of idealization. In the example of the cableway we assumed, without giving it much thought, that the curve of the rope can be described by a mathematical formula. Thus we assumed that the rope has only a length, but not a thickness. This idealization didn't lead to an unacceptable discrepancy from reality because, after all, we didn't require that the formula describes reality with an accuracy of one millimeter.

Graphic representations of structure models are a perfect basis for efficient communication among all people who in one way or another are interested in the system considered. It is possible to point to nodes or edges of the graphs and thus

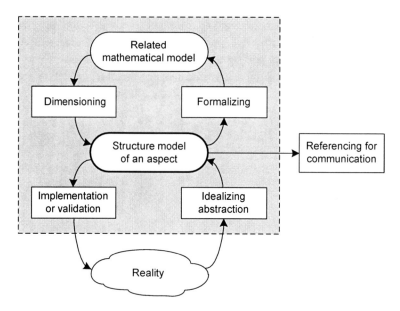

**Fig. 13.5**  Central position of a structure model in the toolbox of engineers

to ensure that everybody knows what is actually being discussed. And since the graphs can always be entirely overlooked, there is no danger of losing the connection between the details and the big picture.

While graphic representations are made for human eyes, mathematical models can be turned over to computers. Mathematical models can be used for deriving dimensioning values, i.e., information about how long or thick a body should be made, what voltage should be chosen, etc. Dimensioning always concerns properties of components or powers of sets which already occur in the structure model. The transition from the model to reality is either an implementation, i.e., building the system according to the specifications from the model, or a validation, i.e., testing the predictions of the model against reality.

Whenever a medical injury leads to a claim, a question concerning the liability of physicians or pharmaceutical companies must be determined. Similarly, a question concerning the liability of engineers comes up in the case of claims in connection with damages and injuries caused by failures of technical systems. In 1998 a bad train accident occurred in northern Germany. It led to the deaths of over 100 people and was caused by the tread ring of a wheel which failed to stand the stress while running over switches at a speed of more than 200 kilometers per hour and broke apart. In such cases, the questions which a judge asks his consultants about the design of the wheel can be characterized by referring to Fig. 13.5. Had reality been abstracted appropriately? Were the mathematical models consistent with the structure models? Were the computations on the basis of the mathematical model correct? Were the consequences from the models correctly transferred into reality? Were the results sufficiently validated by experiments?

We now shall consider a type of system which is called an "input-output-system." Systems of this type are characterized by providing something at their output which is produced from what they get at their input. Even the cableway system can be viewed as a system of this type since, at the mountain station exit, the system delivers people who are "produced" by adding potential energy to the people who enter the system through the entrance at the bottom station. The general model of this type of system is shown in Fig. 13.6. Fig. 13.7 shows what the unions of input and output channels are in the case of a telephone system.

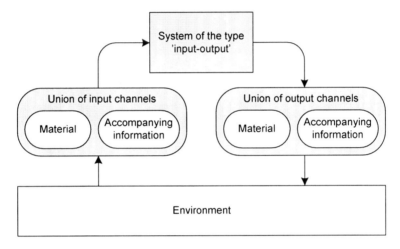

**Fig. 13.6**  Input-output system embedded in its environment

In the top-most view of a system, the system kernel is seen as a so-called *black box* which hides the internal structure. In order to characterize more specific system types, we have to look into the black box. Fig. 13.8 shows the general structure of a so-called *closed-loop control system*. This is one of the most important system models. It is applied not only for designing technical systems, but also for explaining processes in biological and sociological systems, i.e., processes in living creatures, in biotopes, in economics and in politics.

A very simple example of an application of this model is controlling room temperature by using a heating system. The controlled system consists not only of the furnace, but also of the room to be heated. The environment contains the people who expect to get a comfortable room temperature, but it also contains everything else besides the furnace which affects the room temperature. This includes the outside temperature determined by the weather, and the states of the doors and

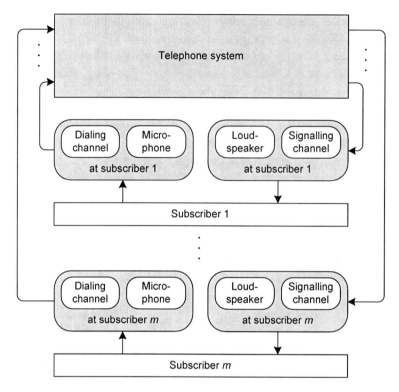

**Fig. 13.7**  Telephone system as an example for Fig. 13.6

windows which might be open or closed. Since the temperature normally is not exactly the same at every point in a room, the structure in Fig. 13.8 contains the two variables $X$ and $X_O$ which both correspond to a temperature in this example. $X$ corresponds to the temperature measured by the controller, while $X_O$ corresponds to the temperature which is sensed by the people in the room and determines whether or not they feel comfortable. The controlling value $Y$ determines the heat production of the furnace. The people in the room have access to a thermostat which allows them to express their wish concerning the desired temperature $X_T$, for example, by turning a knob or by entering digits on a keypad. By comparing the two values $X$ and $X_T$, the controller can decide whether it must change the value $Y$ in a certain direction or leave it alone. Besides the performance of the furnace, all effects which have an influence on the actual room temperature come from the environment and are captured by the variable $N$ which refers to the word *noise*.

The problem of designing a satisfactory controller for a given system to be controlled is solved by using control theory, a subject which is taught in both the departments of mechanical and electrical engineering at a university. Since values

System Models 361

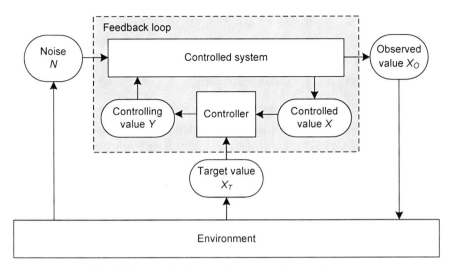

**Fig. 13.8** Feedback control system with its environment

of the noise $N$ and the observed value $X_O$ cannot be captured mathematically, the behavior of the system to be controlled can be only roughly described mathematically. However, this does not apply to the controller, since all three values $X$, $X_T$ and $Y$ which are required to express the behavior of the controller are measurable physical quantities, and the function $Y = f(X, X_T)$ can be expressed by a formula. Although the controlled system cannot be accurately described by a formula, it nevertheless is necessary to model this system component. In order to design an adequate controller, at least some rough information about the system's response time is required. Everybody knows that the room temperature does not jump to the desired value at the moment the furnace is turned on. In this case the response time lies at best in the range of minutes. But there are other cases where the response time of the controlled system is below one millisecond. It would be absurd to design a system where it takes a couple of minutes before the front wheels of a car respond to the turning of its steering wheel.

Since the connection of the controlled system and the controller via the values $X$ and $Y$ constitutes a feedback loop, oscillations will occur if the response times of the controlled system and the controller do not correspond. Such oscillations can also occur in cases where the controller is not a technical system, but a human individual. I very well remember a boat cruise on the "Canal du Midi" in the South of France, a cruise I once took with some friends. There you are permitted to steer a boat without having a boat driver's license. The agents who rent the boats to tourists are convinced that it is sufficient to submit their customers to a fifteen minute practice session, and then let them go on their own. In this example, the boat is the system to be controlled while the person sitting at the steering wheel is both the controller and a part of the environment. Here, the controlled value $X$ and the observed value $X_O$ are identical and correspond to the position of the boat relative to the sides of the canal. The target value $X_T$ is not a measurable physical quantity but comes from the vision of the boat's pilot concerning where he wants the boat to go. The pilot, in the role of the controller, combines his observed actual position of the boat with his vision about where it should be, and from this he determines the angle $Y$ by which to turn the steering wheel. To this point, my description would also apply to steering a car, but there is a significant difference between a car and a boat. While a car responds almost immediately to the turning of the steering wheel, the boat doesn't show any reaction at all for a few seconds. An inexperienced pilot will conclude from this that the angle by which he turned the wheel was not large enough, and he will continue turning the wheel. Only later will he realize that the steering angle was already much too large, because then the boat will turn abruptly by an angle which far exceeds the desired value. In response to this behavior of the boat, the pilot will become panic-stricken and immediately turn the steering wheel in the opposite direction. But again this will not cause an immediate reaction of the boat. A couple of seconds later, the boat will swing in the opposite direction with an angle which again is much larger than desired. Of course, these training sessions are performed in a wide inner harbor where there is enough space between the boat and the edges of the basin. Otherwise, the boat would hit the banks or collide with other boats and cause a lot of damage. Usually it doesn't take more than fifteen minutes before a person learns how to handle the steering wheel adequately in order to avoid all undesired moves of the boat. Then the cruise can begin.

Human individuals are intelligent enough to draw far-reaching conclusions from the behavior of the technical systems they are using, i.e., they can learn from their experience. Ordinary technical controllers, however, are not built for learning; they work correctly only when the system to be controlled behaves according to the assumptions made by the engineers when designing the controller. Nevertheless, it is possible to build so-called *adaptive controllers* which can adapt their behavior to unexpected changes in the controlled system and the environment. The design of such controllers requires much more effort, and therefore they are not used in standard applications.

Basically, a controller is nothing but an information processing system which gets information about measured values of physical quantities at its input. From these, by applying specific formulas, it computes the values which it provides as its outputs. How such information processing system can be designed and built will be explained in Chapter 14.

Each physical quantity which appears in a composition model of a system must be classified according to whether its value domain is continuous or is a finite set. In the latter case, the values are discrete. An example of a quantity having a

System Models 363

continuous value domain is the temperature $X$ to be controlled. While this is given by reality and cannot be changed by the designer of the system, the controlling value $Y$ and the target value $X_T$ can be either continuous or discrete, depending on the design of the system. The domain of the controlling value could be restricted to the finite set {furnace on, furnace off}, but it also could correspond to the continuum of the positions of a valve determining the flow of heating oil. And if the target value $X_T$ corresponds to the position of a knob on a thermostat, the position of this knob could be continuously adjustable, or it could be restricted to a finite set of discrete angles. If the domains of all quantities occurring in a system are discrete, the system is said to be discrete. The simplest case of a discrete system is a *sequential system*. A sequential system is characterized by the fact that its operation is a sequence of steps where each step is associated with exactly one input element and one output element. Although you probably didn't think about their being sequential systems, you certainly have often used systems of this type. Each machine selling soft drinks, candies, cigarettes or bus tickets is a sequential machine.

Fig. 13.9 represents a graphical protocol describing the sequence of steps in the process of buying a bottle of Coke from a vending machine. I assumed that the price of a bottle of Coke is $1.20. When a potential buyer approaches the machine, he sees a prompt for choosing a type of beverage on the display of the machine. Coke can be chosen by pressing a certain button. Pressing this button leaves the output container empty and only changes the message on the display. Now the machine knows that Coke has been ordered and that no payment has yet been made. Therefore the display tells the customer that he must insert $1.20 to get a bottle of Coke. Now the customer inserts a one-dollar bill. As a consequence of this, the output container stays empty and the message displayed says that 20 cents remain to be paid. By inserting a quarter, the customer finally has paid more than the Coke costs, and this causes the machine not only to place a bottle of Coke in the output container, but also to return a nickel. The machine is now in the same state as it was before the whole process began.

The graphical pattern of the protocol in Fig. 13.9 applies not only to this special case of a vending machine, but to all kinds of sequential machines. The actual specific case affects only the naming of the nodes with rounded borders. Each vertical column of such nodes corresponds to a certain type of information. The left column belongs to the input elements; in the next column to the right you find the column of the internal states of the machine; the two columns on the right side belong to the output elements with the right-most column assigned to the supplies, while the display messages are represented to the left. Supplies must not necessarily be material objects like bottles, tickets or coins. A supply may also be a "package of energy" which is supplied when the machine says "Good morning" or "Your input is invalid."

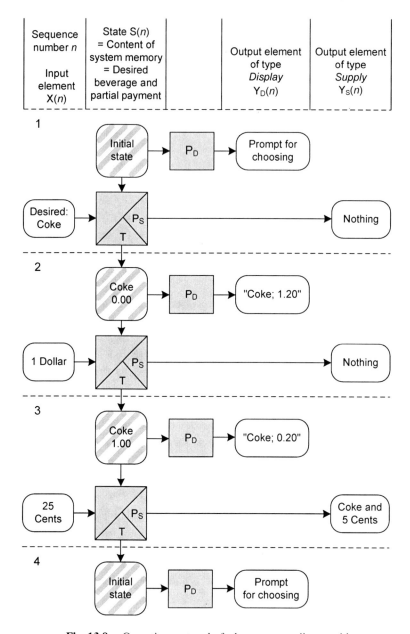

**Fig. 13.9** Operation protocol of a beverage vending machine

In mathematics it is common practice to use the letter $x$ for the input variable and the letter $y$ for the output variable in functional expressions of the form $y=f(x)$. Therefore it is quite appropriate to use the letter $X$ for the input elements and the letter $Y$ for the output elements of a sequential machine. However, the behavior of a sequential machine cannot be described by simply assigning a corresponding output element Y to each possible input element X, since the output does not depend entirely on the actual input. The actual state S of the machine, i.e., the actual content of its memory, must also be taken into account. In the example of the beverage vending machine, memory is required for storing the information about which kind of beverage has been chosen and which payments have already been made.

In order to describe the relationships between the elements occurring in the protocol in Fig. 13.9 in mathematical terms, we have to introduce three functions, the display function $P_D$, the supply function $P_S$ and the state transition function T. The results of the display function and the supply function are output elements of the sequential machine. I used the letter P to symbolize these functions in reference to the words product, presentation and provision. The information which is displayed is completely determined by the state. This is expressed by the equation

$$Y_D(n) = P_D[S(n)].$$

The discrete variable $n$ represents the sequence number which determines the row in the protocol (see Fig. 13.9). This variable $n$ is analogous to the variable $t$ which represents a point in time in formulas describing the behavior of continuous systems. Therefore, $n$ can also be called the discrete time variable. The actual supply $Y_S(n)$ depends on both the input element $X(n)$ and the actual state $S(n)$:

$$Y_S(n) = P_S[S(n), X(n)].$$

The knowledge of what the actual state and the input element are is sufficient to predict what the next state will be:

$$S(n+1) = T[S(n), X(n)].$$

The relationships which we just expressed by formulas are shown again in Fig. 13.10 in form of a composition model. This is a black box model, since it doesn't provide any information about how the state S and the three functions $P_D$, $P_S$ and T are implemented.

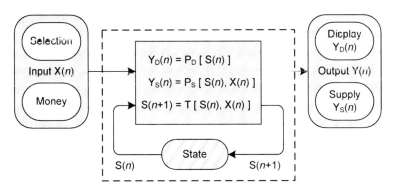

**Fig. 13.10** Black-box model of a vending machine

We shall now consider this problem of implementing our beverage vending machine. As a starting point, we again need an appropriate system model. The model we are looking for can no longer be a black box model as shown in Fig. 13.10, but the model must now show the components within the black box. We shall begin with a model which consists of only two components whose internal composition will be considered later. Fig. 13.11 shows such a composition model having only two components, and this model represents an *instruction execution system*. In my comment concerning the model of a feedback-control system in Fig. 13.8, I said that this model applies not only to technical systems but to many systems outside of the world of technology. The same is true for the model shown in Fig. 13.11.

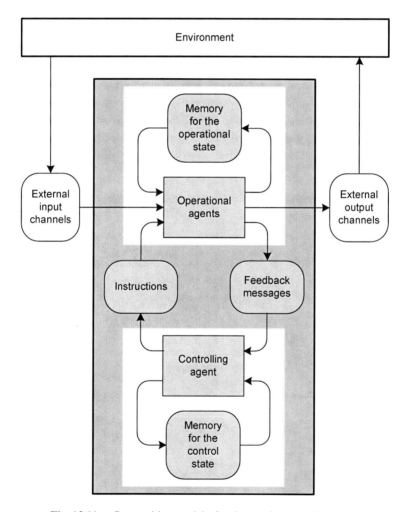

**Fig. 13.11**   Composition model of an instruction execution system

# System Models

In an instruction execution system we always find a set of so-called *operational agents*. Each of these agents is able to perform specific single-step operations without knowing at which time each operation is to be performed, and to which goal his own operations and those of his fellow agents shall lead. Thus, the operational agents need someone to tell them when to execute which operation. As a first example, let's make the absurd assumption that the musicians in an orchestra could not read musical notes and would not know anything about the piece of music which shall be played. But assume each of them would be a master of his instrument and could play any note at any intensity and length. These musicians need a conductor who at each point in time provides the information about which note must now be played, and at which intensity and length. This conductor would be the only one who knows the piece of music being played, and he would know which instruction he must give to each musician at every point in time, either from his memory or by looking at the score. Although the idea of such an orchestra is totally absurd, we need only to go a short step to find a real system which corresponds exactly to this orchestra. Every former merry-go-round organ and every historic orchestrion which you may find at a museum operate exactly as described. Each individual instrument can produce its specific sound which can be turned on and off by controlling a valve or an electromagnetic switch. In most cases, the "score" is a long tape with holes at specific positions which correspond to specific valves. The system which "reads" this tape by unwinding it and forcing air through the holes corresponds to the conductor reading the score and giving instructions to the musicians.

While there is no information flow back from the musicians to the conductor in this example of a conductor giving instructions to the musicians, there are other systems where the controlling agent requires some information which the operational agents must provide. Think of a grandpa sitting in a wheel chair and giving instructions to his two grandsons about where to drill holes in a living room wall for hanging paintings with heavy frames. Suppose he wants a hole at a position where the concrete wall doesn't allow a hole to be drilled. In this case information will flow from the grandson who tried but didn't succeed to his grandpa who gave the order. As a consequence, the grandpa will come up with an alternative position.

While the controlling agent has to deal only with information, this must not be the case with the operational agents. Suppose the purpose of the system is the production of chocolate; in this case at least some of the operational agents must be able to handle matter and energy. All agents are restricted to handling information only if the purpose of the system is pure information processing.

Of course, at the beginning of the design of an instruction execution system, the purpose of the system is well known. But this purpose does not necessarily determine a specific sequence of design steps. The design process begins either with writing the "score," or it begins with determining the set of operational agents and the paths between them. In the case of our beverage vending machine, it is advisable to first design the so-called operational part, and later to decide the order in which the operational agents perform their specific operational steps. Fig. 13.12 represents the result of my design efforts. I tried to come up with a rather symmetrical graphical structure, since this makes it easier to virtually to walk through and to keep it in mind. At the beginning of this chapter, I considered the difference between architects and engineers, and said that esthetics were of minor importance for the work of engineers. Now, I must slightly moderate my former statement because a good engineer will always try to make his drawings meet the esthetical expectations of the viewers. This applies especially to drawings which do not represent real objects drawn to scale, but to drawings which visualize abstract functional relationships. The format of such drawings is totally open concerning the form and size of the graphical symbols, and where they are located. It would have been easy for me to make a drawing with no graphical similarity to Fig. 13.12, but which still represents the same information concerning the existence of specific operational agents and their interconnections. When a good engineer designs such a drawing, he is quite aware that the result of his efforts has a great influence on the quality of the communications among the people involved in the project. Drawings having a layout which has been optimized with respect to communication can easily be kept in mind, and everybody immediately knows what someone is talking about when he refers to certain positions on the drawing. The role of these drawings is comparable to that of maps which people refer to when they are planning a trip.

Although the naming of the nodes provides a rough understanding of the structure shown in Fig. 13.12, it is still necessary for me to give a brief comment. The three nodes at the left represent the channels through which the user and the machine communicate. You may wonder why both the keys and the money channel not only have arrows leading into the machine, but also have arrows leading in the opposite direction. This indicates that the machine can block the keys and close the money channel. As long as no beverage has been selected, the insertion of money should not be possible, and once a beverage has been selected, the selection keys should be blocked. The only key which never will be blocked is the key for aborting the process. Pressing this key will cause the return of the previously-paid money. The four small circles with checkered shading represent the channels through which the controlling agent sends its instructions to the corresponding operational agents. In the case of the "key stroke receiver and analyzer," the corresponding instruction channel is used for requesting the blockage of the keys being turned on or off. The four grey-shaded nodes represent the channels through which the corresponding operational agents can provide the information needed by the controlling agent for its decision about the instructions to be given next. Thus, the controlling agent must be informed when the abort key is pressed, and it must

System Models 369

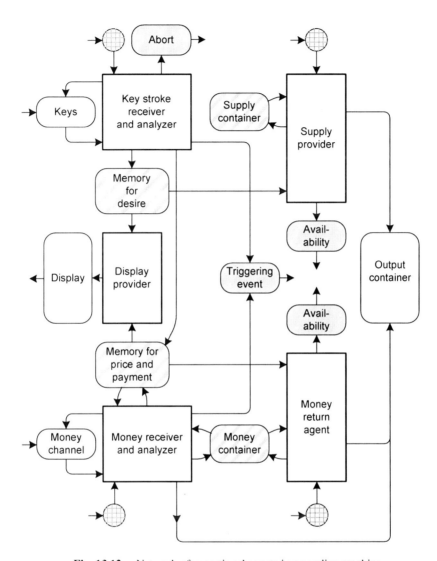

**Fig. 13.12** Network of operational agents in a vending machine

know whether there is at least one bottle of the ordered beverage still available in the supply container. It needs information about whether there is still enough capacity in the money container for inserting additional coins, and whether the coins which are required for returning overpaid amounts are available. The channel in the middle of the drawing is used either by the key stroke receiver or by the money analyzer for sending the message that an event has occurred – either a key has been pressed or money has been inserted. The path which leads directly from the money analyzer to the output container is used for returning a coin or a bill which has been classified as invalid.

The diversity of the operational agents in Fig. 13.12 indicates that the implementation of such a system requires a high degree of division of labor. We can easily identify three subsystems which surely will be designed by different experts. One of these subsystems handles the money. Complete assemblies for this purpose are provided by specialized companies who don't care what will be sold by the machine. The second subsystem is the bottle container with the corresponding output mechanism. It is completely irrelevant to the designer of this subsystem how the process of selecting and paying the beverage is implemented. The third subsystem consists of the keyboard and the display; these are functional units which can be implemented by adaptation of universal components. Such aggregates are required not only in vending machines, but in a wide range of electronic systems.

A machine for selling bottled beverages certainly is not a technical system whose design and operation requires the mastering of a high complexity. I discussed this example only because it is well-suited for introducing the concept of sequential machines and the model of an instruction execution system. Although a lot of pertinent know-how is required for designing and implementing the subsystems in Fig. 13.12, this know-how is not part of the fundamentals of engineering. When talking about know-how, we usually think of all the knowledge an expert accumulates over a long period of time. The source of this knowledge is not only his experience, but also the knowledge about solutions which have been found earlier by others who had to solve similar types of problems.

Once the structure in Fig. 13.12 is found, this does not mean that the design process has been completed. Each component in the structure must now be classified according to whether it can be implemented by applying a well-known method, or whether the structuring process must be continued. The descent over a hierarchy of structures must be continued until finally all components are such that the expert designer can either take them off the shelf or immediately see which method he must apply for their implementation. The great variety of tasks which can be performed by technical systems leads to a corresponding variety of operational agents. Each such operational agent is a processor which receives matter, energy or information through its input channels and provides matter, energy or information at its output channels. There are two basic categories of processors depending upon whether or not they have a memory. The output of a processor with memory depends not only on what is actually provided at its input channels, but also on what has been provided there in the past.

There are two types of operational agents which can be found in many systems, and they sometimes are mistaken for one another by non-experts. Therefore, I now shall consider these two types more closely. Agents of the one type are called

# System Models

*converters*, and agents of the other type are called *amplifiers*. In both cases, the "substance" which flows through the input channel is of the same kind as that flowing through the output channel. This substance is either a specific kind of matter or of energy, or of information. Here, I must remind you of what I said in my comment about Fig. 12.1 where the difference between the purpose of a flow and what is really flowing is emphasized. Here we are interested only in the purpose. For example, information flowing into an information representation converter as energy in the form of an acoustic wave may leave the converter as matter in the form of text on paper.

The table in Fig. 13.13 shows some examples of converters and amplifiers. It is quite natural to wish that an energy converter provides as much energy at its

| Type of operation | Type of flow | Situation at the input | Situation at the output | Type of system |
|---|---|---|---|---|
| Conversion | Energy | Torque times rotation speed $T*N$ | $(f*T)*(N/f)$ | Gear |
| | | Voltage times current $v*i$ | $(f*v)*(i/f)$ | Transformer |
| | | Tons of coal per hour | Tons of hot steam per hour | Steam generator |
| | Information | 8 Megabit per second coding A | 8 Megabit per second coding B | Code converter |
| | | 8 Megabit per second 8 bits parallel | 8 Megabit per second all bits sequential | Parallel-to-sequential converter |
| Amplification | Energy | Torque times rotation speed $T*N$ | $f*(T*N)$ | Power steering |
| | Information | Signal received at the antenna $v(t)*i(t)$ | Output signal $f*(v(t)*i(t))$ | Signal amplifier |
| | | Printed text low contrast | Printed text high contrast | Contrast amplifier |

**Fig. 13.13** Examples of converters and amplifiers

output as it receives at its input. But this wish can never be completely fulfilled because a converting processor always consumes a fraction of the energy from its input in order to function, and therefore the energy coming out at any time will always be somewhat less than the energy coming in. The energy consumed by the converter will cause a rise of its temperature, and this will cause energy to be dissipated into the environment by radiation. However, this dissipated energy is not considered a regular flow through an output channel, but as an undesirable loss. In any case of a conversion, the form of the flowing substance at the output differs from the form at the input. There are special cases of energy conversions and of information conversions where the flowing substance can be expressed as a product of two factors. In these cases, the conversion consists of changing these factors such that the result of the product is left unchanged. Let's consider the first example in Fig. 13.13 which describes the conversion performed by a gear. At the gear's input, the engine provides a certain torque $T$ which causes the axle to run at the rotational speed $N$. The product $T*N$ corresponds to the energy which enters the system per time unit. At the gear's output, the torque differs from the input torque by a factor f and has the value $f*T$. But since the product of torque and speed must be the same at both ends of the gear, the rotational speed of the axle must now have the value $N/f$. This value is approximate because of energy loss in the gears.

The same situation exists in principle in the case of electrical transformers. Such transformers are based on the law of induction which is applied to convert the form of flowing electrical energy, electricity. Both at the input and at the output of such a transformer, the energy per time unit appears as a product of a voltage and a current, both of which are time-dependent sine functions. If the amplitudes of these sine functions at the input are v and i, the corresponding amplitudes at the output are approximately $f*v$ and $i/f$. Thus, the product of the amplitudes is approximately the same at both ends of the transformer. This means that almost as much energy is leaving the transformer as is entering at its input. On its way from the generator in the power plant to the appliances in our home, electrical energy passes through a chain of transformers. The voltage at the generator's output lies in the range of thousands of volts, while the voltage on the overhead power lines should be much higher in order to minimize the loss of energy. This loss occurs in the form of the cables getting warm because of their ohmic resistance, and the loss is proportional to the square of the value of the current. When a certain amount of energy must be transmitted, the current should be as low as possible, and this requires that the voltage be chosen as high as possible. The voltage on many overhead power lines is 380,000 volts. Naturally the voltage at the power outlets in our homes must be much less, because otherwise we would live in danger of injury or death. In many countries the standard voltage for home appliances is either 110 or 220 volts. From these considerations, it follows that there must be at least two transformers on the way from the power plant to our home, one for transforming the generator's voltage to the much higher voltage of the overhead transmission line, and another one for transforming the high transmission voltage down to the voltage in our home. However, it would not be feasible to perform the second transformation in one step, and therefore the transformation downward is implemented by two transformers. First, the high voltage of the long distance transmission line is transformed down at an electrical substation to a medium voltage which is used to distribute the electricity into smaller geographical regions such as neighborhoods, and then this medium voltage is transformed down to the voltage for our power outlets. This last transformation step is performed by transformers which are located near our homes.

In the two examples in Fig. 13.13 concerning information transmission, I assumed that a stream of one megabyte per second is flowing through a converter. The term *byte* was introduced by IBM at about 1965 to denote an ordered set of eight bits. Long before, the term *bit* was introduced to denote one piece of binary information, i.e., for an element of a binary set such as {no, yes} or {0, 1}. In the first example, the converter converts only the coding, but leaves the number of binary channels unchanged. I assumed that the converter has eight binary channels on which all eight bits of a byte can appear at the same time, both at its input and at its output. The next byte is provided every one-millionth of a second. There are $2^8$=256 different combinations of eight bits, and each of these bytes can have its own specific meaning. In the example given, it is assumed that the purpose of the system is the transmission of these meanings and that the converter therefore shall not change the sequence of the meanings. Converting the coding indicates that the meaning which is associated with the byte arriving at the input gets a different byte assignment at the output.

In the next example, the converter does not change the association between byte and meaning, but changes the number of binary channels. It is assumed that there are eight binary input channels on which one byte arrives every millionth of a second, and that the eight bits of this byte can appear only one after the other at the output, since there is only one binary channel coming out.

Now we come to the examples of amplifiers in Fig. 13.13. In the case of conversion, a quantity could be increased only by a certain factor if, at the same time, another quantity was decreased by the same factor. In the case of amplification, however, something is increased without having something else being decreased at the same time. Of course, this requires a specific view, since in general it is impossible for more matter or energy to be delivered at the output of a system than has been flowing into its input. What this specific view is can be best illustrated by an example which has nothing at all to do with the world of technology. Imagine that you went to a farmer's fair where you won a piglet weighing only a few kilograms. Since you live in a city, you certainly cannot care for feeding the piglet yourself, and therefore you try to find a farmer to whom you can give the piglet for care. After about one year, you come back and pick up an adult pig. The farm obviously can be considered a weight amplifier since it got a low-weight piglet at its input and delivered a heavy pig at its output. Of course, no miracle occurred, since besides the input channel through which the piglet entered, the system had another input channel for the swill to feed the pig. In general, an amplifier always has two input channels, one for the substance to be amplified, and the other for the "food" which is required for the amplification. In the case of technical amplifiers, of course, the substance to be amplified is not the weight of animals, but the amount of formed energy or matter. A small amount of formed energy or matters enters at the input of such an amplifier. At the output, much more energy or matter, which has the same form as the energy or matter provided at the input, is delivered. The "food" which is required for the amplification is unformed energy or matter. In the case of hydraulic systems, the unformed energy is provided by a pump, and in the case of electrical systems, the unformed energy is provided by a direct voltage source. The examples given in Fig. 13.13 may help you to understand what I mean by the terms "formed" and "unformed" energy or matter.

A power steering system is similar to a set of gears, and therefore the two systems are well-suited to illustrate the difference between conversion and amplification. In both cases, mechanical power flows into the system and is produced at its output as the product of a torque and a rotational speed. Before power steering, there were gears directly between the steering wheel and the front wheels of the vehicle. I very well remember that, in those days, bus and truck drivers had to apply very strong forces and needed more than one full turn of the steering wheel just to turn the front wheels of a bus or truck by a small angle. All of the energy required for turning the front wheels had to be provided by the driver. Nowadays, drivers themselves must provide only a small fraction of the required energy, with the rest taken from the engine or the battery.

When they read or hear the word "amplifier," most people immediately think of consumer electronics such as radios, TV sets or CD players. Each radio and TV set must contain at least one amplifier, since the antenna provides only a very small signal which is by far insufficient for feeding the loudspeaker or the monitor. In these cases, the information is contained in the form of the time dependency of a voltage or a current. You'll read more about the relationship between information and its forms in Chapter 14.

The last example in Fig. 13.13 shows that amplification is not restricted to energy, but can also be found concerning matter. In this example, "unformed matter" corresponds to the black powder, sometimes called "toner," which is used in a printer to produce the letters and other patterns on paper. As long as this powder is held available in a container for future use, it is unformed, but when the letters and patterns appear on paper, the same powder has been given its form.

# System Models

I have now spent a rather long time on the subject of conversion and amplification. I did this for a particular reason. In all sciences including engineering science, it is a common procedure to search for essential characteristics of different problems and systems, since this may lead to a unified language for talking about the problems and systems, even though the technological differences between them are extreme. If it is possible to talk about different problem fields using the same terms and concepts, it is very likely that the essence of a solution which has been found in one field can be transferred to another field.

The beverage vending machine was introduced as an example of a sequential machine. In the view of engineers, most sequential machines are rather simple systems. Most of the more complex systems they have to deal with are not sequential machines, but are characterized by concurrency. In the case of concurrency, a system can be decomposed into subsystems of sequential machines, but their cooperation cannot be described by a protocol for sequential operation. When I began working on this book, I made a habit of analyzing my everyday contacts with technical systems by asking the question about whether or not they might be suitable for use as examples in this book. Once, on a long railroad trip, I used headphones to listen to some music and, lost in thought, I watched the rotation of the compact disc which I could see under the clear cover of my portable CD player. All of a sudden I became aware of the fact that I had never asked myself how the designers of this player had avoided the kind of whining noise which I had become used to from record players whose rotation speed varies significantly. Could they really control the rotation speed of the disc so precisely that the unavoidable speed variation was not audible, or did they apply other methods? Instead of trying to find out how today's CD players are actually designed, I asked myself how I would solve this problem, and I came up with a straight-forward solution. Later, I was told that this really does correspond to the solution used in today's commercial systems. The basic idea is that the music must not be synchronous with the reading of the information from the disc. The generation of the sound is based on the constant frequency of a quartz-controlled electronic oscillator which is not at all influenced by the varying speeds of any moving mechanical parts. The sound generator gets its input information from an electronic buffer memory which temporarily stores information from the disk which can be read at a constant rate. Of course, what is read from this buffer must be the same information as that which is stored on the disc. An agent is required to transfer the information step-by-step from the disk into the electronic buffer, and this transfer agent must make sure that there is always enough information available in the buffer for the sound generator to read. The amount of buffered but not yet read information must not fall below a certain threshold. For a control system engineer, it is not a difficult problem to design and build a feedback controller (see Fig. 13.8) which keeps the amount of buffered information approximately constant by making the disc run slightly faster or slower when needed.

Such a CD player cannot be modeled as a sequential machine, but must be considered a composition of two sequential machines. Both the subsystem which copies the information from the disc into the electronic buffer and the subsystem which reads from the buffer and generates the sound are sequential machines. They are coupled by using the same buffer memory, but their cooperation cannot be described in the form of a sequential operation protocol.

I shall now present an example which at a first glance has nothing to do with the CD player considered, but soon you'll see that the essence of this new example is exactly the same as in the case of the CD player. Imagine a speaker who realizes half an hour before he must give his speech that he forgot and left the text of his speech on his desk at home, 500 miles from where he actually is located. Fortunately his secretary, who can type spoken text perfectly, has come with him. Therefore, the speaker calls his wife on the phone and asks her to read the speech into the phone so that his secretary can hear the speech and type it. The reading and typing begins about a quarter of an hour before he must step behind the speaker's podium. This guarantees that at least the first page of his manuscript is available at the beginning of his speech. While he is giving his speech, the typing process in the background must run fast enough for providing the next page whenever the speaker needs it. Quite obviously, his wife's reading speed will not be exactly the same as her husband's talking speed.

The CD player and the corresponding speaker system are good examples for introducing the term *steady state* which plays a major role in the world of engineers. In both examples, a so-called start-up-phase is required, either for providing the first buffer content before the sound generation begins, or for providing the first page before the speech begins. At the end of this start-up-phase, the system is in a steady state which is characterized by a process which could continue indefinitely – the wife is reading, the secretary is typing and the husband is speaking. Once I visited a company which produces newsprint, and I was impressed by a huge machine which was about 120 meters long. Chunks of wood were thrown in at one end, and paper 7.3 meters wide was wound onto a spool at a speed of approximately 25 meters per second at the other end. It is understandable that the company must try to keep this machine in operation 24 hours a day and seven days a week, because it is very complicated to start it up and get it into the steady state. While nothing changes at all in a system which is in a static state, changes happen all the time in a steady state, where only certain values which characterize the process stay constant. In the case of the newsprint machine, examples of such constant values are the weight of the wood which is thrown in every hour, the humidity and temperature of some intermediate products and the speed at which the paper comes out and is wound onto spools.

Before powerful computers became available, many things which engineers would have liked to try couldn't be tried because the effort required and the costs were unaffordable. In those days, the only way to find out whether a system behaved as planned was to build the system. A convincing example of the progress in engineering is the design of cars with respect to safety. Engineers always wanted to design the bodies of cars so that, in the case of an accident, as much energy as possible is consumed by deforming the metal body, while not much energy remains for causing injuries to the passengers. Whether or not a body design met this requirement could be tested only by running a real car into a wall. Afterwards, the dummies which represented the passengers could be analyzed in order to see whether unacceptable damage had occurred. Today a model of the whole car and its passengers is described in form of a huge amount of data. This data is then fed into a computer which contains programs containing formulas necessary for describing the dynamics of the car and the dummies according to the laws of physics. Instead of doing experiments with real systems, the experiments are simulated. This is of great advantage. The experiments are very cheap and many experiments with slightly modified data can be performed within a short period of time. Thus, much more useful results become available for optimizing the design than could be obtained from actual experiments.

The requirements for some systems can be described only by referring to probability distributions. Consider a telephone switching system for a big city. The number of subscribers who initiate a call each hour because they want to talk to someone can be estimated only on the basis of past data. On the one hand, the phone company doesn't want a high percentage of call attempts to be rejected because of the switch being overloaded. On the other hand, it is much too expensive to build a system which will not send a busy signal even when half of the city's population decides to call someone from the other half, all at the same time. Before the decision is made about how the system is finally designed, many load cases must be simulated on a computer.

## How the Sine Function Makes the Jobs of Engineers Easier

When I introduced the two functions $\sin(x)$ and $\cos(x)$, I said that the cosine function has the same shape as the sine function and differs from it only by a shift to the left of $\pi/2$ on the x-axis. Whenever the position of the curve relative to the x axis is of no interest and only the shape of the curve is important, the curve is said to be a sine curve, although it would also be correct to call it a cosine curve. The title of this section is meant in this sense.

You probably noticed in my discussion of mathematical considerations that there were only a couple of functions which were mentioned repeatedly. These are the functions which are of greatest importance for both physicists and engineers, namely the exponential function $e^x$ and the trigonometric function $\sin(x)$. The amazing connection between these two functions was shown in Figures 3.22 and 13.3. The sine function is of extremely high importance in various areas of physics and engineering, which at a first glance seem to have nothing at all in common. These areas are characterized by the terms rotation, spectrum, resonance and linearity.

Wherever something is rotating, the sine function cannot be far away. This is an immediate consequence of the definition of this function (Fig. 2.15). You were previously introduced to the concepts of spectrum and resonance, as you can see in Figures 11.2 and 11.6. Resonance is the basis of broadcast systems like radio and television systems, since the effect of resonance can be used for amplifying or eliminating specific frequencies from the spectrum of an electromagnetic wave arriving at an antenna. When you tune your radio to a specific radio station, you are actually changing the resonant frequency of an electronic oscillator so that its frequency matches the transmission frequency of the radio station.

Linearity is something which every engineer and physicist hopes to find in order to avoid complicated computations. A linear relationship between two physical quantities means that doubling or tripling one quantity leads to the doubling or tripling of another quantity. Such a simple relationship is found not only in physics and engineering, but linearity is a quite common effect in everyday life. When we buy twice or three times as much cheese or meat as we bought yesterday, we are not at all surprised that we must pay twice or three times as much. Because of the fact that computations are so simple in the case of linearity, physicists and engineers are always looking for ways to introduce linearity, even in cases where the original problems are actually nonlinear. Now consider Fig. 13.14. Here, the relationship between $x$ and $y$ is obviously not at all linear because the curve describing the function $y=f(x)$ is much different from that of a straight line. But we must ask whether it is necessary, under all circumstances, to take the entire curve into account. It could well be that actually the range of $x$ of interest can be restricted to values of $x$ within the small interval shown. In this interval, the curve which describes the function $y=f(x)$ can be approximated by a straight line near this point of operation. With this view, the values of $\Delta y$ are proportional to the values of $\Delta x$, i.e., $\Delta y = \text{factor} * \Delta x$, where the factor comes from the slope of the line.

As long as linearity applies only to the relationship between $x$ and $y$, the sine function does not come into play. Linearity will lead to the sine function only if it also applies to the derivatives of the function $y=f(x)$. Fortunately, this is the case with many technological systems. A simple example is shown in Fig. 13.15. You always experience such a system when you are traveling in a car or a train. In such vehicles, there are rotating wheels which are attached to some kind of structural assembly and the car or train body containing the passengers is located above them. The block with the mass m represents all parts of the vehicle which can move up and down relative to the wheels. If this mass block were rigidly fixed to the wheel structural assembly, the slightest unevenness in the road or railway would be felt by the passengers as unpleasant bumps. Long ago, mechanical

The Role of the Sine Function

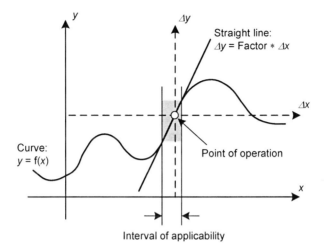

**Fig. 13.14** Linearization around a point of operation

engineers found how the connection between the body and the wheel suspension must be designed so that unpleasant bumps are isolated from the passengers as much as possible. This goal is achieved by placing a spring and a damper in

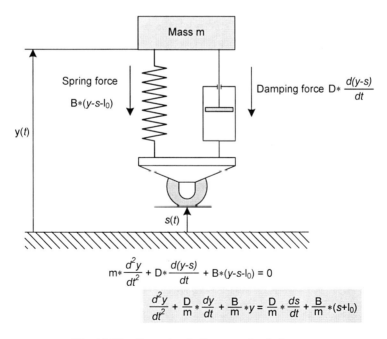

**Fig. 13.15** Example of a linear mechanical system

between the wheel structural assembly and the body. Usually, a damper consists of a cylinder filled with an appropriate liquid wherein a piston can move back and forth. In order to enable this motion, the piston must have holes or valves through which the liquid can get from one side to the other. Such dampers are commonly called "shock absorbers."

The actual state of the vehicle is determined at a time $t$ and at a position $x$. In the case of a moving vehicle, the values of the two variables $t$ and $x$ can be obtained from each other, so only the variable $t$ is considered in Fig. 13.15. Now imagine that the mass block is moving upward, i.e., in positive y-direction. This will lengthen the spring and cause a force pulling downward on the mass block. The longer the spring is compared to the length in its stable state, the greater will be the pulling force. In the stable state neither the wheels nor the mass are moving. In this state the values of y and s stay constant over time and their difference (y-s) has the value $l_0$. Thus, the force of the spring is proportional to (y-s)-$l_0$.

While there is a linear relationship between the force of the spring and the distance y, this does not apply to the damping force because the damping force does not depend on the actual position of the piston in the cylinder. The damping force is determined by the speed at which the piston moves through its liquid. This speed determines how fast the difference (y-s) changes over time, and thus corresponds to the derivative d(y-s)/dt. The sum of the pulling force of the spring and the damping force determine the actual acceleration of the mass block, and this is expressed in form of the differential equation in Fig.13.15. The first version of the differential equation has a form which shows the combination of three forces, with their sum being zero. The second version, which is shaded grey, has a form which is more suitable for mathematical treatment. Only the function y($t$) and its derivatives occur on the left side of this second version, while the so-called stimulation function s($t$) and its derivative occur on the right side. In this situation, the stimulation is based upon the unevenness of the road or railway and the speed of the vehicle.

Electrical engineers must also deal with such linear differential equations, since the voltage between the two terminals of a coil is proportional to the rate at which the current changes its value, and the current through a capacitor is proportional to the rate of change of the voltage between its two terminals. The proportionality factor is symbolized by the capital letter L in the case of coils, and by the capital letter C in the case of capacitors. Fig. 13.16 shows an electric circuit consisting of a coil, a capacitor and an ohmic resistor. The left side of the differential equation which describes the behavior of this circuit has exactly the same formal structure as that of the differential equation in Fig. 13.15: $v_1$ corresponds to $y$, C to m, L to 1/B and R to 1/D.

Although electrical engineers are not afraid of being confronted with mathematical problems, they naturally prefer not to be forced to solve differential

The Role of the Sine Function 381

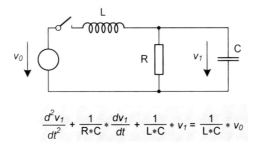

**Fig. 13.16** Linear electric circuit which formally corresponds to the mechanical system in Fig. 13.15

equations every day. In the early days of electrical engineering, it seemed as though alternating current (AC) systems - systems where both current and voltage change their values periodically over time according to sine functions – could be introduced only by forcing engineers to spend much of their time solving differential equations. Many of the components of AC circuits are coils and capacitors, and therefore the formulas describing the relationships between the currents and voltages in these circuits are differential equations. Then why did engineers want to introduce AC systems? It is because power plant generators, which have rotating magnetic fields, provide time dependent voltages and currents whose time dependency is described by sine functions. This time dependency is favorable for power transmission, since it allows the use of transformers – as I discussed in reference to transformers in connection with Fig. 13.13.

Fortunately, the differential equations which describe AC systems are linear, and this is very helpful in this situation where the stimulation functions on the right sides of these equations are sine functions. Because of the linearity of the differential equation and the stimulation being a sine function, it follows that the steady-state solution of the differential equation is also a sine function which has the same frequency as the stimulation. Even though all of this was well known, it still required a genius to conclude that this situation is the basis for the existence of a method which avoids the painful conventional way of solving such differential equations.

By playing around with such differential equations, the German mathematician and engineer Karl Steinmetz (1865-1923) came up with the idea that imaginary numbers must play a significant role in the solution of the problem. His goal was to find a method for computing voltages and currents in AC circuits consisting of resistors, coils and capacitors. He believed the method shouldn't be formally different from the corresponding method for direct current (DC) circuits consisting of only resistors. He actually succeeded and found a method which has become a

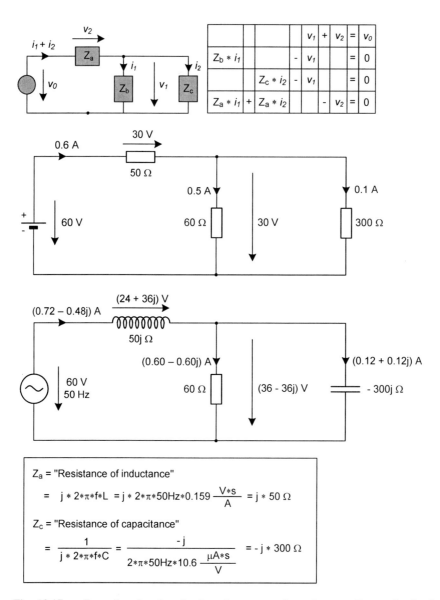

**Fig. 13.17** Examples showing the formal correspondence between linear circuits for direct and alternating currents and voltages

fundamental subject in every electrical engineering curriculum. When Mr. Steinmetz presented his method to the German engineering community, no one at first realized the far reaching consequences of his results. The only people who showed any interest in him were government and police officials who were after him because of his socialistic activities. He first escaped to Switzerland, and then finally emigrated to the United States where he called himself Charles P. Steinmetz and made major contributions to the success of the General Electric Company.

I shall now explain his method for AC circuit analysis using Fig. 13.17. I think the simplicity of this method will surprise you, as it surprised me when I was first introduced to it. The upper left corner of the figure shows a circuit which is composed of a voltage source and three other components, each with two connections. Each of these three components is assumed to be linear in that its voltage is obtained by multiplying its current by a constant factor. With the voltage $v_0$ being given, there are four unknown quantities, the two voltages $v_1$ and $v_2$ and the two currents $i_1$ and $i_2$. To the right of this circuit you see the four equations which describe the circuit. Solving this system of four equations provides the values for the four unknowns. This procedure was well-known before Mr. Steinmetz was born; it had been introduced by the German physicist Gustav Robert Kirchhoff (1824-1887). But originally, it applied only to circuits where all the components are resistors except for voltage sources. If a circuit is of this type, the standard symbols for voltage sources and resistors can be used; this is the case for the circuit diagram shown in the second row, just below the top circuit in the figure. Here, the voltage source is a battery. I chose the resistor values $Z_a$=50 Ω, $Z_b$=60 Ω and $Z_c$=300 Ω which led to the values shown for the voltages and currents. The Greek letter Ω (omega) was introduced previously as a substitute for V/A in Fig. 9.18.

Now we consider the AC circuit in the third row of the figure. It has exactly the same structure as the circuit in Fig. 13.16. The voltage source is assumed to provide a sinusoidal voltage whose amplitude is 60 V and whose frequency is 50 Hz (50 cycles per second). The ohmic resistor has the same 60 Ω value as in the circuit above, but the values of $Z_a$ and $Z_c$ now express the fact that the corresponding components are a coil and a capacitor, respectively. The physical unit for $Z_a$ and $Z_c$ is still Ω, but their numbers are imaginary – the positive imaginary value 50i Ω indicates that the component is a coil, while the negative imaginary value -300i Ω indicates that the component is a capacitor. You'll notice that in Fig. 13.17, the letter j is used instead of i as the symbol for the square root of -1. This is a consequence of the fact the letter i is used as the variable for electric current in electrical engineering.

The Ω-values for coils and capacitors are not constant properties assigned to these components by their design, but are computed using certain other constant properties of the components and the frequency of the sinusoidal voltage source. A coil is described by the constant value L, the factor by which its voltage is obtained from the rate of change of its current. Thus, the physical unit of L must be V/(A/s)=V*s/A= Ω*s. A capacitor is described by the constant value C, the factor by which its current is obtained from the rate of change of its voltage. Thus, the physical unit of C must be A/(V/s)=A*s/V=s/Ω. From this it follows that the

Ω-values for coils and capacitors exist only in the case of a constant supply frequency. This applies to all AC energy supply systems. This constant supply frequency is 50 Hz in Europe, 60 Hz in the U.S., and 16 2/3 Hz or one third of 50 Hz in the German railroad system.

In the box in the bottom section of Fig. 13.17, you find Mr. Steinmetz's formulas for computing the Ω-values for coils and capacitors from the values of L, C and the frequency f. I really don't expect you to be interested in these formulas; I present them only to show you how useful imaginary numbers can be.

The numbers we get for the four unknowns when we solve the system of four equations, are all complex numbers. Electrical engineers are used to calling them "complex amplitudes," which immediately leads to the question about how they should be interpreted. Mr. Steinmetz provided the required interpretation rules. Fig. 2.6 shows how complex numbers can be represented as points in a plane, and Fig. 2.15 illustrates the connection between complex numbers and the sine function. Since Charles Steinmetz knew all of this, it is no wonder that he represented complex voltages and currents as points in a "complex plane." Fig. 13.18 shows such a plane with the points for $v_0$, $v_1$ and $v_2$. The corresponding sine functions on the right side are obtained the same way the sine function in Fig. 2.15 was obtained: it is assumed that the "complex arrows" rotate counterclockwise at the constant speed $2\pi*f$, and that the vertical coordinates of these arrows define the values of the sine functions. The time needed for one full rotation of an arrow is 1/f which is 20 milliseconds when f is 50 Hz. In this interpretation, the length of an arrow corresponds to the amplitude of its corresponding sine function, and the angle determines the position of the sine function on the horizontal time axis.

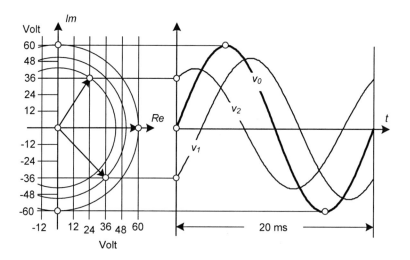

**Fig. 13.18**  Interpretation of the complex voltage amplitudes from Fig. 13.17

Perhaps you noticed that there is a switch in the circuit of Fig. 13.16. As long as this switch is open, no current flows through the voltage source and the coil, and the capacitor which originally might have been charged becomes discharged by a current flowing through the resistor. Finally, the circuit will be in a state where the voltage $v_1$ is zero and no current flows. In this state, neither the coil nor the capacitor contains any stored energy. Now look at Fig. 13.18. If we assume that the switch is closed at time t=0, it is impossible for the voltage $v_1$ to jump to -36 volts and the voltage $v_2$ to +36 volts while $v_0$ is still zero. In reality, both $v_1$ and $v_2$ will still be zero immediately after the switch has been closed and they will have significant non-zero values only after some time. Does this mean that Fig. 13.18 is wrong? No. It only means that t=0 cannot be the moment when the switch is closed. In sections above, I emphasized the difference between the startup phase and the steady state of a system, and this difference must be considered here. Fig. 13.18 shows the behavior of the circuit in its steady state which is reached when the start-up phase is over, i.e., when the switch has been closed for a long time before time t=0. The start-up phase begins when the switch is closed, and it is over when the time variations of all currents and voltages are sine functions. Applying the Steinmetz method is an easy way to get the steady-state behavior of an AC circuit, but this method doesn't help at all to obtain information about transient processes. Fortunately, in 99 percent of cases, only the steady state is of interest, and therefore the painful task of computing the transient behavior of a circuit by actually solving differential equations is not required very often.

I mentioned previously that sinusoidal voltages and currents result from rotations within generators at power plants. The rotation in such a generator involves the rotation of a coil in a magnetic field and this, according to Faraday's law of induction, causes a voltage between the ends of the coil. Thus, the two ends of the coil constitute the voltage source from which electrical energy is fed into the power transmission system. But it would be very difficult, if not impossible, to transfer large amounts of energy from the ends of a coil if these ends were rotating. Therefore, in modern generators, it's the magnetic field which actually rotates, and not the coil. Of course, there is still another coil which actually rotates. This coil is needed to create the rotating magnetic field. But the energy flowing into this rotating coil is only a small fraction of the energy flowing away from the static coil which sends energy to the power transmission system.

It didn't take long from the time the first generators were built until some engineers got the idea that, instead of using only one static coil, they could use more coils. These coils are to be placed at different angles on the circumference of the generator. Thus, they invented the so-called *three-phase-power* systems, and this led to the typical configurations you see of the way high voltage transmission lines are hung on their towers (Fig. 13.19). Perhaps you didn't realize that the number of transmission lines hanging from the insulators on these towers is, in most cases, a multiple of 3. The electrical energy which you get from the outlets in your home is provided in form of a single sinusoidal voltage, and for this you need only two wires. But in a three-phase-power system, the energy is transmitted in the form of a triple set of sinusoidal voltages, and this requires three transmission lines which are labeled R, S and T. The tower in Fig. 13.19 carries two such three-phase systems. Voltages are present both between any two of the three lines, and between any one of the lines and the ground. The voltage between any two of the lines is higher than the voltage between a line and the ground by a factor of √3. Where does this factor come from? It follows from the geometry shown in Fig. 13.20 which is similar to Fig. 13.18, since it also shows complex amplitudes in a complex plane. The three sine functions in the grey shaded area on the right side of the figure represent the voltages between each one of the three lines R, S and T and the ground. The sum of these three sine functions is zero. The corresponding complex amplitudes are represented as arrows on the left side. There is an angle of 120 degrees between any two of these arrows. The arrows can also be viewed as illustrations of the positions of the three coils in a three-phase generator. There is a fourth arrow for (R-S) which corresponds to the voltage between the two lines R and S; it represents the difference between the two sine functions belonging to R and S. The factor by which the arrow for (R-S) is longer than any of the arrows for R, S and T can be obtained by applying the law of Pythagoras to one half of the grey-shaded equilateral triangle.

Although the factor √3 plays a major role in three-phase systems, the factor √2 is also of great importance in the world of sinusoidal voltages and currents. You might have obtained the impression from my comments that electrical engineers are primarily interested in the amplitudes of the sine functions. But actually, the

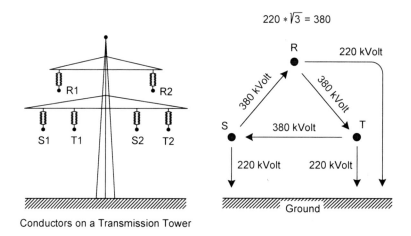

**Fig. 13.19** High voltage transmission of three-phase systems

# The Role of the Sine Function

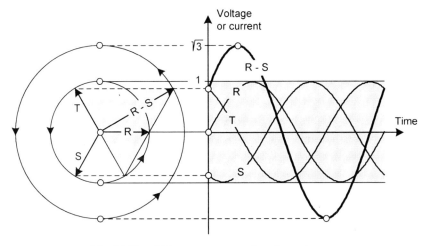

**Fig. 13.20** Voltages or currents in a three-phase system

amplitude values are not the main values of interest. In fact, the transmitted AC power is of much greater importance than the amplitudes. A sinusoidal current flowing through a resistor causes the resistor to get warm or hot. This means electrical energy flowing into the resistor is being converted to thermal energy. It is now interesting to consider which equivalent constant (DC) current would cause the same amount of heating, i.e., what value of direct current would supply the same amount of power to the resistor as an alternating current with a particular amplitude. The value of a direct voltage or current, which produces the same power consumption as its corresponding AC voltage or current, is called the *effective value* or *root-mean-square value* of the AC signal. The root-mean-square value, abbreviated as the rms value, refers to the method of computing this value from the corresponding sine function. This method is illustrated in Fig. 13.21.

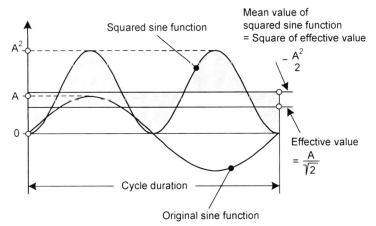

**Fig. 13.21** Effective value of a sine function

The formula for electric power, *power = voltage\*current*, was given in Chapter 9. It was also pointed out there that the voltage across a resistor is proportional to the current flowing through it. Therefore the power, i.e., the energy per unit time, which is flowing into a resistor, is proportional to the square of its voltage or current. When we square a sine function with amplitude A and frequency f, we get a function which is the sum of $A^2/2$ and a sine function with amplitude $A^2/2$ and frequency 2f. The mean value of this resulting sine function is $A^2/2$, and the square root of this, $A/\sqrt{2}$, is the effective or rms value of the sine function. Thus the effective value of a sine function is obtained by dividing the amplitude by $\sqrt{2}$. Now, I would like to note that the values of the voltages included in Fig. 13.19 are not amplitude values, but effective values. Electrical engineers normally use the effective values of AC quantities, and not the amplitude values. The table in Fig. 13.22 contains all four kinds of values which occur in a three-phase system.

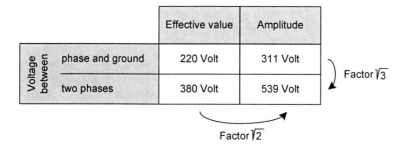

**Fig. 13.22**  Amplitudes and effective values of a three-phase system

Three-phase voltages and currents result from the specific design of the rotating generators in power plants. But rotation occurs not only at the source of the power transmission, but also at the load end where the power is consumed. Many devices that are large consumers of power are electric motors. Such motors are very common in factories. Motors can have a much simpler design if their power is provided by a three-phase system, and not by a single-phase system. Engineers always dreamed of using three-phase motors in electric locomotives, but with only one power line suspended above the track, it seemed impossible for this dream to ever become a reality. However, the development of modern semiconductor components made it possible to transform the incoming one phase power system into a three-phase system. Thus, most electric locomotives today have three-phase electric motors.

# Chapter 14
# Everything becomes Digital – Really Everything?

When Mrs. Miller complained, "In our area, cable TV service was converted to digital, and therefore we had to rent a converter box," she probably believed that digital technology is something quite new which didn't exist twenty years ago. And if I had asked her what she knows about digital systems, she might have answered, "That's a kind of electronics with only zeros and ones." Although this answer is actually correct with respect to over 99 percent of today's digital systems, it's not a correct definition, nevertheless. A digital system does not necessarily need to be implemented electronically or to operate with zeros and ones.

## What Zeros and Ones Have to Do with Digital Systems

The English word "digital" comes from the Latin word *digitus* which means both a finger and a toe. Therefore it is no wonder that this English word refers to the elementary symbols of the decimal number system. In the technological world, the word "digital" is used to indicate that information is transmitted or processed in the form of symbols. The ability to symbolize information exists only if the informational items are elements from a finite set. Examples of such sets are the three colors red, yellow and green from traffic lights, the keys of a computer keyboard, the set of telephone subscribers and the floors which can be reached using an elevator.

Technical systems which handle only informational items from finite sets have been in existence for hundreds of years. Before electronics became the standard technology for implementing computers, mechanical desk calculators containing over 10,000 precision parts were built. The most advanced of such machines could not only execute the four arithmetic operations addition, subtraction, multiplication and division, but they could even compute square roots. The operands which had to be keyed in were decimal numbers of finite length. Today, we find such machines only in museums.

But even today, there is a certain kind of digital system which contains only mechanical parts. The mechanical pendulum clock is such a system. The pendulum which periodically swings back and forth determines the points in time when the discrete state of the clock is changed. Many such clocks have a cycle time of one second, and twice within each period – at the moments when the pendulum changes its direction – the state of the clock is changed. Since, after 12 hours, the clock reaches the state it had 12 hours earlier, it must pass sequentially through a total of 12*60*60*2=86,400 different states. These states are defined by specific positions of the cogwheels within the clock. The number of different states gets even higher if we take into account that the mechanical energy source also passes through many different states. Such a state can be the spring force or the position of a weight. Although it seems as though these energy states would change continuously, this is not the case. The downward movement of the weight is a stepwise movement, because energy is needed only for changing the positions of the cogwheels, and this happens only when the pendulum changes its direction. In the case of chiming clocks, additional energy, which is usually provided by a spring or by the downward movement of a second weight, is required. Thus, it is clear that a mechanical clock is a digital system. The first systems of this kind were built about 400 years ago.

The counterpart of digital technology is analog technology which is used to transmit or process physical quantities with continuous time dependency. Consider the problem of transmitting spoken language over a long distance from one location to another. The speaker produces continuous variations in air pressure which is converted by a microphone into an electrical current whose time variation is the same as the air pressure. The time dependent current flows to a distant loudspeaker which uses electromagnetic forces to convert the current variation back into air pressure variations. The term "analog technology" refers to the fact that the time dependency of a physical quantity at the output of a converter is *analogous* to the time dependency of a (usually different) physical quantity at the input. In the example of the long-distance audio system mentioned above, the current is used as an analog of the air pressure. Later, I shall introduce you to "tricks" which make it possible to apply digital methods to transmitting and processing information which is originally continuous.

The pair {continuous, discrete} which belongs to the world of mathematics corresponds to the pair {analog, digital} which belongs to the world of technology. The two pairs should not be confused, i.e., you should not use the word analog as a synonym for continuous or the word digital as a synonym for discrete. The Latin word *discretus* is used to characterize something as being easily distinguishable from anything else, and that's exactly the characteristic of something which can be assigned a natural number. The opposite of the discrete world is the world of quantities which can change continuously. We are used to representing the variation of such quantities as continuous curves in appropriate coordinate systems, and we know that the points on such a curve cannot be mapped to natural numbers. Thus, the words discrete and continuous basically distinguish between natural and real numbers. In contrast to this, the words analog and digital distinguish between two methods of handling information in technical systems.

In technical systems, elements of a discrete world can be handled by assigning a so-called "technical symbol" to each of the original elements. The simplest technical symbols are binary. Each binary symbol represents a well-defined counterpart situation. Think of a switch for turning the light in the living room on or off. Each of the two possible states – on or off – has exactly one counterpart in the switch position. Whenever engineers want to talk or write about binary symbols without considering an actual technical implementation, they refer to 0 and 1 as the two possibilities of a situation. That's what Mrs. Miller meant when she defined digital systems as systems dealing only with zeros and ones. Mrs. Miller's definition is wrong because digital is not synonymous with binary. But, of course, there must be a reason why almost all modern technical systems dealing with information use binary symbols exclusively.

Railway control systems for setting the binary states of signals and track switches, and telephone switching systems for connecting two subscribers according to the number dialed, are examples of binary systems from the early days of computing. An old telephone system contained both a digital subsystem and an analog subsystem; the sound was transmitted in analog form, while the creation of the electrical path connecting two subscribers was a digital task, since it required only that a specific set of switches be closed.

In 1963 I attended in a lecture from which I expected to learn how to design telephone switching systems. The lecturer was a department head in the state agency which owned and operated the German telephone system in those days. It might well be that the lecturer was a competent department head, but as a teacher he proved to be didactically incompetent, although he was very friendly to us students. He darkened the classroom and projected a slide showing a section of a switching circuit containing about 50 relays, each controlling two to three switches. To us, the wires which connected the contacts of the switches and the relay coils appeared to be drawn at random, but the lecturer moved along these wires with his pointer trying to explain which switch would be the next one to be opened or closed. It's no wonder that nobody learned anything from this lecture, and at least half of the students soon stopped looking at the screen and took a nap. When I was in this lecture, I didn't yet know how the subject should be taught, but I knew from the very beginning that the way this lecturer taught it was certainly not the right way.

Today, the majority of digital systems are implemented using semiconductor switches, i.e., transistors, and the use of relays is restricted to very specific applications. Nevertheless, the principles and methods for designing binary systems have not really changed. Therefore, it is still reasonable to introduce the principles of binary technology by using relay circuits because these are much more illustrative than the corresponding transistor circuits. The basic components of all binary systems are elementary logical circuits, the so-called *logical gates*. An example of such a gate is shown in Fig. 14.1.

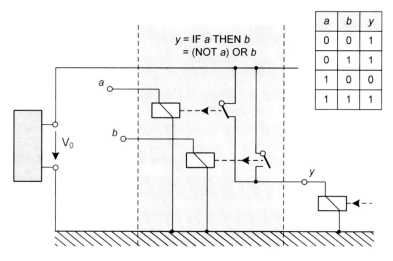

**Fig. 14.1** A logical gate using relays to implement the function "IF . . . THEN . . ."

The physical quantities which are interpreted as binary symbols, are the voltages at the points $a$, $b$ and $y$ relative to the horizontal wire at the bottom of the circuit. The value of the voltage at point $y$ can obviously be either zero or $V_0$, since either there is a vertical connection which brings the voltage $V_0$ from the upper horizontal line to the point y, or such a connection does not exist and y does not get a non-zero voltage. The three rectangular symbols with a diagonal line running through them represent the coils of relays which consist of copper wires wound around iron cores. When a current passes through such a coil, it creates a magnetic field which produces a force. This force pulls the corresponding switch in the direction of the dashed arrow. Thus, when a current flows through a coil, the corresponding switch, which was originally open, will be closed, and vice-versa. Looking at a logical gate, we must always assume that the voltages at the input contacts of the gate are elements of the same binary set to which the output voltage belongs. Thus the set $\{0, V_0\}$, which we found as the set of possible output voltages, is also assumed to be the set of possible input voltages. The relay following the point $y$ does not belong to the gate being discussed. It was added only in order to show that an output point of one gate can be used as an input point of another gate which follows the first one.

The binary situation at $y$ is completely determined by the binary situations at $a$ and $b$. The functional relationship between $(a, b)$ and $y$ can be described by the table in the upper right corner of Fig. 14.1. Here, the set $\{0, 1\}$ of binary symbols is substituted for the binary set of voltages $\{0, V_0\}$. There is only one row in this table which has a zero in the y column. This row corresponds to the combination of the switch $a$ being opened by a force and the switch $b$ being left in its original open state. If you now go back and look again at Fig. 4.6, you'll see that the pattern from the table in Fig. 14.1 is exactly the same as the pattern in the table on the left side of Fig. 4.6 which describes the logical function "if $a$ then $b$". This is a consequence of the fact that the function of the circuit in Fig. 14.1 can be described by the following statement: "If there is a force which opens the a-switch, then there must also be a force which closes the b-switch in order to bring the voltage $V_0$ to the point $y$."

By appropriately connecting the switches of two relays, each of the functional patterns from Fig. 4.7 can be implemented. This simple relationship between logical functions and electrical circuits is the basis of all modern technical information processing systems. I previously indicated the possibility of cascading logical gates where the outputs of one gate layer are used as inputs to the next gate layer. This was done by connecting an additional relay to the output $y$ in Fig. 14.1. There is no limit to the number of gate layers. Thus it is possible to implement any required functional mapping of the type $(y_1, y_2, \ldots, y_n) = f(x_1, x_2, \ldots, x_m)$ where $x_j$ and $y_k$ are binary symbols. Circuits which implement such functions don't contain a memory; they are called *combinatorial circuits*. However, more complex information processing systems must have a memory and cannot be implemented as a mere combinatorial circuit. Therefore, I'll now show you how a binary memory cell can be implemented.

Binary memory cells are often called *flip-flops* in reference to the sound of switches being moved back and forth between their two possible positions. The desired behavior of such a cell is described in Fig. 14.2. The actual content of the memory is either 0 or 1 and is displayed on the output line Q. The rising edges of the clock signal $c(t)$ determine the points in time when a new binary value enters the memory. The value to be stored is obtained by sampling the binary input function $D(t)$ in the grey-shaded time intervals. Within the sampling intervals, the value of $D(t)$ is kept constant; outside of these intervals, the D-values are irrelevant. The input value D from the sampling interval does not become visible at the output Q before the falling edge of the clock appears. In the intervals between two such falling edges, the displayed value of Q never changes; this indicates that the cell really is a memory cell.

The set of curves of binary functions $c(t)$, $D(t)$ and $Q(t)$ as shown in Fig. 14.2 is not the only way to describe the behavior of the binary memory cell. The same behavior can also be described by a so-called *state transition graph* (Fig. 14.3). This is a graphical structure containing two types of nodes, rectangles and ovals, which are connected by arrows. The first example of this type of graph was presented in Fig. 6.5; further examples were introduced in Chapter 13. However, its structural type does not determine the interpretation of the graph. The graphs in Chapter 13 were interpreted as representations of systems composed of components, while the graph in Fig. 6.5 was interpreted as the structure of a process. The state transition graph in Fig. 14.3 also represents a process, a sequential process

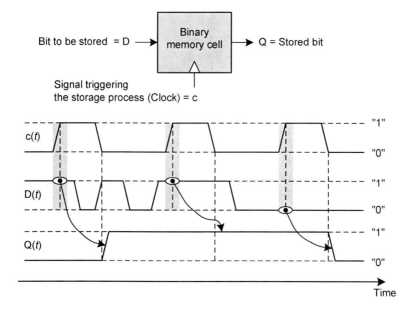

**Fig. 14.2**  Behavior of a clocked binary memory cell

where each step is followed by a next one. The rectangular nodes represent the steps while the oval nodes represent the static states in between the steps. Each step corresponds to a transition from one state to a different next state. In this case, the steps are associated with specific edges of the clock signal. The labels on the nodes of the graph in Fig. 14.3 should enable you to see the correspondence to Fig. 14.2.

While the system specification in Fig. 14.2 doesn't give us any hint about how we could implement the memory cell, the approach in Fig. 14.3 is a very good starting point for finding an implementation. The following considerations will lead us to the composition of the structure of the memory cell. Since all values occurring in the system should be binary, the states must be encoded as binary words, i.e., as sequences of bits. The six oval nodes in the graph in Fig. 14.3 correspond to six different states. If the binary words assigned to these states all have the same length – which is a reasonable requirement – the words must have a length of at least three bits. You can see which words I assigned to the states, although I could have made quite different assignments. The only condition I had to observe was that the six words assigned had to be unique. But I decided to observe two additional conditions, (1) that the first bit of each word be equal to the displayed value Q of the actual content of the memory, and (2) that each state transition changes only one bit of the state code. The reason for this last condition will soon become clear.

# Zeros and Ones

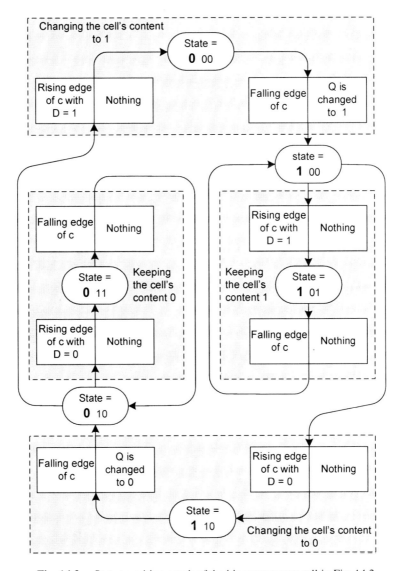

**Fig. 14.3** State transition graph of the binary memory cell in Fig. 14.2

Fig. 14.4 illustrates the way to move from the state transition graph to an implementing circuit. The different situations which must be considered are determined by the actual state ($S_1$, $S_2$, $S_3$) and the actual values of the input variables ($c$, $D$). Such a situation is either stable or unstable, meaning that it will either last until the value of c is changed, or not last and cause a state transition. An example of a stable situation is the combination of the state 000 with c being 1. An example of an unstable situation is the combination of the state 000 with c being 0 since

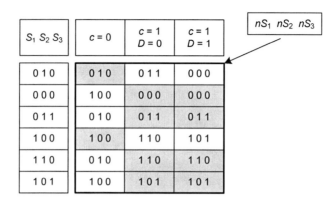

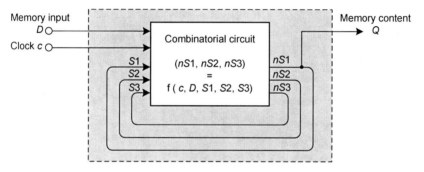

**Fig. 14.4** Function table corresponding to the state transition graph in Fig. 14.3 and its implementation

this will cause a state transition from 000 to 100. The table in Fig. 14.4 shows all relevant situations which are defined by the state transition graph in Fig. 14.3. Whether or not a situation is stable can be deduced by comparing the actual state ($S_1$, $S_2$, $S_3$) in the leftmost column with the next state ($nS_1$, $nS_2$, $nS_3$) which can be found at the intersection of the row of the actual state and the column of the actual input values. If these two states differ, the situation considered is unstable. Only the grey-shaded cells of the table correspond to stable situations. The table defines a logical function ($nS_1$, $nS_2$, $nS_3$) = f ($c$, $D$, $S_1$, $S_2$, $S_3$) which can be implemented by a combinatorial circuit as shown in Fig. 14.4. By connecting the output ($nS_1$, $nS_2$, $nS_3$) of this circuit to the corresponding input ($S_1$, $S_2$, $S_3$) we get a feedback loop. Whenever a state transition occurs, each bit by which the actual state differs from the next state will cause a positive or a negative binary transition edge running from the right to the left. If two or more of such edges occurred at the same time, this would constitute a conflict with an unknown outcome. Because I wanted to prohibit such conflicts, I encoded the states in Fig. 14.3 in such a way that two neighboring states differ by only one bit.

When I specified the behavior of the binary memory cell in Fig. 14.2, I made sure that the output signal Q(t) never changes its value within a grey-shaded sampling interval. This allows us to connect the output Q with the input D of the same memory cell or another memory cell of the same type. The simplest structure of connected memory cells is shown in Fig. 14.5. The output $Q_i$ of each memory cell which has a right neighbor is connected to the input $D_{i+1}$ of this neighbor. This circuit is called a shifter since each clock pulse causes the actual content of the memory chain to be shifted one position to the right with the new content of the leftmost cell coming from outside.

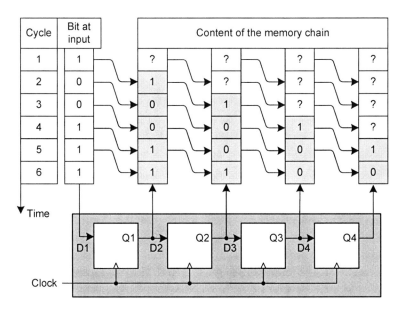

**Fig. 14.5**  Shifter composed of binary memory cells

The system in Fig. 14.5 is a simple special case of the general structure of the clocked sequential circuit shown in Fig. 14.6. All input values $D_i$ which are sampled at the rising edge of the clock signal are generated by a combinatorial circuit from the external input values $x_k$ and the actual contents $Q_i$ of the memory cells. The same combinatorial circuit also provides the external outputs $y_j$. Since the numbers of inputs, memory cells and outputs are finite, the function of the combinatorial circuit can be specified in the form of a function table. Efficient methods have been developed which lead from this table to an implementation of the specified function in form of a network of logical gates. Even the most complex binary information processing systems have the general structure shown in Fig. 14.6.

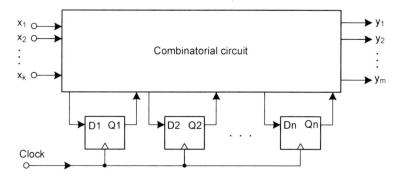

**Fig. 14.6** General structure of a clocked sequential circuit

However, you should bear in mind that in many of today's systems, the number of logical gates and binary memory cells is much greater than one million. Therefore, special methods for handling such complexity are required.

## Why Engineers Want to Digitize as much as Possible

If information from the discrete world is to be transmitted or processed, digital technology is the only choice. Correspondingly, until some decades ago, analog technology seemed to be the only way to transmit or process information from the continuous world, e.g., sound or temperature. But analog technology has an intrinsic deficiency which engineers have been unhappy about from the very beginning: there is no way to protect analog information against noise, since it is practically impossible to detect which part of the received information corresponds to the information provided at the input, and which undesirable part was added as noise along the way from the transmitter to the receiver. Think of a situation where you are listening to someone who called you on the phone, and assume that this person sounds to you as if he has a cold. In this case, you cannot know whether he really has a cold, or whether the particular sound you hear is the result of noise being superimposed on the sound of his voice in the telephone transmission system.

Digital technology doesn't have this deficiency. There are both simple and sophisticated methods for regenerating the original information from the information received. Remember that digital technology means that information is represented in form of symbols. You know from graphical symbols that they can still be recognized even when they are blurred or smeared over with dirt. If the transmitted information consists of sequences of only zeros and ones, e.g., low voltage representing the 0 and a higher voltage representing the 1, the receiver can still decide what sequence of zeros and ones entered the channel as long as the superimposed noise stays below a certain threshold. If, however, the noise exceeds this threshold, the channel can be divided into shorter sections, each of which adds a noise below the threshold. Regenerators can be inserted between the sections which receive the noisy information at the output of the preceding section and provide the regenerated information at the input of the next section. Thus it is possible to transmit binary sequences over very long distances even though the channel used is rather noisy.

However, there are types of noise which cannot be compensated by the method just described. In these cases, an additional method must be applied. This method consists of adding sufficient redundancy to the original information in order to determine from the received information whether or not it corresponds to the information sent. Adding redundancy means that additional bits are appended at one end of the original binary sequence. If enough redundancy is added, it is even possible to reconstruct the original binary sequence from a sequence which has been detected as not being the original one. The possibility of detecting and possibly correcting binary transmission errors is explained by referring to Fig. 14.7.

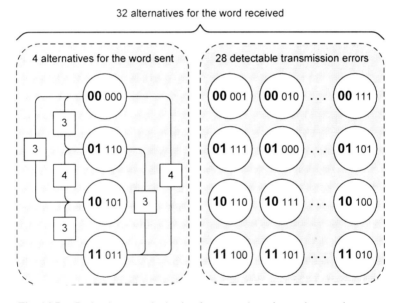

**Fig. 14.7** Redundancy as the basis of an error detecting and correcting system

Assume that the original binary sequence to be transmitted consists of only two bits. In the figure, the four possible original sequences are printed using two bold digits. To each of these four original sequences, redundancy is added in the form of three more bits. These three bits are chosen such that the differences between pairs of the four new sequences of length five are maximized. On the left side of the figure, you see not only the four alternative sequences which can be sent, but you also see how many bits are different with respect to any two of the four sequences.

There are four pairs where the difference is only 3 bits, while the remaining two pairs differ in 4 bits. The number of bits by which two binary sequences of equal length differ is called the *Hamming distance* of the two sequences in honor of the American mathematician Richard W. Hamming (1915-1998).

While the transmitter must select the sequence to be sent from the set of four sequences on the left side of Fig. 14.17, any of the $2^5=32$ possible sequences of five bits may be received. If one of the four sequences from the original set is received, the probability is rather high that this sequence actually had been sent, but there is still a small probability that the received sequence resulted from a different input sequence where the differing bits were changed. If the received sequence does not correspond to any of the four input alternatives, the receiver knows with certainty that a transmission error has occurred. In this case, the receiver must decide how to react. Either the receiver asks the transmitter to send the same sequence again, or the receiver assumes that the sequence sent was the one which differs least from the received one. Both alternative reactions are well known from our human communication processes. When we have conversations via telephone, it sometimes happens that we hear a spoken word or sentence which we don't understand. In such a case, we either guess what might have been said or we ask the partner at the other end to repeat what he just said.

In Fig. 14.7, the so-called *redundancy rate* is 60 percent since three redundant bits are added to the two bits representing the non-redundant information. There are different methods for adding a certain percentage of redundancy to given binary sequences, and some of these methods are very sophisticated. What the redundancy rate should be, and which method for finding the redundant bits should be chosen, depends strongly on the properties of the particular transmission channel. In the case of a space probe which takes pictures from the planet Mars or Jupiter and sends them back to us, much noise will be superimposed onto the signal before it reaches the antenna at the space center on earth. The information about the original pictures can be extracted from such a noisy signal only if a very high redundancy rate is used. Compared to this case, the noise on conventional communication channels on earth – e.g., using cables or point-to-point-radio – is quite low and can be compensated by redundancy rates in the range of 20 to 30 percent.

Having a way to compensate for the superimposed noise and to regenerate the original signal is one essential reason why engineers began to search for methods to represent continuous functions as binary sequences. However, there are two more reasons to prefer digital technology. If originally continuous information is transmitted as a stream of bits, information about discrete facts can be transmitted easily over the same channel by inserting additional bits into the stream. When your telephone rings, you quite naturally expect to see the number and possibly even the name of the caller on the telephone's display. The number and name represent digital information which is transmitted over the same channel used for transmitting the voice of the caller.

The third reason for preferring digital technology for transmitting originally continuous information is the opportunity to get much higher utilization of the transmission capacities of the channels used. Similar to a pipe which has a maximum throughput measured in liters per second, an information channel has a maximum throughput measured in bits per second. This maximum throughput is called the *channel's capacity*. The capacity says how much original information can be transmitted per unit time if the redundancy rate and the modulation, i.e., the conversion of the bit stream into a continuous signal at the channel input, are optimally chosen. It was the American mathematician Claude E. Shannon (1916-2001), an employee of AT&T Bell Labs who, in 1948, published his communication theory which is based on the definition of the capacity of information transmission channels.

Today, it seems to be quite natural to measure information in bits. A memory is said to have a capacity of so many million (Mega) or billion (Giga) bits, and a channel is said to have a capacity of so many thousand (kilo) or million (Mega) bits per second. Although we can symbolize elements of finite sets by finite sequences of binary symbols, it does not necessarily follow that the binary symbol is an adequate unit for measuring quantities of information. In fact, Shannon's theory does not refer to the binary representation of information at all. He was aware that information can be characterized by a specific probability. We may say that information is the difference between two states of knowledge. But how can we measure a state of knowledge? We have knowledge about a certain subject if we can answer questions concerning this subject. If we don't know all there is to know about the subject, we nevertheless can guess what might be the correct answer to a certain question, and our guess could be correct with a certain probability. The value of this probability characterizes our actual knowledge about the corresponding subject. We now assume that someone tells us something about a particular subject. If we already know all he is telling us, the probability will not change. But, it is also possible that we actually receive new information, and this will cause the probability of making correct guesses to increase. There is a maximum amount of information we can receive, since the probability of making correct guesses cannot exceed the value 1.

The concept of entropy was introduced in Chapter 10. There is a great formal resemblance between the concept of entropy and the concept of information. In both cases, pairs of states $(S_1, S_2)$ are considered, with each state $S_i$ having a probability $p_i$ assigned to it. Here, the states are states of knowledge. In both cases, we are interested in finding a function f(S) such that the difference $f(S_2)-f(S_1)$ is determined by the two corresponding probabilities, i.e., $f(S_2)-f(S_1) = F(p_2, p_1)$. In the case of entropy, we had the additional requirement that the function F must be determined by the ratio of the two probabilities, i.e., $F(p_2, p_1) = G(p_2/p_1)$, and this could be satisfied only by choosing the logarithm as the function G. Couldn't it be that it makes sense to introduce the same requirement in our search for an adequate formal definition of the quantity of information? I started this discussion with the statement that information is the difference between two states of knowledge with each state being characterized by the probability of guessing a correct answer to a question concerning a particular subject. Wouldn't it make sense for the information which transfers the state $S_1$ to the state $S_2$ to correspond to the result of the function $\log_B(p_2/p_1)$? The greater the ratio $(p_2/p_1)$, the more information is needed to change the state of knowledge from $p_1$ to $p_2$. Let's consider the two situations, $(p_2, p_1) = (1, 0.5)$ and $(p_2, p_1) = (0.2, 0.1)$. In both situations the ratio $(p_2/p_1)$ has the value 2. Is it reasonable to say that in both cases the same amount of information has been provided? The first situation corresponds to a question which has only two possible answers, yes or no. In the state $S_1$ the person has no idea what the correct answer is, and a guess will be correct with a probability of 50 percent. If this person is now told what the correct answer is, the probability of making a correct guess will rise to 1. The second situation corresponds to a question which has ten possible answers. Someone who has no idea what the right answer is will have to guess, and the probability that the guess is correct is 10 percent. Now let us partition the set of ten possible answers into two sets, each containing five answers, and the question becomes which of the two sets contains the correct answer. This corresponds to the first situation, since now the question has only two possible answers. The answer will reduce the number of possible answers to the original question from ten to five, and the guess about which of the five answers left is the correct one will be correct with a probability of 20 percent. It really makes sense to say that a certain amount of information is required to halve the number of possible answers, and that this amount does not depend on how many answers we had originally.

These ideas lead us to Shannon's definition: "If the probability of guessing the correct answer to a given question is p, the amount of information provided by the correct answer is $\log_B(1/p)$." The base B of the logarithm can be chosen arbitrarily. Usually the base 2 is chosen, and instead of writing $\log_2$, the two letters ld are used. The reason for preferring the base 2 lies in the fact that the implementation of digital systems is based on binary symbols.

I have given you the three reasons engineers prefer digital technology. But we all know that goals are not automatically fulfilled. The goal of digitizing the transmission of continuous information can be fulfilled only if two conditions are satisfied. One condition requires that a method be found for converting a continuous function into a sequence of bits. The second condition requires that digital technology is not more expensive than the analog technology which is to be replaced. With figures 14.1 and 14.4, I showed you that logical gates are the basic components of today's digital systems. Substituting digital technology for analog technology requires that we perform extremely complex information processing tasks. Typically, the number of gates needed for a satisfactory performance easily exceeds one million, and the reaction time of the gates must be in the range of billionths of a second or less. Gates implemented with relays as shown in Fig. 14.1 are absolutely inappropriate as components of such systems because their reaction time is in the range of milliseconds. This is too long by a factor of one million, and relays are so big that a million of them would require the space of many huge cabinets. The space available comes from the requirement that the entire system must fit into the application such as a TV set or a cell phone. From this it follows that the substitution of digital technology for analog technology could begin only after the development of microelectronics had reached a sufficient stage. The driving forces behind the development of microelectronic components originally came from the computer systems industry, since such components were needed to build faster and more powerful computers.

The first computers were built shortly before the beginning of the Second World War. It's an amazing fact that mathematicians and engineers both in Europe and in the United States, at nearly the same time and without knowing each other, began to design and build their first computers. In the United States, it was the mathematician Howard H. Aiken (1900-1973) who, at Harvard University, developed a computer which later got the name Mark I. He did this between 1939 and 1944. He used almost the same technology as the German civil engineer Konrad Zuse (1910-1995). This technology included electromechanical components such as relays, electric motors and cog wheels used together with punched tapes for storing programs. Zuse's first machine, which was finished in 1938, didn't work well because of the lack of precision of its mechanical parts. His third machine, finished in 1941, was called the Z3. This machine is considered the world's first fully-functional computer.

In my comments about Fig. 14.1, I told you that switching electrical currents on and off can be done not only with relays, but also with electronic devices such as transistors. At first, the relays were replaced by electron tubes where a current flows from one electrode through a vacuum to a second electrode. This current can be interrupted by applying a negative voltage to a grid-shaped third electrode which is located between the other two electrodes. This negative voltage can stop the flow of electrons because of their negative charge. The use of electron tubes didn't reduce the space required in comparison to relays, but the speed of operation was increased by factors of about one thousand. In 1947, the transistor was invented and it soon replaced the electron tube in most applications. In Chapter 11, I told you that a transistor is a sandwich consisting of three layers of semiconductor material. Today, most transistors are made of silicon. When a transistor is used as an individual component, the piece of silicon is enclosed in a small container from which three wires connect the three layers to the external world. Around 1960, physicists began to implement more than one transistor on the same piece of silicon. These transistors could be connected together internally without using any wires. That's the principle of integrated circuits, the so-called *chips*. Today, integrated circuits which contain many millions of transistors are made with a piece of silicon no bigger than a quarter. It was the availability of such powerful integrated circuits which finally made it possible to replace almost any analog technology by digital technology.

From all the conditions which had to be satisfied before digital technology could be used for transmitting continuous information such as sound, there is only one left which we did not yet consider in detail. This condition requires a method for converting a continuous function into a sequence of bits. As early as 1933, the Russian mathematician Wladimir Kotelnikow (1908-2005) deduced a mathematical relationship which later was called the *sampling theorem*. Fifteen years later, in 1948, the American mathematician Claude E. Shannon – the same man who introduced the exact definition of the quantity of information – deduced the same sampling theorem, not knowing that it had been deduced 15 years earlier. For many years, the professional world was convinced that the sampling theorem was first deduced by Shannon, and therefore it was taught in engineering courses as "Shannon's sampling theorem". Since this theorem is of extreme relevance for today's communication technology, I shall not present just the final result, but also guide you step by step along the path to showing how it was deduced.

We start by assuming that a time-dependant physical quantity is given, and that its time dependency corresponds to a sine function whose cycle time or period T is known. But we don't know the amplitude and the points in time when the function value is zero. We want to know the minimum number of measurements necessary to determine the missing information. A single measurement certainly will not be sufficient since we do not know where this measurement is located within one period of the sine function. Thus, the sampling point could accidentally be at the location of the maximum value, the amplitude, or it could also be very close to a zero point of the sine function which would mean that the measured value is only a small fraction of the amplitude. If a single measurement is not sufficient, it might be that two measurements are sufficient. The distance $\Delta$ between these two measurements should not be an integer multiple of half the cycle time, i.e., $\Delta \neq n*T/2$, because otherwise we would get either two identical values or two values which differ only with respect to their signs. If, however, the sampling distance is made less than half the cycle time, the two measured values will be independent from each other.

Fig. 14.8 shows an example. Here, the distance between the two sampling points is 0.3 T. In order to represent a sine function in the form of a mathematical formula, the position of the zero point on the x-axis must be determined. The fact

Advantages of Digitizing 405

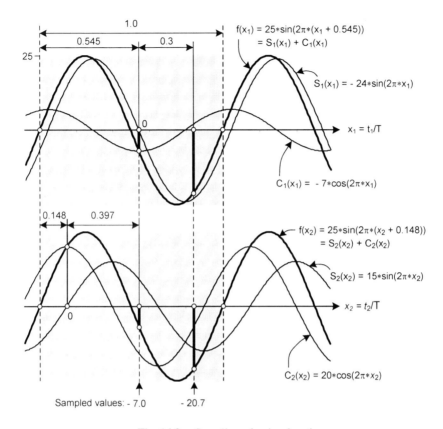

**Fig. 14.8** Sampling of a sine function

that this position can be chosen arbitrarily is illustrated in Fig. 14.8 where the same sine curve is described by two different expressions, one for each of the two positions of the zero point on the x-axis. Once the position of the zero point on the x-axis has been determined, a sine curve can be described alternatively in the forms $A*\sin(2\pi*(x+x_0))$ or $[S*\sin(2\pi*x) + C*\cos(2\pi*x)]$. In each case two values, either $(A, x_0)$ or $(S, C)$, are required to characterize the actual sine function. The three numbers A, S and C are related by the law of Pythagoras: $A^2 = S^2 + C^2$. The two diagrams in Fig. 14.8 show two sampled functions, $f(x_1)$ and $f(x_2)$, and their corresponding sine and cosine functions characterized by $(S_1, C_1)$ and $(S_2, C_2)$, respectively. In Fig. 14.8, the value of the amplitude A is 25. Once the coordinates of the sampling points are given, the values of S and C can be easily computed. However, I shall omit the corresponding formulas. The coordinates of the two sampling points are $(x_1, f(x_1)) = (0, -7.0)$ and $(0.3, -20.7)$ in the case of the upper diagram, and $(x_2, f(x_2)) = (0.397, -7.0)$ and $(0.697, -20.7)$ in the case of the lower diagram. The corresponding sine and cosine values are $(S_1, C_1) = (-24, -7)$ and $(S_2, C_2) = (15, 20)$, respectively.

Since we know that two sampling measurements are sufficient for deducing the amplitude and the time axis position of a sine function with known cycle time, we may guess that four sampling measurements will be sufficient for deducing all missing information about a curve which is composed of two sine functions with known cycle times. Fig. 14.9 shows such a curve; the frequency ratio of its two components is $f_1:f_2=1:1.2=5:6$. The corresponding cycle lengths of the two components are 1 and 1/1.2, and these determine the value 5 of the cycle length of the sum. How can this be explained? The explanation must refer to the findings of Mr. Fourier who proved that any periodic function can be expressed as a sum of sine functions whose frequencies are integer multiples of the basic frequency $\Delta f$. $\Delta f$ is the reciprocal of the cycle length of the periodic function. In our example, $\Delta f$ is 0.2, since the frequencies of the components are 5∗0.2 and 6∗0.2, and from this it follows that the cycle length of the curve which corresponds to the sum is $1/0.2 = 5$.

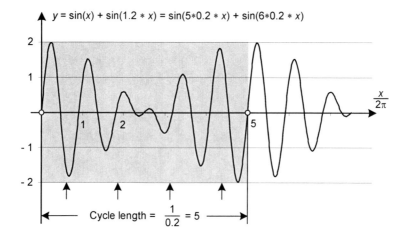

**Fig. 14.9**  Periodic function resulting from adding two sine functions

When sampling a single sine function, the two sampling points must be positioned in such a way that their distance is less than half the cycle length. From this we may conclude that the four sampling points which are required in the case of sampling a sum of two sine functions must be positioned in such a way that the distance between any two of them is less than one quarter of the cycle length. Fig. 14.9 shows one way to position the four sampling points.

# Advantages of Digitizing

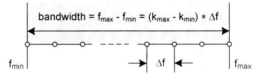

**Assumptions:**

The cycle length of the function to be sampled is $\frac{1}{\Delta f}$

The frequencies $f = k*\Delta f$ of the spectrum are restricted to the interval

$k_{min}* \Delta f = f_{min} \leq f \leq f_{max} = k_{max}* \Delta f$.

---

The length of this interval is called the *bandwidth*.

bandwidth = $f_{max} - f_{min} = (k_{max} - k_{min}) * \Delta f$

Within the bandwidth (including its two limits)
$(k_{max} - k_{min} + 1)$ spectral frequencies can be placed at equal distance $\Delta f$.

---

For each frequency $k*\Delta f$ the amplitudes $S_k$ and $C_k$ of a sine and a cosine function must be determined. This requires two sampling values. Thus, the total number of required sampling values is

$2 * (k_{max} - k_{min} + 1)$.

If the function is sampled at equal distances within its cycle length, the distance between two neighboring sampling locations is

$$\frac{1}{\Delta f * [2 * (k_{max} - k_{min} + 1)]} = \frac{1}{2 * (bandwidth + \Delta f)}$$

---

The so-called *sampling theorem* says:

The distance between two neighboring sampling locations must be less than the reciprocal of the doubled bandwidth.

**Fig. 14.10** Deduction of the sampling theorem

From this point, we need only one more step to reach the universal sampling theorem which can be applied to a non-periodic function. Fig. 14.10 is obtained by generalizing the comments I made about Fig. 14.9. The five variables $\Delta f$, $f_{min}$, $f_{max}$, $k_{min}$ and $k_{max}$, which are introduced in Fig. 14.10, have the values 0.2, 1.0, 1.2, 5 and 6 in Fig. 14.9. In the example in Fig. 14.9, the bandwidth which is defined in Fig. 14.10 equals the basic frequency $\Delta f$, but this doesn't correspond to reality. In reality, the signals to be sampled aren't periodic at all. With a non-periodic function, the cycle length is infinite and $\Delta f$ is zero, since it is the reciprocal value of the cycle length. Therefore, $\Delta f$ doesn't occur in the final version of the sampling theorem. The upper limit for the distance between two neighboring sampling points is determined only by the bandwidth. Thus, a continuous signal can be digitized only if its bandwidth is known and finite. This is the case for all signals of technical relevance. The bandwidth of acoustic signals such as voice or music is determined by the lowest and highest frequencies a person can hear. The lowest frequency is approximately 20 Hz. Humans cannot produce tones of such low frequency, but organ pipes can if they are long enough. The highest frequencies humans can hear are in the range of 20,000 Hz, but only young people can hear such high tones. The maximum frequency a person can hear decreases with advancing age. The required frequency for sampling an acoustic signal with full bandwidth is 40,000 Hz, i.e., the distance between two neighboring sampling points is 25 microseconds. However, this high sampling frequency is required only for transmitting or storing music with the highest quality. Adequate speech quality can be obtained by assuming 4,000 Hz as the upper limit of the bandwidth. This corresponds to 8,000 sampling points per second. All kinds of continuous signals which are to be stored or transmitted by technical systems have been analyzed with respect to their bandwidth. In comparison to the bandwidth of sound signals, video signals have a much higher bandwidth; it is in the range of Megahertz, i.e., millions of cycles per second.

Sampling a continuous signal provides a sequence of numbers. Representing these in binary form (Fig. 4.4) completes the conversion of the continuous signal into a sequence of zeros and ones. The number of bits used for each sampling value is usually minimized experimentally. The number of bits is repeatedly reduced until the reverse conversion from the binary sequence back to the continuous signal no longer provides an acceptable result.

In order to avoid any unnecessary effort, many tricks are applied to reduce the number of bits which must be transmitted per unit time. In fact, it is possible to further reduce the bit stream which results from sampling. For example, a person speaking doesn't speak continuously, but inserts short breaks between words and phrases. Such breaks need not be stored or transmitted using over 50,000 bits per second, but can be captured by specific binary sequences of much shorter length. Besides this simple case, there are many more ways that are used to reduce the length of a bit stream which is the result of sampling a continuous signal. All these methods are called "source oriented encoding," since they use information about specific properties of the source generating the original continuous signal.

A summary of all my explanations concerning possible ways to digitally transmit continuous information is given in the system shown in Fig. 14.11. This figure shows all the components of a system which digitally transmits a continuous signal from a source to a sink. The analog-to-digital converter samples the continuous signal provided by the source and encodes the samples as binary numbers. Thus the output of this converter is a stream of bits. Besides providing a continuous signal, the source may also provide information about discrete facts such as the phone number of a caller. This information is inserted into the bit stream. The resulting bit stream then enters the source-oriented encoder which eliminates as much redundancy as possible. Next, the resulting reduced bit stream is converted

## Advantages of Digitizing

into a continuous signal which enters the transmission channel. When it reaches the exit of the channel, the signal will most likely have been distorted by noise, but the regenerator will nevertheless be able to extract the original binary sequence with a very high probability. This bit stream now has to be enriched again by the redundancy which had been eliminated on the left side. From the resulting bit stream, the bits which belong to the discrete information can be immediately interpreted by the sink, while the bits which correspond to the sampled values enter the digital-to-analog converter which generates the corresponding continuous signal.

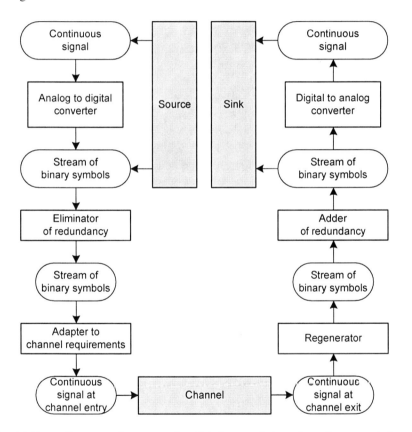

**Fig. 14.11** Structure of a system for digital transmission of continuous and discrete information

Since a telephone is used both for sending and receiving information, all components shown in Fig. 14.11 except the channel must be contained in the telephone. Without the tremendous progress in microelectronics which made it possible to integrate a million or more logical gates on a silicon wafer the size of a quarter dollar, using digital technology to implement telephone systems would still be a utopian dream. In a CD-player, you find only the components from the right side of Fig. 14.11. The channel and the components on the left side are replaced by the system reading the compact disc. The bit stream which is stored on the CD contains not only the sampling information about the sound signal, but also the discrete information which appears on the display, i.e., the number of pieces of music and their corresponding playing times.

While the system shown in Fig. 14.11 contains all the components required to explain the principal function of fixed-line or wired digital telephone systems, CD-players and digital broadcast systems such as radio and television, it is not sufficient to explain how cell phone systems work. In the case of broadcast systems, everyone is entitled to receive the broadcasted information such as radio and television programs. The selection of a specific program requires only that the receiver be tuned to the transmitter's frequency. In the case of telephone systems, however, the person placing the call expects that the information sent is received only by the individual whose number was dialed. This can be guaranteed easily in the case of fixed-line or wired networks, where the signal flows only along the path which ends at the receiver. But in the case of mobile telephone systems, the signals are broadcast, and therefore they can be received by anyone who has the appropriate receiver. Thus, it is necessary to distinguish between receiving a signal and receiving the information contained in the signal. How can this be done?

The overall area of a mobile telephone system is divided into so-called cells whose sizes depend on the local subscriber density, i.e., the number of subscribers per square mile. The diameters of the cells vary from some hundred meters in big cities up to kilometers in rural regions. The power of the electromagnetic waves radiated from the fixed antennas used must be such that the signals can be received clearly even at the borders of the cell, but fade away before they reach cells at greater distances. Of course, waves can be received in cells which are just beyond the border of the cell where the antenna is located. To make sure that, even in this case, the corresponding senders can be distinguished, different transmission frequencies are used in adjacent cells. Thus, the receiving telephone can be adjusted to the frequency which is actually wanted. Perhaps you remember travelling in a car and listening to a radio program. If you experienced fading of the radio signal, you needed to change your radio to a different radio station carrying the same program. It is quite common for certain radio programs to be transmitted at different frequencies in different regions. While a radio program is transmitted without considering whether anybody is actually listening, this does not apply to mobile telephone signals. Therefore, the operation centers which control the transmitting and receiving antennas must always register which subscribers are actually making or receiving a call, and the cell in which they are presently located.

Since a great number of telephone conversations take place at the same time, it is impossible to assign an exclusive frequency to each subscriber. Nevertheless, subscribers using the same assigned frequency can still be technically separated. This is done by periodically dividing time into as many short intervals as there are subscribers to be separated. Thus, each subscriber of the group gets his own time slot. According to a standard actually used in Europe, eight subscribers are grouped together and get assigned two frequencies, one for receiving and another one for transmitting. The distance between the two frequencies is 45 Megahertz. The time slots have durations of half a millisecond, and each subscriber gets 250 slots per second for receiving and another 250 slots for transmitting. The interval between the end of a slot for receiving and the beginning of the next slot for transmitting is one millisecond.

Using different frequencies and time slots guarantees that different conversations don't interfere, but this does not yet solve the problem of how to avoid having someone build an appropriate receiver and listen in to the conversations by selecting the corresponding frequencies and time slots. Fortunately, the system uses digital technology where all information transmitted is a sequence of zeros and ones. This makes it possible to assign to each subscriber his own procedure for scrambling a binary sequence in such a way that the transmitted sequence cannot be unscrambled by someone who doesn't know the individual procedure. The scrambling and unscrambling methods are individualized by using information which is known only to the transmitter and the receiver. Each cell phone contains a memory chip where individual information is stored. This can be used only if the subscriber enters his personal identification number.

## Computer Hardware:
## How Digital Systems Which Execute Programs Are Built

It's certainly pointless to consider the question of what Socrates would find more miraculous: the mobile telephone or the computer. Today, people seem to be more impressed by the functionality of computers than by that of mobile phones, at least as long as these are used only for making phone calls. Of course, now there are more advanced mobile phones that incorporate some computer functions. Using a typical mobile phone doesn't differ much from using a fixed-line telephone, and this makes the users forget the great differences between the implementations of the underlying systems, if they are even aware of these differences. The know-how required for using a telephone is much less than the know-how required for using a computer. How much a potential computer user has to learn depends greatly on whether he wants to write new programs or use existing programs. Decades ago, I read the following appropriate characterization: "A computer is a half-witted maid with extremely high job performance." The high job performance is the consequence not only of the high speed of operation, but also because a computer doesn't forget anything which was stored in the past, unless it is explicitly told to forget. When a new computer is brought into service, it "knows" only what has been stored in its memory in the process of its production, and this primary knowledge enables it to communicate with its environment. It can acquire additional information via the keyboard and possibly other input devices, and it can provide output information via its monitor and possibly other output devices. Although acoustic communication is possible – think of a Global Positioning System which talks to you with the voice of a friendly lady, or of a computer assistant in a car which accepts acoustic commands from a disabled driver – this is not yet the standard form of communication between computers and humans. Today's standard is communication via written text and graphics.

Almost all of the fundamental findings on which the building of computers is based were presented in previous chapters of this book. Since a computer is a purely digital system, it can be implemented exclusively by applying the concepts introduced in figures 14.1 through 14.6. But before the process of designing a computer can begin, we must ask what the function of a computer actually should be. It is quite evident what a computer is not expected to do, e.g., wallpaper the living room or prepare a Thanksgiving dinner. Its tasks should be exclusively informational, i.e., it should assist us in processing information. In the days when the first computers were built, the tasks the designers had in mind were only mathematical tasks, i.e., calculating new numbers from given numbers. The design of a computer is always guided by the model of an instruction execution system which is shown in Fig. 13.11. When this model is applied to computers, the controlling agent is given a very specific structure which I shall now develop. Let's begin with Fig. 14.12 which shows the graphical representation of a program. Although this program cannot be executed by a computer because it refers to matter and energy and not to information, it nevertheless has the characteristic structural properties of any program. The graph in Fig. 14.12 belongs to the same class of graphs as the graph in Fig. 6.5 which shows the protocol for using a telephone system. The program in Fig. 14.12 is so simple that you'll understand it without an explicit explanation. But what I want to point out is the fact that this program contains all three types of structural elements found in sequential programs. The grey rectangles represent three different activities which are to be executed one after the other in the described order and from top to bottom. Each activity must be described as being elementary or structured. The first two grey activities are structured while the third one is elementary. The internal structures of the first two grey activities differ significantly. The first one contains a distinction of cases which means that only one of the two activities represented by the rectangles within the grey activity will be executed. Which one will be executed is decided by the actual situation, and only

Computer Hardware 413

one of the two assigned conditions can be true. The structure of the second grey activity is a conditional repetition which requires that the activity in the loop be repeated until the exit condition of the loop is satisfied.

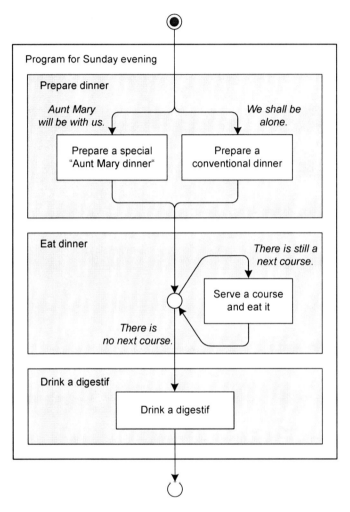

Fig. 14.12  Example of an imperative program

The program in Fig. 14.12 is a so-called *imperative program*, i.e., a program consisting of instructions. All the text in the elementary rectangles can be interpreted as instructions. By looking at the program, we cannot predict how many instructions will actually be executed as steps in the execution of the entire program. The only thing we know for sure is that, in any case, the last instruction will be executed exactly once.

The restriction that a sequential program be composed of only three types of structural patterns (the sequence, the distinction of cases and the repetition) makes it easy to convert the graphical form of the program into an equivalent text. But before I discuss this conversion, I shall stay with the graphical representation in my next discussions since it better illustrates the program. Of course, we now assume that the instructions no longer refer to matter or energy, but only to information. Fig. 14.13 shows the structure of a system which is suited for executing such programs. There is the container for the program in the lower left. You may imagine that the drawing of the program graph is kept here, and it can be viewed by the controlling agent. The container underneath the controlling agent is labelled "program marking." This is the information about how far the program execution has proceeded, and is nothing more than the identification of the small circle in the program graph which actually contains the token. When the execution of a program is started, the token is placed in the circle at the top of the graph. Fig. 14.12 shows the token in this position. Once the entire program has been executed, the token is located in the circle at the bottom of the program. If we compare the program to the score of a conductor, the marking may be thought of as the position of the conductor's finger which indicates the next note to be played.

By having the program and its marking in different containers, this structure guarantees that the program cannot be changed, but only read. The controlling

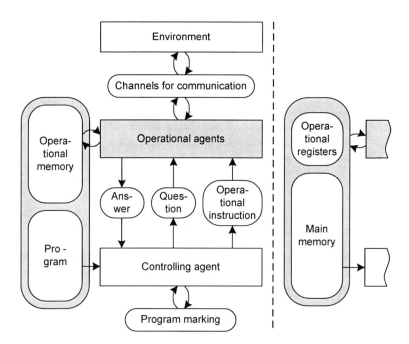

**Fig. 14.13** System for the execution of imperative programs

Computer Hardware                                                                 415

agent, which reads the program, must not only give instructions to the operational agents, but in certain situations, it also must ask questions. The answers are required for distinguishing cases and for finishing repetitions, i.e., for deciding where the token should be placed next.

In the structure on the left side of the dashed line in Fig. 14.13, the operational memory is strictly separated from the program container. The operational memory contains the data items to which the instructions and questions of the controlling agent refer. In the structure shown, the controlling agent cannot request changes in the program by giving appropriate instructions. In fact, at the beginning of the computer era, nobody saw any reason to change a program after its execution had been started. But the Hungarian chemist and mathematician John von Neumann (1903-1957), who came to the United States in 1929 and remained there, soon had the brilliant idea to structure the system's memory as it is shown on the right side of the dashed line in Fig. 14.13. The former operational memory, to which only the operational agents had access, is now reduced to so-called *operational registers*. However, there is still another memory called *main memory* which the operational agents can access for reading and writing, and this memory is also used for storing the program. As before, the controlling agent can only read from this memory and has no writing access. There is no longer a specific container reserved for storing the program, i.e., the program can be stored anywhere within the main memory. The memory cells from which the controlling agent actually must read are determined by the specific marking information.

All nodes with curved borders are locations for information represented by binary sequences. The node represents either a channel, in which case information is flowing, or it represents a memory, in which case information is stored. How a binary sequence must be interpreted cannot be decided by just looking at it. The rule for interpretation follows from the actual situation, i.e., from the actual location of the sequence to be interpreted, and concurrently from the contents of other cells. If you could look into the main memory, you would see billions of zeros and ones, but you would not know what they mean. But if the controlling agent looks into a main memory cell which is identified by the contents of the marking container, the sequence read must be an instruction or a question. Which of these two it actually is can be determined by looking at the sequence itself.

But why did John von Neumann suggest the structure on the right side of the dashed line in Fig. 14.13 in the first place? He was the first to realize that operational data could become program instructions. The binary form of the program which the machine requires for executing the instructions is not suitable for being read by humans. In the early years of the computer age, programmers had to punch the program instructions as patterns of holes in tapes which could be read by the machine. This required programmers to be familiar with the technical details of their computers. Computer engineers soon began to search for ways to eliminate this tight connection between programming and digital technology.

| | |
|---|---|
| Instruction in natural language | Compute the function f with the arguments from the cells $x_1$ through $x_k$ and store the result into the cell y. |
| Instruction in formal language | $y := f( x_1, x_2, x_3, \ldots x_k )$ |
| Example | in programming language $P_A$:     ADD x1, x2, y |
| | in programming language $P_B$:     y := x1 + x2 |

**Fig. 14.14**   Readable form of instructions of the type "compute and store"

Wouldn't it be convenient if programmers could write instructions in a form illustrated by the examples in Fig. 14.14? In this case, the instructions could be fed into the computer by typing them on a keyboard which provides a specific binary sequence for each of its keys at its output. A typewriter keyboard has a maximum of 60 keys, most of which stand for two symbols, lower and upper case letters. Thus, 120 different symbols must be encoded, and this can be done by using binary sequences containing 7 bits, since there are $2^7=128$ different binary sequences of length 7. The chains of binary coded symbols provided by the keyboard cannot be interpreted directly by the machine according to the programmer's intention. However, the machine can take these chains as input data to a program which converts them into a sequence of instructions, and these instructions can be executed by the machine. A program which tells the computer how to convert a sequence of binary coded symbols written by a programmer into a sequence of machine instructions is called a *compiler*. In the view of a compiler, the binary sequences which it provides as its output are still data. In order to enable the controlling agent to read this information and interpret it as a program, the operational agents which execute the compiling program must write their output data into the same memory from which the controlling agent is reading.

There are two different types of machine instructions which can be distinguished by the controlling agent, operational instructions and branching instructions. Operational instructions are not interpreted by the controlling agent but are transferred to the operational agents for execution. The examples shown in Fig. 14.14 are of this type. In such an instruction, a memory location into which a new content is to be stored is identified. This new content is also specified in the instruction, either by providing it directly as a part of the instruction, or by specifying an operation which provides the new content as its result. While the programmer may assign arbitrary names to the memory locations he is referring to, e.g., y, x1, radius or weight, the machine requires binary sequences as identifiers for memory locations. For this purpose, the main memory is considered a

Computer Hardware 417

long chain of small memory cells which are enumerated, and a cell can be identified by providing the number which corresponds to its position in this enumeration. In most cases, the size of such a memory cell is one byte, i.e., 8 bits. While the capacity of the main memories of early computers was in the range of a few thousand bytes, today's computers have main memories with capacities usually exceeding one billion bytes.

If an operational instruction identifies a function to be computed, a specific binary sequence must be interpreted by an operational agent as the function's name. Simple functions such as addition can be implemented as combinatorial circuits, while more complex functions such as the square root require instruction execution systems of their own which are contained within the corresponding operational agents. Since a computer can contain only a finite number of operational agents, the number of different functions which can occur in operational instructions must also be finite. The selection of an appropriate set of functions is an important decision the computer designer has to make. The four arithmetic operations addition, subtraction, multiplication and division are selected in most cases. Other functions which are also required are the so-called logical functions. A logical function requires two bytes at its input and provides one byte at its output. The output byte is obtained by applying one of the logical functions from Fig. 4.7 to each pair of corresponding bits of the input bytes. Another function which also doesn't interpret the binary sequences is the shift function. This function takes an input byte and converts it into an output byte by shifting each bit one position to the left or to the right.

Besides operational instructions, a program usually also contains branching instructions. Such an instruction is not transferred to an operational agent, but is executed by the controlling agent itself. The execution of a branching instruction means that the content of the program marking is changed as specified explicitly in the instruction. Since the program is a sequence of instructions, the instructions can be enumerated and the program marking is nothing but the number which corresponds to a specific position in this enumeration. Once an operational instruction has been executed, the next instruction to be executed is found by increasing the program marking by one. Therefore, the operational instructions do not contain any explicit information about where the next instruction comes from. This does not allow any deviation from the strict sequence given by the order of the instructions in the memory. However, programs may contain distinction of cases and conditional repetitions (Fig. 14.12), and these require deviations from the sequence given by the order of the instructions in memory. It is the purpose of branching instructions to make such deviations possible. A branching instruction contains two kinds of information, namely the identification of a memory location where the binary state of the actual condition can be found, and the number of the instruction which is to be executed next in the case the condition is true. If the condition is false, the program marker is increased by one which means that the next instruction will be the one at the position following the actual branching instruction in memory. Thus, a branching instruction can always be read: "If the content of the binary location $c$ is 1, the number of the instruction to be executed next is $n$." An unconditional branching instruction results if $c$ identifies a location which always contains a 1.

While computer hardware design necessarily must restrict the condition in branching instructions to the content of a binary memory cell, such a restriction is not acceptable with respect to writing reasonable programs. In a program, it must be possible to write "If the number in cell $x$ is greater than 15 and less than 20, then get the next instruction from position $n$." Converting such a conditional statement into an equivalent sequence of instructions executable by the machine can be performed by the compiler. The resulting sequence of instructions must request the computation of the expression $(15-x) \bullet (x-20)-1$ since the result will be positive only if the given condition is satisfied. The location of the resulting sign in the memory corresponds to the location $c$ which must be specified in the corresponding branching instruction.

As I mentioned, the inventors of the computer originally did not plan to build a machine for processing every kind of information; they only wanted to have a machine for calculating with numbers. It was not until a decade later that users of these machines became aware that any kind of information could be represented as a sequence of zeros and ones, and that any functional relationship between such binary sequences could be implemented by a computer program. Today, everybody knows that a computer is a universal information processing system. It is evident that its performance strongly depends on how the information to be processed is represented by sequences of zeros and ones. This dependency was realized by the first computer engineers who encoded only numbers. At the very beginning, the German inventor Konrad Zuse recognized and ingeniously solved a problem which at first was completely overlooked by inventors in the U.S. Mr. Zuse realized that there are two different kinds of applications which require different ways of encoding the numbers. One kind of application requires that the numbers be represented with absolute precision. A typical example for such an application is in the area of business and banking where the amounts of money must be represented to the last cent. In such applications it is always possible to determine how many positions on the right side of the decimal point are required. The corresponding numbers are called *fixed point numbers*.

The situation is quite different when engineers or physicists deal with numbers. Their numbers, which always belong to physical quantities, can never be determined with absolute precision. Therefore, engineers and physicists are used to indicating the relative precision of their numbers by saying how many of the leftmost non-zero digits are relevant. Although the numbers might be represented with more digits, the digits to the right of the relevant digits need not be taken into account. In the course of an engineering calculation, very high numbers and very small numbers may occur simultaneously, e.g. 3,72?,??? and 0.0000372?. By my using question marks instead of digits at the right of the first three non-zero digits, I indicated that only the first three highest-weighted digits are to be considered. At the beginning of performing an engineering calculation, the range of the numbers which will occur in the course of this calculation cannot be predicted. Therefore, a number representation must be found whose length is independent of the position of the decimal point, and depends only on the number of relevant digits.

Mr. Zuse solved this problem by separating the number representation into two sections, one containing the information about the relevant digits, and the other containing the information about the position of the decimal point. The way to split up the information becomes obvious when the two numbers which I considered above as examples are written as follows: $372 \bullet 10^4$ and $372 \bullet 10^{-7}$. In this case, the first section of the number representation contains the positive integer 372, and the second section contains the integer 4 or -7, respectively. The section which contains the relevant digits is called the *mantissa*, and the second section which identifies the position of the decimal point is called the *exponent*. Numbers which are represented in this way are called *floating point numbers*. Although the numbers in my examples are represented as decimal numbers and the exponent belongs to the base 10, you should be aware that in computers the numbers are represented as binary numbers and the base is 2.

The principle of floating point representation is illustrated in Fig. 14.15. Here, the lengths of the mantissa and the exponent are naturally much shorter than the lengths actually used in computers. Typical lengths in computers are 24 bits for the mantissa and 8 bits for the exponent. In Fig. 14.15, the numbers are represented with a relative precision of three binary digits. The leftmost bit of the mantissa is needed for representing the sign. The table contains all numbers which result from varying the mantissa and the exponent in the intervals $-8 \leq m \leq +7$ and $-4 \leq exp \leq +3$. You will notice that some of the numbers occur in the table more than once. An example of such a number is 2 which can be represented as $1 \bullet 2^1$, $2 \bullet 2^0$, or $4 \bullet 2^{-1}$. In order to simplify the implementation of the arithmetic operators using combinatorial circuits, only one representation is selected for each number, and all others are excluded. In the table in Fig. 14.15, the excluded representations are shaded grey. From the three possible representations for the number 2, the form $4 \bullet 2^{-1}$ is selected and the other two are excluded.

Floating point representation has a characteristic effect on the distance between two adjacent numbers, as is illustrated at the top of Fig. 14.15. As long as the values of the numbers are small, the distances between two adjacent numbers are also small. But the distances increase with increasing values of the numbers. Arithmetic operations on pairs of floating point numbers will not necessarily result

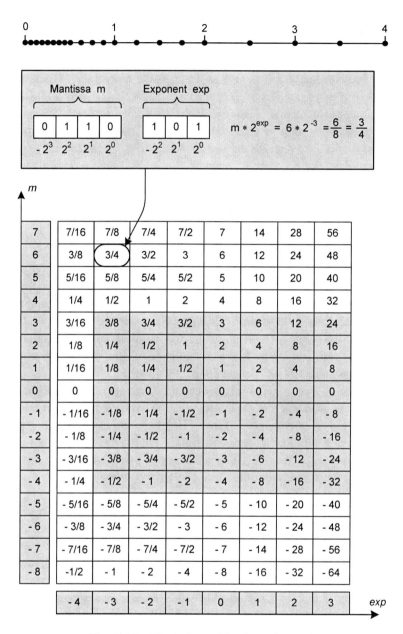

**Fig. 14.15** Illustrations of floating point numbers

in numbers which can be represented exactly by a mantissa and an exponent in the given ranges. Let's consider the product of the two numbers $0.75 = 6 \cdot 2^{-3}$ and $3.5 = 7 \cdot 2^{-1}$ which both occur in the table in Fig. 14.15. Their product is 2.625, and this cannot be found in the table, although it lies within the interval $-64 \leq x \leq 56$ which contains all numbers from the table. The consequence is that this number must be approximated before it can be represented by a mantissa and an exponent. The interval between two adjacent numbers from the table which contains the product is $2.5 \leq 2.625 \leq 3$, and therefore the product is approximated by the number 2.5.

If Mr. Zuse had not invented the concept of floating point numbers when designing his first computer, this concept surely would have been invented by someone else within the next decade. Without it, programming engineering applications would be much more difficult.

## Computer Software:
## How Programmers Can Tell Their Computers What They Expect Them to Do

What I told you up to this point was exclusively about the so-called *computer hardware*. When you once learned the meaning of the word hardware, it was most likely not in connection with computers, but in connection with things like pans, shovels, knives and hammers, which we buy at hardware stores. When the word hardware is used in connection with a computer, it refers to its material structure contributing to its weight. This would include metal, plastics, glass, etc. In contrast to this, the word *software* means encoded information which can be stored in the memory of a computer as sequences of zeros and ones, and which determine the computer's behavior. The word software was created by computer scientists as a complement to the word hardware. When talking about the capability of computers, one should always make clear whether the properties discussed are those determined by the hardware designers or those determined by the programmers. Every now and then, someone who had problems with handling a certain computer application expected me to help him by demonstrating how to interact with the specific user interface of the given application. He would say, "Aren't you teaching university courses in computer technology? Then certainly you know how to use this software for processing video clips. Please tell me what I must do in order to convert this rainy grey sky into a sunny blue one." Those who were convinced that I would be able to help them were always very disappointed when I told them that I wasn't familiar with this particular application of software, and I had no idea how to use it. It's a consequence of the computer being a universal information processing machine that an infinite number of programs can be written for it. Each of these turns the computer into a very specific system having its own individual behavior. In order to understand the hardware of a given computer, it's sufficient to know the ideas implemented by the engineers who designed this specific machine. But if we want to understand the behavior of a system which is determined by specific software, we must find out what the programmers had in mind when they developed this software.

422                                                                                    14. Digital Technology

Until now, I have not presented a single program which actually could be executed by a computer. The example shown in Fig. 14.12 looks like a computer program, but it contains instructions referring to matter and energy and therefore cannot be executed by a computer. When I was looking for a problem suitable for being presented in this book as a starting point for developing a programmed solution, I had to satisfy two conditions. First, the problem should be easy to describe, and second, it shouldn't be so simple that almost everyone could immediately see how it could be solved using a sequence of computer instructions. I finally was convinced that the so-called SUDOKU puzzle satisfies both conditions. For those readers who are not familiar with this kind of puzzle, I shall now give a short explanation. SUDOKU has some resemblance to a crossword puzzle, but instead of filling the empty cells of rows and columns with the letters of words, cells must be filled with digits from the set 1 through 9. A SUDOKU problem is always given in the form of a square array containing 9×9 cells, with some of them empty and others containing digits. An example of a SUDOKU problem is shown in Fig. 14.16. Everything in grey which surrounds the 9×9 square does not belong to

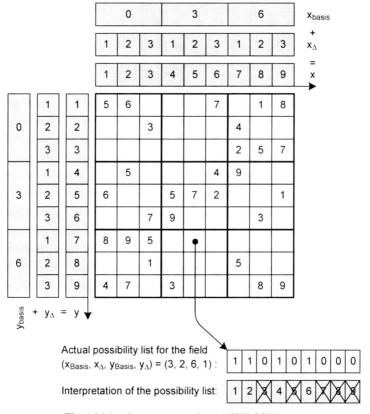

**Fig. 14.16** Data structure for the SUDOKU program

Computer Software

the original problem, but was added as a help for developing the program. The program we now are going to develop enables the computer to find a digit for each empty cell with the restriction that each row, each column and each of the nine 3×3 subsquares contain all nine digits from 1 through 9.

Each of the 9•9=81 cells may be identified using the coordinates x and y. In order to have a simple way to identify each of the nine 3×3 subsquares, I introduced the additional coordinates $x_{Basis}$ and $y_{Basis}$ which are related to x and y according to the equations $x = x_{Basis}+x_\Delta$ and $y = y_{Basis}+y_\Delta$. Using these coordinates, each cell, each row, each column and each subsquare can be identified by two numbers: a cell by (x, y), a row by ($y_{Basis}$, $y_\Delta$), a column by ($x_{Basis}$, $x_\Delta$) and a subsquare by ($x_{Basis}$, $y_{Basis}$).

Making a computer solve a SUDOKU puzzle requires a program which is merely a formal description of our own solving process. For each empty cell we determine the set of digits which remain after we eliminated all digits already contained in the corresponding row, column or subsquare. This is formalized by associating a so-called possibility list to each of the 81 cells. Such a list is a sequence of nine bits, with each corresponding to one digit. The example shown in Fig. 14.16 indicates that only the digits 1, 2, 4 and 6 are left as possible entries for the selected cell, because the other digits from the set 1 through 9 already occur either in the corresponding row, namely 5, 8 and 9, column, namely 7, or subsquare, namely 3.

Before a program is written in a formal programming language, it should be written in natural language, because the program can be understood much easier in this form. If it contains logical errors, these can be detected much faster in natural language. My design of the SUDOKU program is shown in Fig. 14.17. While empty cells can be left empty when a SUDOKU puzzle is printed, cells in a computer memory can never be empty, but must always contain some information. I could have decided to store the word "empty" in those memory cells which correspond to the empty cells of the printed puzzle, but this would not have been an appropriate decision. Since each of the non-empty cells contains a decimal digit from the domain 1 through 9, the digit 0 (zero) was still available, and therefore I chose to use it for characterizing an empty cell. In the course of the program execution, the number of empty cells will hopefully decrease. In order to check whether such a decrease actually occurred, I introduced two memory cells which I called prevCEC and actualCEC. These abbreviations stand for previous and actual count of empty cells. Only if actualCEC is not yet zero, but less than prevCEC, does it make sense to execute the block of instructions in the loop. Otherwise, one of two possible situations occurs. Either the original problem square contained 81 empty cells, or the last execution of the block in the loop had been unsuccessful, i.e., actualCEC was not decreased.

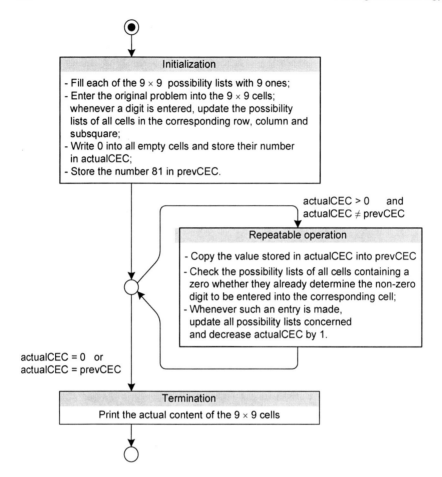

**Fig. 14.17** Program for straight-forward searching for a SUDOKU solution

The most powerful part of the repeatable operation is determined by the instruction "Check the possibility lists of all cells containing a 0 to see if they specify the non-zero digit to be entered into the corresponding cell." There are two possibilities for such a check being positive. The first case can be detected easily, since this case occurs when a possibility list contains eight 0's and a single 1. In this case, the position of the 1 determines the digit to be entered into the corresponding cell. The detection of the second case requires a much greater effort, since this case occurs when a possibility list contains a 1 at a position where all other relevant possibility lists have a 0, either in the row, the column or the subsquare.

Once I had developed the program in the form shown in Fig. 14.17, I had to choose a programming language. Many different formal languages have been developed for writing imperative programs, and the criteria for selecting one are in reference to their specific properties. I selected FORTRAN, although today's computer scientists believe this language is long outdated. It was introduced in 1957 as the first high level programming language for engineering applications. It was developed by a team working at IBM under the leadership of John W. Backus (1924-2007). I didn't want to spend a lot of time developing my program, and so I therefore selected FORTRAN in its least sophisticated form. I believe that learning how to read and understand programs written in this language is not too difficult, but I don't expect my readers to understand it. The program I ended up with has a length of three pages. Fig. 14.18 may give you an impression of the looks of such a program. Careful programmers enhance the readability of their programs by inserting co-called comments which are written in natural language and are not meant for being interpreted by the machine. In the program text in Fig. 14.18, comments are characterized by an exclamation mark (!) at the beginning of the line.

```
! ************ Declaration section ************ !

INTEGER, DIMENSION (1:9, 1:9)      :: cell
LOGICAL, DIMENSION (1:9, 1:9, 1:9) :: possibiliy

INTEGER prevcec, actualcec
INTEGER position
INTEGER xbasis, ybasis, x1, y1, x2, y2, z

! ********* Initialization ********* !

   .
   .
   .

! ******* Repeatable operation ******* !
prevcec = actualcec
DO ybasis = 0, 6, 3
  DO y1 = ybasis+1, ybasis+3
    DO xbasis = 0, 6, 3
      DO x1 = xbasis+1, xbasis+3
        IF (coll(x1, y1) = 0) THEN
          . . .
```

**Fig. 14.18**  A section of the SUDOKU FORTRAN program

Usually, a program begins with the so-called "data declaration section" where the programmer tells the machine how many memory cells are required, which names will be used to identify them and what kind of information will be written into these cells. At first, the 9×9 cells are declared. The declaration says that each cell will be identified by two coordinates, each of which has a value from the domain 1 through 9. Thus, the term cell(2:7) identifies the cell in the second column and the seventh row which in Fig. 14.16 contains the digit 9. The word INTEGER tells the machine that the corresponding cells are used to store numbers of the integer type. The second declaration introduces the binary entries of the possibility lists. Each such entry is identified by the two coordinates of the corresponding cell and the position of the bit within the list. Thus, the term possibility(5:7:2) identifies the second position within the possibility list which belongs to the cell in the fifth column and the seventh row. In Fig. 14.16 this position contains the binary symbol 1. The word LOGICAL indicates that each entry of a possibility list is a binary symbol from the set {0, 1} or {FALSE, TRUE}.

The lower section of Fig. 14.18 shows the program text which corresponds to the beginning of the block in the loop in Fig. 14.17. The four lines which begin with the word DO together with the line following them represent the FORTRAN version of the English text, "For all cells of the 9×9 square which contain a zero do ..."

After having entered the program into my computer's memory by typing it on the keyboard, I started the execution of the FORTRAN-compiler which converted my program text into a sequence of instructions which could be interpreted and executed by the machine. Our daily newspaper provided enough SUDOKU puzzles which I could use to check whether my program worked as I had planned. After the first examples were solved correctly, I had no doubts that the program would also handle all future examples to my satisfaction. But then, to my great surprise, it happened that a "solution" was printed with some of the 9×9 cells still containing 0's. This forced me to conclude that, although my program wasn't completely wrong, it didn't cover all possible cases. Therefore, I had to find a way which would lead from the partial solution it had found to the final solution. I asked myself what I would do if I had to find the final solution myself, and I came up with the following method: I checked the possibility lists of the empty cells to find one which contains a minimum number of 1's. I was lucky and found one with only two 1's. I arbitrarily selected one of these 1's and entered the corresponding digit into the corresponding cell. Now, I considered the modified 9×9 square as if it were an original SUDOKU puzzle, and used it as an input to my program. Since the digit I chose was not the correct one, the program ended with a result where a possibility list had no 1's at all. This was an indication that this modified puzzle had no solution, and that I had better select the other alternative from the list with two 1's. With this alternative, the program actually found the final solution.

The principle of searching a solution by trial and error is a universal principle which can be applied to other problems as well as the SUDOKU problem. This is

Computer Software

a fundamental principle that was developed and programmed in the early days of computer science. The best way to introduce and explain this principle is to present a protocol of a trial and error process in a special graphical form (Fig. 14.19). This protocol illustrates the fact that a trial and error process contains a sequence of alternating steps leading forward and backward. The process begins in the upper left corner with the given problem P which is submitted to a program F, and then F tries to find a solution in a straight-forward way. In the case of our SUDOKU problem, this would be the program in Fig. 14.17. I assumed that this program does not find the final solution of the problem P, but only a partial solution pS. In our SUDOKU example, this partial solution would consist of a 9×9

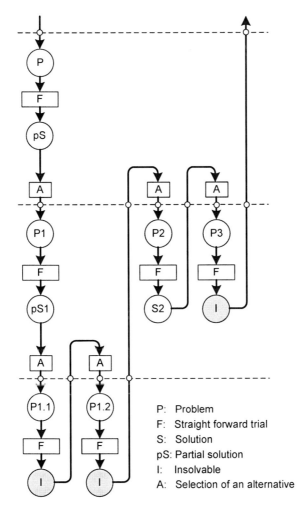

**Fig. 14.19** Protocol for searching for a solution by trying out alternatives

square which still contains some empty cells. The three A's in the row underneath the node pS indicate that there are three alternatives which can be tried. The selection of the leftmost alternative leads to the new problem P1, which is now submitted to the program F. Again this doesn't provide a final solution, but gives the new partial solution pS1. I assumed that this partial solution leaves two alternatives open, and their selection provides the problems P1.1 and P1.2. But the program F cannot find solutions to either of these problems. From this it follows that the problem P1 has no solution and that now the next alternative for pS must be tried. This leads to the problem P2 to which the program F finds the solution S2. This is also a solution of the original problem P. But since it could be that the problem P has more than one correct solution, the third alternative for pS should also be tried. However, I assumed that S2 is the only solution of the problem P, i.e., that the problem P3 has no solution. From the fact that P3 cannot be solved, it does not necessarily follow that the partial solution pS has no further alternatives. Thus, this must be checked, and therefore the process path leads upwards again until it cuts the dashed line at the top. All possible alternatives have been tried only when the process path has returned to the top.

The rectangular nodes in Fig. 14.19 represent activities. F represents the execution of a program for the straight-forward trial, and A represents the selection of an alternative and using it for determining the next problem. The circular nodes represent information that is given or produced. As long as the process path leads downward, the contents of the circular nodes along the way must be kept available, since they will be referred to on the way back. The information below the dashed line may not be deleted before the process path cuts a horizontal dashed line on its way upward. Each time the process path cuts a dashed line on its way downward, new information is added. On the way back, it's deleted again.

If you assume that the process described in Fig. 14.19 does not happen within a computer, but is performed by yourself sitting at a desk, you'll soon realize that the alternation of the adding and deleting of information corresponds to adding and removing documents to and from a stack. Many information processing tasks require the use of such stacks. In programming languages, the instructions for adding and removing elements to and from a stack are PUSH(x) and POP, respectively (see Fig. 14.20). The element lying on the top of a stack is referred to by the word TOP. Only the information contained in the top element can actually be read or changed, not only in the case of a stack of paper documents, but also in the case of a stack in computer memory. In order to get access to an element somewhere below the top, all the elements lying above must first be removed.

While the graph in Fig. 14.19 represents the protocol of a program's execution, and not a program, the corresponding program is represented in Fig. 14.21. In order to check whether an execution of this program can lead to the protocol in Fig. 14.19, you follow the flow of the black token which sits in the top circle of

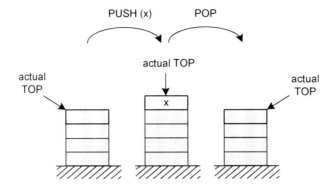

Fig. 14.20   The two elementary stack operations

the diagram in Fig. 14.21 at the beginning. Whenever a branching decision is requested, the protocol in Fig. 14.19 provides the information about which direction to select. PUSH- and POP-operations must correspond to intersections of the process path and the horizontal dashed lines in Fig. 14.19.

The SUDOKU example illustrates a certain type of problem and its solution in the form of a computer program. This type of problem is characterized by the fact that the corresponding programs are formal descriptions of the same methods a human being would apply if he had to solve the problem without using a computer. But this is not the only type of problem which can be solved by computer programming. There are two other types of problems which can be solved using a computer. These are either problems which humans solve without applying any method at all, or they are problems which can be solved only by a coordinated effort of many programmers.

When people first read or hear the term "artificial intelligence," they usually think that this is a kind of intelligence which enables computers to solve problems which are too difficult for humans. But in fact, it's just the other way around. Problems which a computer can solve by applying artificial intelligence methods are those which humans can solve easily without applying any method at all. All of these problems have to do with pattern recognition where the patterns may be either patterns in perceivable communication signals – acoustical or visual – or patterns in electrical signals and data structures. In their early childhood, humans learn to recognize many acoustical and visual patterns, not only spoken and written words, but also the faces of relatives and friends, and the sound of speech associated with a certain person. Later they learn that there are accents, i.e., characteristic words or ways of pronouncing words which depend on the region where a person grew up. In a computer, all such patterns can be represented as sequences of zeros and ones, but the patterns don't provide any hints about how they could

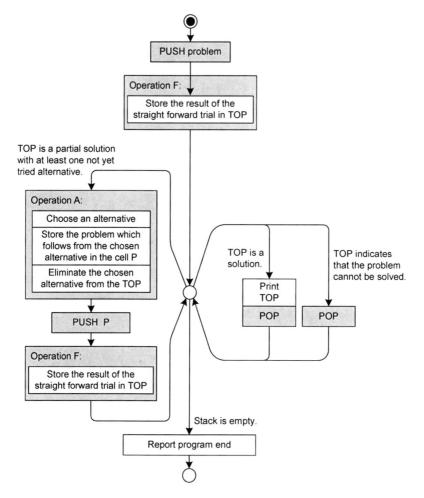

**Fig. 14.21**  Trial and error program using a stack

be detected by analyzing the corresponding binary sequences. As early as 1960, computer scientists began to search for methods which would enable a computer to recognize the kind of patterns described. They came up with many ideas, most of which proved to be useless, but every now and then an idea was born which actually could be successfully applied. Many useful ideas, which became the basis of artificial intelligence were developed in the past approximately fifty years. Today, there are programs which turn a computer into a typist, i.e., you can dictate words of a text into a machine which then types the words on paper like a human secretary would do. There are even programs for detecting characteristic patterns in human voices. Think of a case of blackmail where the police recorded a phone call in which the blackmailer disguised his voice. When a suspect is questioned later, a computer can be used to tell how likely it is that the suspect and the blackmailer are the same person.

My primary concern is to convince you that computers cannot provide meaningful results when programmers cannot explain how the results were obtained. The less someone knows about the way a computer works, the more will he be inclined to believe that computers have superhuman powers. In order to sensitize you to this problem, I shall now tell you what happened to me about forty years ago. At that time, I was a research assistant at the Institute of Information Processing Systems at the University of Karlsruhe. One day, a man came into my office and told me that he was an astrologer, and that he had developed a method based on the constellation of the planets to predict the six numbers of the Lotto game (see Fig. 5.2) one week in advance, before they were officially selected at random by the Lotto machine. The only problem remaining was that his method sometimes provided numbers which had to be added or subtracted in order to obtain the final numbers. As an example, he told me that last week his method provided the numbers {4, 9, 23, 33, 39, 41}, while the Lotto-machine selected the numbers {10, 12, 19, 20, 42, 43} which could easily be obtained by combining his numbers as follows: {10=33-23, 12=41+4-33=39-23-4, 19=23-4, 20=33-9-4, 42=33+9, 43=39+4}. He was convinced that our computer could easily find the right combinations once the major task of deriving the six numbers from the constellation of the planets was done. It was quite clear to me, from the very beginning, that this gentleman had no idea about how a computer must be programmed before it can solve a given problem. But I didn't want to humiliate him, and therefore I told him how much time and effort I would have to invest before our computer could go to work on his problem. I also told him that our institute would have to charge him for this. Of course, he said that he certainly hadn't expected that my efforts would be free, and that it would be only a question of the amount he could afford. I named a price of 10,000 German Marks which corresponded to about 5,000 U.S. Dollars, since I was sure that this was far above what he had expected to pay. He said, "For heaven's sake, that's far too much. I had expected something in the range of 200 Marks." This reaction showed me that the gentleman was not really deeply convinced about either his method or of the capabilities of our computer, since he could easily pay the sum I had requested after he had won one million Marks or so using our computer with his method.

The execution of a program which provides the solution to a given problem corresponds to the computation of a function with given input information. In analogy to

*product* = multiplication *(first factor, second factor)*

we may write

*solution* = execution(*program, data*).

This kind of computing requires that both the data and the program are given as formally described information. Although the programs for such kinds of problems may be very sophisticated, developing such a program is not really an engineering task because the program can be developed by a single programmer and doesn't require a division of labor. At the beginning of the computer age, all computer programming tasks were of this type, and the computer design according to Fig. 14.13 is appropriate for such tasks.

It was around the year 1960 when computer engineers had the idea that it would be useful to modify the design of computers so that running programs could be interrupted. From then on, the execution of a program could be interrupted by certain events. This means that the execution of the actual program was stopped and its continuation was postponed to a later point in time in order to execute a more urgent program first. We are all used to the fact that we might get interrupted, no matter what we are doing. If, for example, the phone rings while we are eating our breakfast, we'll most likely stop eating breakfast in order to answer the phone call. After the phone conversation is over, we'll resume our breakfast. This, of course, requires that the situation on the breakfast table remains as it was when the phone began to ring. Correspondingly, the contents of the memory cells which were used by the interrupted program must still be available when the execution of the program is resumed. However, some of these cells will also be used by the program which is executed rather than the one that is interrupted. As an example, think of the cell which contains the actual program marking; necessarily, this cell is used by all programs. Therefore, when a program is interrupted, the contents of the cells to be shared must be "saved," i.e., stored in cells of the main memory which were reserved for this purpose. Before the execution of the interrupted program can be resumed, these contents must be brought back into the cells from which they had been previously saved. There is no reason why the program which is executed, rather than the interrupted one, could not also be interrupted. Therefore, the contents of the cells to be shared must be piled up in a stack. The top of this stack contains the information belonging to the program which was interrupted most recently. The execution of this program will be resumed when the more urgent programs have terminated.

Interrupts are triggered by events. In the example of being interrupted while we were eating breakfast, the interrupting event was the ringing of the phone. Interrupting events can generally be classified into two different categories, since an event is either a synchronous consequence of the last activity of the interrupted process, or it is not such a consequence and its occurrence is asynchronous to the interrupted process. While it is possible to ignore an event of the second type, an event of the first type always requires a reaction. Obviously, the ringing of the phone which interrupts the process of eating breakfast is an asynchronous event which is independent of this process and could be ignored. But our eating breakfast

Computer Software 433

could also be interrupted by an event which we generate ourselves. Assume that you accidentally knock over your coffee cup and the coffee spills across the table. In this case, the event cannot be ignored, and requires an appropriate reaction.

The same two categories of events also exist in the case of interrupting the execution of a computer program. In the course of executing a program, it can happen that an instruction requires a division where the denominator has the value zero. This is a synchronous event which cannot be ignored, since the execution cannot be continued as if the event had not occurred. The execution of the incorrect program must be stopped and an appropriate reaction program must be executed instead. An appropriate reaction would be to inform the user about the incorrect instruction. In contrast to synchronous events which are always triggered by instruction executions, asynchronous events are generated by the environment (Fig. 14.13) to attract the attention of the processor executing the programs. Almost any device which is connected to a computer can request an interrupt. For example, the keyboard attracts the attention of the processor whenever the "enter" key is pressed, or the printer may want to inform the system about the fact that it ran out of paper. For each different event, there must be an appropriate reaction program stored in the main memory. The execution of such a program can be started whenever the corresponding event triggers an interrupt. Each computer system contains an electronic clock which, as a component of the environment, can generate interrupt events. This clock can generate periodic events which correspond to the chiming of a church clock, although the cycle time of the computer clock is not one hour, but can be set to very short values in the range of fractions of milliseconds. The computer clock can also be used as an alarm clock which forces the system to execute specific programs at preset times.

The possibility of ignoring asynchronous interrupt requests requires that the computer system contains a dedicated memory cell which stores all information concerning interrupt requests. Whenever an interrupt request occurs, the system must look into this cell in order to determine whether it should accept or ignore the request. This corresponds to the situation where a department head says to his secretary, "This morning, I don't want to be interrupted unless it's the president or my wife."

The feature of interruptability is also a prerequisite for establishing communication networks for computers which exchange messages. The internet wouldn't exist if the arrival of a message couldn't cause the receiving computer to react by executing a specific program. Today's computers can be interrupted over a thousand times per second, since they can execute many millions of instructions per second. Thus, a computer can change the role it is playing many times every second, and this gives its users the impression that the computer is simultaneously playing different roles. It's appropriate to consider such a computer like a company with many employees who cooperate and communicate. In the first years of the computer era, a computer always executed one program exclusively from its beginning to its end, and the user knew exactly which program was actually running and would provide the results. Today, a computer user can be compared to a manager having multiple telephone sets by which he can communicate almost simultaneously with different partners who may speak different languages. The windows on a computer monitor can be considered like telephone sets, with each connected to a different agent.

The number of agents which act simultaneously within a modern computer may very well be greater than 100, and therefore it's obvious that the programs which determine the behavior of these agents can no longer be developed by a single programmer. Today, there are software systems which consist of more than 100,000,000 lines of program code, and to which more than 1,000 software developers have contributed. For example, consider today's operating systems, or enterprise resource planning systems which support a variety of business activities within huge corporations. The effort required for developing such complex software systems easily exceeds 10,000 developer years, e.g., 2,500 developers working 4 years or 5,000 developers working 2 years. When you consider the following, it may give you an impression of the size of a software program which is running on your laptop computer. About 60 lines of program code can be printed on one side of a piece of letter-size paper. A package of 500 pieces of paper has a height of 5 cm. Thus, the printout of 200 million lines of code would require a paper stack having a height of about 333 meters, exceeding that of the Eiffel Tower in Paris.

## An Engineering Job Which Is Not Yet Adequately Done

Until now, I strictly refrained from expressing any personal opinions, but now I shall leave the field of proven facts and present a personal conviction which is definitely not shared by most of my professional colleagues. Those who know me well can tell you that, for over 30 years, I have been like a missionary who uses every occasion to present his sermon. I am convinced that basic problems concerning the quality of software systems will not be solved unless the responsibility for software engineering is taken away from computer science and is established as its own discipline.

In the world of software there are two kinds of problems which have practically nothing in common. On the one hand, there are the problems whose solutions can be obtained by computing the results of specific functions – like our SUDOKU example or a program which provides the printed version of spoken text. On the other hand, there is the problem of mastering the complexity of systems which consist of many sophisticated interacting components. These components are developed by different software developers who never get to know each other, and who never communicate directly. Nevertheless, it must be guaranteed that the whole picture is never lost and that all parts interact as desired. From the very beginning of computer science as an academic discipline, its focus was directed towards problems of the first kind, and actually the new discipline succeeded in turning the originally amateurish way of programming into a serious professional activity based on logic and mathematics. Solving this kind of problem by developing a program which provides the result of a function requires that human conceptions are turned into appropriate formal descriptions, i.e., that human ideas are communicated to a machine. In contrast to this, mastering the complexity of huge software systems requires that individuals adequately communicate their conceptions to other individuals. For this kind of problem, formalisms are of no help. Although mastering the complexity of systems characterizes the jobs of engineers, mastering the complexity of software systems by using a high degree of division of labor isn't even mentioned in computer science education. For decades, I studied computer science curricula at a large number of colleges and universities. I came to realize that, with few exceptions, the criteria which guided the teaching and research of the professors in computer science were not those of engineers, but of mathematicians. It is therefore no wonder that almost all methods proposed by these scientists as contributions to the problem of unmastered system complexity completely missed the point. A few years ago, I experienced a situation which made it clear what the problem is. A professor who had been hired to teach courses in software engineering was talking to his assistants, and he emphasized that he was proud to be a scientist, and not an engineer.

It is not reasonable to have professors in the same academic department who do not share a common view about problems and methods to solve them. The best example of a good solution is the peaceful coexistence of the physics department on one side, and the engineering departments on the other side. In most cases, it is quite clear which problems belong in which department. The professors in the physics department can easily accept that their physics courses for engineering students are considered a service within the engineering curriculum. A similar relation should be established between computer scientists and software engineers.

Of course, most computer scientists are opposed to my recommendations. They fight for keeping the responsibility for software engineering within the computer science departments. Isn't it great to have the reputation of being responsible for everything which somehow has to do with computers? But the consequences of this are deficiencies which cannot be overlooked. There is no other engineering discipline where the ratio between effort, time and money on one hand and the quality and mastery of the systems built on the other hand is as poor as in the area of software engineering.

Software is not visible, and therefore, the general public does not yet realize that, long ago, software became a technology which is as important for our civilization as steel and electricity. Most people believe that as long as they don't touch a keyboard or view a computer monitor, software doesn't play any role in their lives. They are not aware that almost no technical devices, not even their car or their microwave oven, would function correctly if the software hidden inside of these devices had any deficiencies. The computers which you actually can see represent less than two percent of all computers in operation. Telephone switching systems, filling equipment at a dairy, railroad signal and track switching controllers, fuel-saving engines in cars and airplane navigation systems are all based on software. The spreading of software throughout all technical areas makes it impossible for any engineering discipline to keep its sovereignty unless the discipline of software engineering also gains its own sovereignty. To achieve this should become a strategic goal of economics and politics.

# Concluding Remarks

Surely you remember that, in Chapter 1, I asked the question about how someone who had died hundreds or thousands of years before our time would view today's world. Considering this question led me to assume that the Greek philosopher Socrates would come and ask me for explanations about our technology-based world. This assumption was my guideline for both the selection of the subjects in this book and for my didactical efforts. Of course, I never really expected Socrates to show up one day in my study or lecture hall. But I am convinced that if my explanations would have satisfied Socrates, then my readers also would appreciate these explanations. This conviction is based on historical reports about Socrates which say that he could easily detect whether a person who explained a difficult subject really had understood the subject himself. You may imagine how embarrassed his dialogue partners were when he pointed out to them that they did not know or understand what they were talking about. I hope that, after reading this book, you are in a state of knowledge which helps you to avoid such embarrassments when you talk about technological subjects.

Not only did the bad habit of using words or expressions without knowing exactly what they mean exist at the time of Socrates, but also this bad habit is a phenomenon of our time. In order to sensitize my students about this habit and help them to avoid it, I invented the following story:

Socrates once became aware of the fact that many people around him often used the expression *xyz*, while he had no idea what that meant. (Here I am using *xyz* as a substitute for the words or expressions I actually used with my students, expressions such as "object orientation", "artificial intelligence", "political correctness" or "social justice.") Socrates knew that an acquaintance of his, named Polimaikes, often used this expression. Therefore, Socrates visited him and said, "My dear friend, people throughout all of Athens say that you are a wise and understanding man and, in particular, that you are an expert concerning *xyz*. I have no idea what *xyz* means and I would really like to know. Therefore, I would greatly appreciate if you would explain to me what *xyz* is." Polimaikes felt very flattered by this request, and he immediately declared his readiness to provide his help. He offered to host Socrates in his home for several weeks, and to be available night and day for discussions until Socrates came to know all there is to know about *xyz*. With great gratitude, Socrates accepted this offer. After six weeks, Socrates said to Polimaikes, "My dear friend, you cannot imagine how thankful I am for everything you have done for me in the last few weeks. I lived like a prince in your house. The meals were the best I ever had, and the wine was out of this world. You listened to all my questions with extreme patience, and answered them in great detail. Unfortunately, I am sorry to have to tell you that I still don't know what *xyz* is. But there is one thing I now know for sure – that you don't know either."

No wonder Socrates finally was sentenced to drink the cup of poison!

# Acknowledgments

Without the help of many devoted and competent people, this book would never have been written and published. The contributions of those who helped me prepare the German edition have already been acknowledged there. But two persons remain to whom I want to express my special thanks here. One is my friend and former colleague, Dr. James H. Burghart. He converted the rough English text, which I was able to provide, into the text which you now find in this book. The other person I want to thank is my wife Ursula who had hoped that, after the German edition was published, her husband would become mentally present again after having been absorbed for such a long time while brooding over his manuscript. I certainly know how hard it was for her to live two more years with a retired husband who, though physically present, was mentally absent most of the time. Finally, I would like to thank all the people from Springer Verlag for their decision to publish this English edition. It was a pleasure to see how much effort they put into giving it a form of such high quality.

# References

Explicitly referred to in the chapters:

[BOR]   Born, M.: Die Relativitätstheorie Einsteins. 5. Auflage. Springer, Berlin (1969); (quotes from pages 301 and 292)
[BR]    Der Brockhaus multimedial 2002. Bibliographisches Institut & F. A. Brockhaus, Mannheim (2001)
[CH 1]  Chargaff, E.: Unbegreifliches Geheimnis. Klett-Cotta, Stuttgart (1988); (quote from page 183)
[CH 2]  Chargaff, E.: Über das Lebendige. Klett-Cotta, Stuttgart (1993); (quote from page 33)
[DES]   Descartes, R.: Discours de la Methode. Bobin & Le Gras, Paris (1668)
[EIN]   Einstein, A.: Über die spezielle und die allgemeine Relativitätstheorie. 21. Auflage; Vieweg, Braunschweig (1969); (quote from page 64)
[EU]    Euler, L.: Vollständige Anleitung zur niedern und höhern Algebra. Erster Theil. G. C. Nauk, Berlin (1796); (quote from page 71)
[FA]    Faraday, M.: Experimental Researches in Electricity. Taylor, London (1839)
[FOU]   Fourier, J.B.J.: Théorie analytique de la Chaleur. Didot, Paris (1822)
[FU]    Fuld, W.: Die Bildungslüge. Argon, Berlin (2004); (quote from page 116)
[HA 1]  Hawking, S.W.: Eine kurze Geschichte der Zeit. Rowohlt, Hamburg (1988)
[HA 2]  Hawking, S.W.: Einsteins Traum - Expeditionen an die Grenzen der Raumzeit. Rowohlt, Hamburg (1996); (quote from page 77)
[HEG]   Hegel, G.F.W.: Enzyklopädie der philosophischen Wissenschaften. Zweiter Teil: Die Naturphilosophie. Werkausgabe. Suhrkamp, Berlin (1970); (quote from §323 (Zusatz) page 278)
[HEI]   Heisenberg, W.: Der Teil und das Ganze. Piper, München (1969); (quotes from pages 59, 63 and 86)
[HIL]   Hilbert, D.: Grundlagen der Geometrie. Teubner, Leipzig (1922)
[HUE]   HÜTTE - Das Ingenieurwissen. 32. Auflage. Springer, Heidelberg (2004); (quote from page A50)
[HUN]   Hund, F.: Geschichte der Quantentheorie. 2. Auflage Bibliographisches Institut, Mannheim (1975)
[LA]    Biographie: Lavoisier. Spektrum der Wissenschaft, Heidelberg (1999)
[LEI]   Leibniz, G.W.: Schriften zur Logik und zur philosophi schen Grundlegung von Mathematik und Naturwissenschaft. Insel, Frankfurt/Main (1992); (quote from page 257)
[LI]    Liessmann, K.P.: Geistige Selbstkolonialisierung. In: Forschung und Lehre 2007 Heft 1. Deutscher Hochschulverband, Bonn (2007); (quote from page 28)

| | |
|---|---|
| [LO] | Lorenz, J.F.: Euklid's Elemente. Buchhandlung des Waisenhauses, Halle (1825); (quote from page 1) |
| [MA] | Maxwell, J.C.: A Treatise on Electricity and Magnetism. Clarendon Press, Oxford (1873) |
| [ME] | Meyers Enzyklopädisches Lexikon. Bibliographisches Institut, Mannheim (1974) |
| [OYG 1] | Ortega y Gasset: Mission of the University. Routledge & Kegan, London (1952); (quotes from pages 69 and 47) |
| [OYG 2] | Ortega y Gasset: Betrachtungen über die Technik. in Signale unserer Zeit. Europäischer Buchclub, Stuttgart (1965); (quote from page 478) |
| [RB] | Renneberg, R.: Biotechnologie für Einsteiger. Spektrum Akademischer Verlag, Heidelberg (2006); (quote from page 16) |
| [SCH] | Schopenhauer, A.: Sämtliche Werke, Band 6. Brockhaus, Mannheim (1988); (quote from page 520) |
| [SCR] | Scriba, C.J., Schreiber, P.: 5000 Jahre Geometrie. Springer, Berlin (2003); (quote from page 476) |
| [SPI] | Philosophische Bibliothek, Band 41: Descartes' Prinzipien der Philo- sophie, geometrisch begründet von Benedict von Spinoza. Heimann, Berlin (1871); (quote from page 56) |
| [SPO] | Spoerl, A.: Pachmayr - Lebenslauf einer Leiche. Ullstein, Frankfurt/M (1973) |

---

Books from which I got valuable information on subjects outside of the scope of my professional competence:

Brock, W.H.: Viewegs Geschichte der Chemie. Vieweg, Braunschweig (1997)

Bruss, D.: Quanteninformation. Fischer Taschenbuchverlag, Frankfurt/Main (2003)

Demtröder, W.: Experimentalphysik. Springer, Heidelberg (2004)

Fischer, E.P.: Am Anfang war die Doppelhelix. Ullstein, München (2003)

Fritzsch, H.: Elementarteilchen. C. H. Beck, München (2004)

Ingold, G.-L.: Quantentheorie. C. H. Beck, München (2003)

Jacob, F.: Die Logik des Lebenden. Fischer Taschenbuchverlag, Frankfurt/Main (2002)

Ridley, M.: Genome. Harper Collins, New York (2000)

Zankl, H.: Genetik. C. H. Beck, München (1998)

Zeilinger, A.: Einsteins Schleier. C. H. Beck, München (2003)

# Name Index

Achilles  40
Aiken, H.  403
Al-Hwarizmi  44, 87
Ampère, A. M.  209, 236
Aristotle  88, 147
Avery, O.  323
Avogadro, A.  252

Bateson, W.  322
Becquerel, A. E.  272
Becquerel, A. H.  259
Bernoulli, D.  249
Berthelot, M.  314
Berzelius, J. J.  314
Bohr, N.  283
Boltzmann, L.  217, 254
Boole, G.  91
Born, M.  202, 283
Broglie, L. V. de  281
Buchner, E.  332

Cantor, G.  29
Cardano, G.  22
Celsius, A.  153, 251
Chargaff, E.  4, 309, 323, 343
Clausius, R.  257
Compton, A. H.  275
Coulomb, Ch. A. de  214
Crick, F.  323, 328
Curie, M.  259

Dalton, J.  243
Darwin, Ch.  317
Democritus  242
Descartes, R.  28, 311
Dirac, P.  262
Doppler, Ch.  176

Einstein, A.  7, 173, 272, 302
Empedocles  242
Euclid  53
Euler, L.  20, 43, 78

Faraday, M.  209, 214, 244
Fert, A.  302
Feynman R.  5
Fourier, J. B. J.  270, 275, 406
Franklin, B.  207
Franklin, R.  329
Fraunhofer, J. von  268
Frege, G.  91
Fuld, W.  7

Galilei, G.  148
Galton, F.  110, 121
Galvani, L.  208
Gauss, C. F.  120, 126, 199
Gell-Mann, M.  263
Grossmann, M.  190
Grünberg, P.  302

Hahn, O.  189
Hallwachs, W.  272
Hamming, R. W.  400
Hawking, S.  7, 196
Hegel, G. W. F.  6
Heisenberg, W.  278, 283
Heraclitus  242
Hermite, Ch.  296
Hertz, H.  240, 273
Hilbert, D.  97, 190, 291
Hund, F.  273
Huygens, Ch.  267

Johannsen, W.  322
Joule, J. P.  236
Joyce, J.  263

Kekulé von Stradonitz, F. A.  244
Kelvin, Lord  252
Kepler, J.  153
Kopernikus, N.  148
Kotelnikow, W.  404
Kronecker, L.  14
Kühne, W. F.  331

Langdon-Down, J.  339
Lavoisier, A. L. de  243, 249
Leibniz, G. W.  72, 243
Liessmann, K. P.  6
Linné, C. von  246
Lorentz, H. A.  179, 227

Malus, E. L.  278
Marconi, G.  240
Maxwell, J. C.  216, 224
Meitner, L.  189
Mendel, G.  317
Mendelejew, D. I.  246
Meyer, L.  246
Miescher, F.  323
Minkowski, H.  196
Muller, H. J.  317

Neumann, J. von  415
Newton, I.  155, 161

Oersted, H. Ch.  208
Ohm, G. S.  211, 239
Onnes, H. K.  301
Ortega y Gasset, J.  8, 9, 352
Ostwald, W.  332

Pascal, B.  114
Pauli, W.  263

Peano, G.  93
Planck, M.  271
Protagoras  151, 241
Pythagoras  54

Roemer, O.  239
Röntgen, W. C.  275
Roscoe, H.  244
Rutherford, E. Lord  260

Schleiden, M.  311
Schopenhauer, A.  7
Schrödinger, E.  283
Schwann, T.  312
Shannon, C.  401, 404
Siemens, W. von  228
Socrates  2, 437
Spinoza, B. de  243
Steinmetz, Ch. P.  381
Strassmann, F.  189

Thompson, B.  249
Thompson, W.  252

Vinci, L. da  352
Virchow, R.  311
Volta, A.  208

Watson, J.  323, 328
Watt, J.  236
Wheatstone, Ch.  228
Wöhler, F.  314
Zenon of Elea  40
Zuse, K.  403, 418

# Subject Index

AC circuit 381
acid 324
action quantum 272
addition 14
adenine 324
algebra 44
algorithm 87
alpha-particle 259
amino acid 337
Ampere (unit) 234
Ampere's circuital law 226
amplifier 371
analog technology 390
analog-to-digital converter 408
angular acceleration 166
angular momentum 166
angular velocity 165
antibiotics 315
antiparticles 262
argument (of a function) 31
arithmetic 14
artificial intelligence 429
associative law 49
atom 243, 261
atomic weight 244
Avogadro number 253
axiom 95

bandwidth 407
base (chemical) 324
base (of a power) 19
base pair 327
battery, electrical 208
beta-particle 259
binary number 86
biochemist 313
bio-information scientist 339
bipartite graph 356
bit 303, 373, 401

black box 359
board of Galton 110
Boltzmann constant 254
byte 373

cableway 353
calculus, differential and integral 64
calculus, logical 95
capacitor 211
Cartesian product 28
catenary 354
CD player 375, 410
cell (living) 311
cell division 334
centigrade 153
CERN 265
channel's capacity 401
Chargaff's rules 325
chemical energy 255
chemistry 243
chemistry, inorganic 314
chemistry, organic 314
chip 404
chromosome 322
circuit, electrical 209
clock signal 393
closed function 47
closed-loop control system 359
close-range effect 217
combinatorial circuit 393
combinatorics 112
communication 131
communication theory 401
commutative function 33
commutative law 48
compiler 416
complement (of a set) 27
complex amplitude 383
complex number 21

composite symbol   133
composition model   356
computer   403
concurrency   143, 375
conic section   154
conservation of energy   254
context-sensitivity   139
control theory   360
converter   371
coordinate system   57
correlation factor   121
cosine-function   36
Coulomb (unit)   235
Coulomb's law   214
current, electrical   209
cytosine   324

damper   379
DC circuit   381
decimal number   85
degree (of a polynomial)   35
denominator   15
derivative   71
desoxyribonucleic acid (DNA)   322
detector chamber   263
diagonal matrix   286
dielectrical flow   217, 225
differential equation   73
differential quotient   70
differentiation   71
digital broadcast system   410
digital systems   389
digital-to-analog converter   409
diode   302
discrete time   365
disjoint (sets)   26
distributive law   46
divergence   219
division   15
domain   27
dominant property   320
Doppler effect   176
double-clutching   350
Down's syndrome   339

effective value   387
eigendirection   286
eigenfrequency   270
eigenfunction   291
eigenvalue   284, 286
eigenvector   286

electric field   225
electrical charge   209, 233
electrolysis   244
electromagnetic wave   232, 239
electron microscope   318
element (of a set)   26
ellipse   155
encoding   140
energy   157
energy operator   295
energy, kinetic   164
energy, potential   164
engineering science   348
entanglement (of quantum states)   303
entropy   257, 401
enzyme   331
equation   44
Euclidean geometry   53
Euler's formula   78
existential quantification   92
expected value   120
exponent (of a power)   19
exponential function   43

factorial   77
feedback loop   361
field, algebraic   51
fire substance   249
fixed point number   418
flip-flop   393
floating point number   419
fluorescence   260
focus   155
force   157
formal logic   88
formalism   81
FORTRAN   424
Fourier transformation   270
fraction   15
Fraunhofer lines   268
fruit fly   317, 322
function (mathematical)   31
function chain   34

transformation   177
gamma-rays   259
gas theory   249
gas-thermometer   254
gate, logical   391
gear   372
generator, electrical   228

# Subject Index

generator, electrical  322
genetic engineering  338
genetic fingerprint  341
genetic text  326
genetics  322
geometry  53
geometry, axiomatic  97
gradient  222
grammar  135
gravity, law of  160
group theory  51
group, algebraic  49
guanine  324
gyroscope  164

Hamming distance  400
hardware (of computers)  411, 421
heat  249
hormone  332
hydrogen  244
hydrogen bridge  326
hyperbola  155
hyperbolic function  355

imaginary number  21
imaginary time  196
imperative program  413
induction, complete  95
induction, law of  216
inequality  44
infix form (of a function)  48
information, quantity of  401
input-output-system  359
instruction execution system  366, 412
instruction, branching  417
instruction, operational  416
integral  71
integrated circuit  404
integration  71
interaction (of particles)  265
interference (of waves)  267
interrupt (of a program execution)  432
interrupt, asynchronous  432
interrupt, synchronous  432
intersection (of sets)  26
invertability (of an operand)  49
ion  324
irrational number  18
isomorphism  97

Joule (unit)  236
junk DANN  338

Kelvin (unit)  252

language  131
lattice, algebraic  51
life  309
life line  173
life power  314
limit value  42
linearity  378
linguist  133
litmus test  324
logarithm  20
Lorentz transformation  178
Loschmidt number  253
Lotto  112

magnetic field  215
magnetic flow  226
main memory  415
mantissa  419
mass, heavy  191
mass, inertial  191
matrix  58
matrix multiplication  58
Maxwell's equations  224
mechanics  154
median  120
mental state  310
meter  151
microelectronics  403
mobile telephone system  410
model (of a theory)  96
molecular biology  315
molecule  244
moment of inertia  166
momentum  156
mongolism  339
Morse code  140, 228
multiplication  15
muon  185, 264
mutation  317

natural logarithm  79
natural number  14
negative number  16
neutral operand  49
neutrino  263

neutron 260
Newton (unit) 236
noise 360, 398
non-terminal (of a grammar) 134
normal distribution 126
nuclear fission 189
nucleus, atomic 261
number, binary 86
number, complex 21
number, decimal 85
number, imaginary 21
number, irrational 18
number, natural 14
number, negative 16
number, rational 17
number, real 18
number, transcendental 18
numerator 15

Ohm's law 211
operating know-how 3
operational agent 367
operational register 415
operator 291
orbit 155
orchestrion 367
organ (of a creature) 311
organism 311
organization 311
oxygen 244

parabola 155
paradox (of Zenon) 40
Pascal's triangle 114
pattern recognition 429
pendulum clock 389
periodic table (of chemical elements) 246
permeability 224
perpendicular product (of vectors) 62
philologist 133
photoelectric effect 272
photon 272
Planck's constant 272
pneumococcal bacteria 323
polarizing filter 279, 303
pollination, artificial 319
polynomial 34
POP 428
positron 262
potential (field of) 161, 235
power (of a set) 29

power (operation) 19
power engineering, electrical 227
power steering 374
predicate 91
predicate logic 91
probability 110
probabiliy density 125
program (for a computer) 412
propositional logic 91
protocol 141, 3
proton 260
PUSH 428
Pythagoras, law of 54

quantificational logic 92
quantifier 92
quantity, physical 148
quantum 272
quantum bit 303
quantum computing 307
quarks 263

radioactivity 259
random process 110
rare gas 246
rational number 17
real number 18
recursive (function definition) 37
redundancy 399
relay 392
resistor, ohmic 211
resonance 270, 378
reversible function 32
right hand rule 64, 166
root (of a tree) 135
root (operation) 20
root-mean-square value (rms value) 387
rotation (of streaming media) 220
rotation, mechanical 164, 192
Roulette 112

sampling theorem 404
scalar 58
scalar product (of vectors) 59
Schrödinger's cat 291
Schrödinger's equation 295
second 151
semantics 134
semiconductor 302
sequential system 363

Subject Index 449

series 77
set 25
set theory 25
sexual reproduction 334
shifter 397
shock absorber 380
simulation 377
sine-function 36
software 421
software engineering 434
source (of streaming media) 218
source oriented encoding 408
space, n-dimensional 100
spance, bent 196
spectrum 268
spherical coordinate system 299
stack (in a computer) 428
standard deviation 120
start-up-phase 376
state transition graph 393
state, final 83
state, initial 83
statistics 129
steady state 376
structured random event 112
subset 26
subtraction 14
SUDOKU 422
superconductivity 301
symbol 132
syntax 134
system kernel 356
system model 348

team player 351
temperature scale 153
terminal (of a grammar) 134
theory (axiomatic) 96
thermodynamics 249
thermostat 360
three-phase-system 385
thymine 324
time slot 411

tomography 310
TOP 428
topology 55
torque 166
Towers of Hanoi 38
transcendental function 36
transcendental number 18
transformation, isomorphic 99

transistor 302, 403
transuranium 247
trigonometric function 355
tuple 27
turbulence 218

ultracentrifuge 343
union (of sets) 26
universal quantification 92
universe (of elements) 26
urea 314

valency (of an atom) 244, 301
variable 28
variance 120
vector 58
vector product (of vectors) 62
virus 340
vitamin 332
Volt 235
voltage 211, 235

Watt (unit) 236
wave length 267, 273
wave packet 276
wave resistance 239

X-chromosome 333
X-rays 275

Y-chromosome 333

zero 14

CPSIA information can be obtained at www.ICGtesting.com
233786LV00004B/8/P